FIELD THEORY, QUANTIZATION AND STATISTICAL PHYSICS

MATHEMATICAL PHYSICS AND
APPLIED MATHEMATICS

VOLUME 6

FIELD THEORY, QUANTIZATION AND STATISTICAL PHYSICS

In Memory of Bernard Jouvet

Edited by

E. TIRAPEGUI

D. REIDEL PUBLISHING COMPANY

DORDRECHT : HOLLAND / BOSTON : U.S.A.
LONDON : ENGLAND

Library of Congress Cataloging in Publication Data

Main entry under title:

Field theory, quantization, and statistical physics.

(Mathematical physics and applied mathematics; v. 6)
Includes bibliographical references and index.
1. Quantum field theory—Addresses, essays, lectures.
2. Statistical physics—Addresses, essays, lectures. 3. Jouvet, Bernard, 1927–1978. I. Tirapegui, E., 1940– II. Jouvet, Bernard, 1927–1978. III. Series.
QC174.46.F53 530.1′43 80–24425
ISBN-13: 978-94-009-8370-0 e-ISBN-13: 978-94-009-8368-7
DOI: 10.1007/978-94-009-8368-7

Published by D. Reidel Publishing Company,
P.O. Box 17,3300 AA Dordrecht, Holland

Sold and distributed in the U.S.A. and Canada
by Kluwer Boston Inc.,
190 Old Derby Street, Hingham, MA 02043, U.S.A.

In all other countries, sold and distributed
by Kluwer Academic Publishers Group,
P.O. Box 322, 3300 AH Dordrecht, Holland

D. Reidel Publishing Company is a member of the Kluwer Group

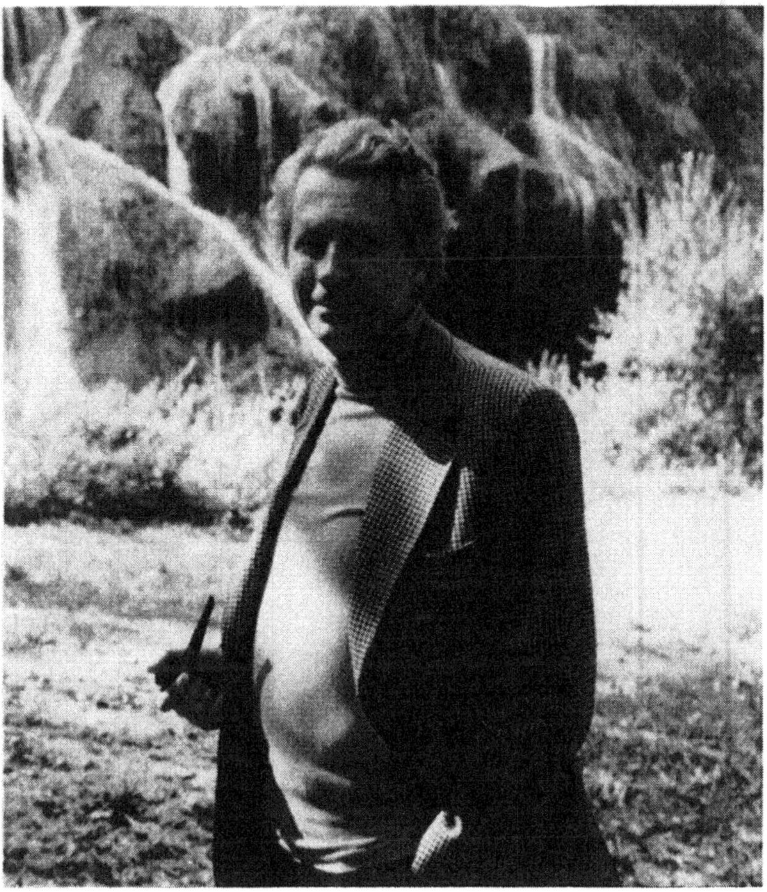

BERNARD JOUVET (1927–1978)

FOREWORD

It is with great emotion that we present here this volume dedicated to the memory of Bernard Jouvet, Docteur ès Sciences, Directeur des Recherches at the Centre National pour la Recherche Scientifique.

The life and the career as a physicist of Professor Jouvet are evoked in the following pages by Professor F. Cerulus, a friend of long standing of Professor Jouvet. The contributions have been written by physicists who were friends, collaborators or former students of Professor Jouvet. I express here my gratitude for their contributions.

I wish also to thank Mrs. France Jouvet for her kind help in the realization of this book. Without her support this would have been impossible.

I am also especially indebted to Professor M. Flato for his constant encouragement and kind cooperation, and to F. Langouche and D. Roekaerts for their generous help in the preparation of this volume.

E. TIRAPEGUI

TABLE OF CONTENTS

PART ONE: FIELD THEORY AND QUANTIZATION

PART TWO: STATISTICAL PHYSICS

BIOGRAPHICAL SKETCH

Bernard Pierre François Jouvet was born on February 5th, 1927, in Lons-le-Saunier, a small provincial town in the Jura hills in the eastern part of France, near the Swiss border. His father, André Jouvet, was a physician and Bernard was the youngest child of three sons and one daughter.

He went to school in his home town and obtained the 'baccalauréat' in 1945. The next two years were spent at the Lycée du Parc in Lyon studying mathematics ('Math. Elem.' and 'Math. Sup.') in preparation for his university studies. These took place in Paris and he obtained his first degree, the 'licence ès science', in 1949. He married in the same year. He had known his fiancée, France Tournier, for many years, already at the Lycée in Lons-le-Saunier; she was to give him a lifelong happy home and three sons.

In 1950 he obtained a fellowship of the Centre National pour la Recherche Scientifique (CNRS), the official French research organisation. He felt inclined to theoretical research and had been noticed by his professors, especially by A. Proca whose research group he joined at the Institut Henri Poincaré.

Amidst this group and at the Proca seminar he learned of Feynman's methods and of the decisive progress which had been made in quantized field theory. Almost all of these discoveries had been made in the U.S.A. during the years 1945–1949 and, because of the war and its aftermath, their development had taken place without any participation by French physicists. Most of the new methods were discovered by fairly young men in research groups with which – even before the 1939–1945 war – the French theorists had few – if any – personal contacts.

The efforts of Jouvet and his fellow-students at the Institut Henri Poincaré were, therefore, exclusively directed towards understanding the printed publications without the benefit of a verbal and intuitive initiation into this new field of theoretical physics but also without the concomitant prejudices. In addition to this were the cramped quarters at the Institut Henri Poincaré and the habits of its members which resulted in Jouvet working mostly at home. As a new idea came to his mind he would examine and develop it by himself and discuss it in the group when it was already fairly well thought-out.

At the beginning of his research activity in 1949 Proca suggested that he should look at the symmetry between space and time which Schwinger had managed to build into his theory of relativistic quantum fields. Jouvet was therefore induced to study seriously relativity theory and its supporting mathematics. He noticed a puzzling analogy between the mechanical equation of motion and Maxwell's equation of the electromagnetic field. By a formal treatment, copying the methods of special relativity, he proposed an extension of the electromagnetic theory; this meant that, for constant fields at least, the contraction of the electromagnetic tensor ($F_{\mu\nu}$ $F^{\mu\nu}$) would be bounded by a constant. He then envisaged a further extension, analogous to the transition from special to general relativity, to Riemannian space for non-constant fields with the hope that this could perhaps provide a cure for the divergence of the electron self energy; this meant, however, a non-linear theory. Proca suggested at this point a comparison with the non-linear electrodynamics of M. Born, and Jouvet followed the hint. His exposé of the Born theory and of his own theory was presented as the required thesis for the degree 'diplôme d'études supérieures' in 1950. In 1951 he obtained a position of research fellow ('attaché de recherche') at the CNRS and spent two months at the summer school of theoretical physics at Les Houches.

Proca arranged for Jouvet to present his theory to W. Pauli when the latter came to Paris that same year. Pauli was apparently favourably impressed for he invited Jouvet to come to Zürich for three months in the spring of 1952.

A few days after his arrival in Zürich Jouvet explained his work in detail to Pauli and Schafroth; the next day Pauli proved that the envisaged generalization could not work and proposed that Jouvet should turn his attention to nuclear physics and try to find the magic numbers in the shell model by using a Thomas–Fermi method for the nucleus.

Jouvet thought the problem over and came up with a novel way of computing by introducing a variational method. Numerically it was very succesful and predicted the spins of quite a few nuclei, some of which were known at the time and others were subsequently verified. He realised, too, that the success of the method depended crucially on an invariance property of his trial wave functions for which he had no physical justification. Pauli appreciated the swiftness Jouvet had demonstrated in this research and the lucid analysis in presenting it, and he invited him for another three months the following spring.

The summer 1952 was again spent at the Les Houches school where its founder, Mrs. C. Morette-de Witt, had managed to bring together some of the best specialists in theoretical physics and some thirty young physicists from France and other European countries. The importance of these gatherings for those students can hardly be overestimated and Jouvet attributed decisive progress in his understanding of field theory to the course of R. Jost.

The spring of 1953 was again spent at the E.T.H. in Zürich, with many discussions and some calculations on the meson states in a Bethe–Salpeter equation. In the fall of that year Jouvet wondered how the idea of L. de Broglie that photons were bound states of neutrino–antineutrino pairs would work out in quantum field theory.

He stated the problem in the formalism of Feynman graphs and established an equivalence between an interacting-field theory of the Yukawa type (with a three-field interaction term) and one of a Fermi type (with a four-fermion field interaction). This equivalence meant that the fine structure constant, i.e. the coupling constant in the Yukawa theory for electromagnetism, was related to the divergence of the vacuum polarization in the Fermi-type theory; it was, in fact, a function of the cut-off used to regularize the diverging integrals. The result, first achieved for mass zero photons and neutrinos, was also extended to massive fermions and massive bosons and meant that mesons could be interpreted as bound states of nucleon-antinucleon pairs.

His doctoral thesis ('Doctorat ès Sciences') was devoted to the electrodynamic part of this theory and presented in 1954. Following this he obtained the position of 'chargé de recherches' at the C.N.R.S.

Proca supported successfully his application to a fellowship at the CERN theory division which was then still at the Niels Bohr Institute in Copenhagen, and he stayed there for two years (1955–1957). In those years the Institute had some fifteen CERN fellows and as many other visitors and was at the focus of European research in theoretical nuclear physics and in field theory. The latter research group was conducted by G. Källén who was then already well known for his work on quantum fields in the Heisenberg representation and the sumrules for the renormalization constants. Other members of the group were H. Lehmann, W. Glaser, S. Deser, A. Peterman and A. Visconti.

The Institute was a lively place and discussions of everybody's research were going on all the time. Källén brought to them an immense enthusiasm and a superior intelligence; he would infallibly spot any error in the many papers submitted to him and would also leave no doubt as to his opinion.

The research interests of the field-theory group went toward the then fashionable problems: the Lee model, ghosts, spectral form of the propagator, dispersion relations, etc. They used of course the most advanced theoretical tools which were in part developed by them.

Jouvet explained his work to Källén and the others but it was not approved of. One reason for this was a plain language problem; Källén did not understand French and the subtlety of an argument would be lost in Jouvet's scientific English. Another reason – Jouvet explained later – was his use of series of Feynman graphs and the regularization technique instead of the compact methods of renormalization theory his opponents were used to; at any rate they did not point out any mistake in his calculations but opposed him with objections on general grounds. They thought it was impossible that a renormalizable Yukawa theory could be equivalent to a non-renormalizable Fermi theory; in addition it seemed that the photon appearing as a bound state in intermediate states only would make it impossible to construct an S-matrix with real incoming and outgoing photons.

Jouvet took up the study of renormalization theory and tried to formulate his theory in this way in order to obtain an answer to the objections. He noticed then that the equation which determined the Yukawa coupling constant was simply $Z_3 = 0$ where Z_3 was the renormalization constant of the composed field. This led to an open conflict with Källén who had proved a few months earlier that Z_3 was identically zero and this forced Jouvet to a critical examination of that proof. He convinced himself it could not be true, by an argument based on the compensation of infra-red divergencies (this argument he never published as such; however, it can be found in his lectures at the South American summer school).

He wrote up the results obtained so far which were published in *Il Nuovo Cimento* during 1956 and 1957. During this second year in Copenhagen the scientific conflict with Källén remained tense but the discussions with the members of the group and the many visitors who came through Copenhagen gave him confidence in his mastery of the methods of field theory and added to his determination to pursue the investigation of composed fields and the many new research possibilities they opened up.

Proca had died in 1956 and coming back to France in 1957 Jouvet looked for a suitable position. He was determined to find a position that would allow him to choose freely his research line even at the price of doing a job unrelated to this. He became one of the two 'resident theorists' with the newly founded experimental group in high-energy physics within the laboratory of F. Perrin at the Collège de France; the promotion to 'maître de recherches' in 1958 made his scientific independence secure.

At the Collège de France laboratory he was to devote his efforts to two distinct activities. On the one hand he studied the phenomenological theories in particle physics and discussed with the experimentalists both their plans and their results, and on the other hand he pursued his investigations in field theory with the many students who came to him. As the years went by, from 1957 onwards, he looked more closely at a great many problems suggested by the hypothesis of bosons composed of fermions. He and his coworkers investigated in detail the mathematical structure of such theories, thereby answering the objections raised in Copenhagen, and extending the scope of the theory. The use of the relation $Z_3 = 0$ as a test of compositeness and its experimental verification was also thoroughly examined. The selected bibliography in the present volume shows the extent of this effort and the names of his collaborators.

He liked very much to be with his family and friends and the Jouvets returned regularly to their native Lons-le-Saunier and its beautiful countryside of hills and lakes. He felt definitely reluctant to go on sales travels in order to promote his ideas. These were slow in being accepted and were in part even rediscovered; his stay in Copenhagen had not contributed to promoting his name onto the list of authors it was advantageous or fashionable to cite. As time went on, his discoveries became gradually known and he noted with pleasure when they were confirmed, used or expanded by reputed scientists.

The job at the laboratory led him to a number of investigations into the theory and the possible experiments with muons (1960). He introduced also the ρ-exchange mechanism in order to explain the $K^+ + p \to K^0 + \Delta^{++}$ reaction and there were several other papers originating from his permanent contact with experimental physics which, however, was naturally more devoted to the informal discussions and counseling in the laboratory than to original research. In 1964 he was promoted to 'Directeur de Recherche titulaire' at the CNRS.

From 1971 onwards he started worrying about the future of his research. He was concerned over his responsibility as the leader of a group of young theorists in the face of changing times which, he thought, would make it increasingly difficult for them to pursue a career in pure research.

Perhaps in order to break away from his past he accepted a visitor's position at Santiago University (Chili). He was then already quite familiar with the South-American continent which he visited regularly since 1967, but this time he devoted more than a year (1971–1972) to lecturing – a pleasure the CNRS position had rarely afforded him – guiding graduate students and collecting mineral specimens during numerous trips.

When he returned, things were changing rapidly: a thorough reorganisation of the high-energy experimental physics in Paris was taking place due to the simultaneous retirement of F. Perrin and L. Leprince–Ringuet and the members of his theory group had to find their way into other universities and laboratories.

He had, in the meantime, been thinking of new research subjects which he felt should be more applied and stable than the fickle fashions in high-energy physics where even the 'established' facts were never quite secure.

He chose to work at the Laboratoire de Physique Corpusculaire et Hautes Energies in Orsay and took up the study of the theory of phase transitions which he applied to problems of solid state physics. He was also attracted to the theory of turbulence and finally found this last subject profoundly challenging to his theoretical gifts and mathematical ability.

Before he could make contributions to it illness seized him; he bore it for months with equanimity and finally passed from among us on January 8th, 1978.

The above biographical sketch is of course very incomplete and passes lightly over the years between 1958 and 1971 when he was guiding his students and producing his better-known research. I thought it more interesting, however, for the reader to go back to his days in Zürich and Copenhagen which I witnessed and which determined his career and explain some puzzling features of his biography. Already then the prominent traits of his character were apparent: wit and caustic humour, a very French taste for good cuisine and a personality distrusting accepted ideas and relying on its own judgment in everything, but which would rather observe events than take part in them.

In the present volume, theorists who were associated with him give, through their work, a last tribute to Bernard Jouvet. This tribute goes to the scientist who did research with a passion but also to the man who did not allow this passion to become narrow and single-minded, who had time for friendship and wide interests in men and things. That so many contributed shows the genuine bereavement we all feel because he is no longer here.

All who knew Bernard Jouvet will be grateful to Enrique Tirapegui who proposed this memorial volume and saw it through all its stages.

Je ne pourrais pas clore ces lignes sans une pensée émue pour Madame France Jouvet; que ce livre soit pour elle le message que l'oeuvre de son mari est vivante et que ses idées font maintenant partie du vaste ensemble des connaissances humaines.

Heverlee, January, 1980 F. CERULUS

LIST OF SELECTED SCIENTIFIC PUBLICATIONS

The following list is not an exhaustive compilation of all the publications of B. Jouvet, which number some seventy items. The selection is his own, being copied from the most recent version of his curriculum vitae.

1. Theory of composite particles and fields

Thesis: 'l'électromagnétisme électro-neutrinien', *Journ. de Math.* **33**, 201 (1954).
'A realistic theory of mesons', *Nuovo Cimento Suppl.* **2**, 941 (1955).
'On the meaning of Fermi Coupling', *Nuovo Cimento* **3**, 1133 (1956).
'Fermi Coupling and Mass and Charge Spectra of Bosons', *Nuovo Cimento* **5**, 1 (1957).
'Etude d'un modèle de champ à constante de renormalisation nulle', *Nuovo Cimento* **18**, 466 (1960) (in collaboration with J. C. Houard).
'Renormalisation d'un doublet de bosons en interaction avec un doublet de fermions', *Nuovo Cimento* **29**, 53 (1963) (in collaboration with J. P. Adam).
'Particules et champs composites résultant d'interactions entre plusieurs courants', *Nuovo Cimento* **29**, 1275 (1963) (in collaboration with J. P. Adam).
'Un test de la nature composite du meson ρ dans les facteurs de forme des nucléons', *Physics Letters* **6**, 307 (1963).
'Renormalisation et propriétés des matrices S, non relativistes et relativistes, résultant de couplages de Fermi', *Nuovo Cimento* **37**, 20 (1965), (in collaboration with A. Jacquemin).
'The composite ρ-meson coupling constant in the H and X models', *Nuovo Cimento* **36**, 287 (1965) (in collaboration with M. G. Clement et A. Jacquemin).
'Diffusion élastique des neutrinos et structure du photon', *Nuovo Cimento* **37**, 951 (1965).
'Particules auto-composées', *Nuovo Cimento* **49**, 667 (1967) (in collaboration with J. C. Le Guillou).
'Dynamique d'un champ autocomposé et la construction de l'espace-

temps', *Ann. Inst. H. Poincaré* **A8**, 191 (1968) (in collaboration with G. Clement).
'Infinitely composite particles and the condition $Z=0$', *Phys. Rev.* **4D**, 3222 (1971) (in collaboration with M. Whippman).
'Behaviour of the function Z_3 (α) in 2nd class QED', *Nuovo Cimento* **28A**, 273 (1975) (in collaboration with G. Bordes).

2. Studies about renormalization constants, gauge invariance, and other field-theoretical problems

'Les invariances de jauge classiques de l'électrodynamique quantique', *Nuovo Cimento* **80**, 28 (1961).
'Utilisation des infinis d'énergie propre du photon pour la détermination de la forme asymptotique de son propagateur canonique', *Nuovo Cimento* **25**, 135 (1962).
'Anomalous properties of the vector field in interaction with Dirac's current', *Nuovo Cimento* **28**, 283 (1962).
'Sur les termes en α^3 de la constante de renormalisation $Z_3(\alpha)$ du champ de photon', *C.R.A.S.* **262**, 163 (1966) (in collaboration with M. Astaud).
'Extension du "groupe" de renormalisation', *Nuovo Cimento* **53**, 841 (1968) (in collaboration with M. Astaud).
'Zero dimensional physics, stairs, continued fractions and convergence problems', *Nuovo Cimento* **53A**, 69 (1968) (in collaboration with E. Tirapegui).
'Renormalization "group" extended to masses. Application to the structure of the renormalization constants Z_3 and m_0', *Nuovo Cimento* **63**, 5 (1969) (in collaboration with M. Astaud).
'An exact content of the renormalization group postulate', *Lettere al Nuovo Cimento* **4**, 413 (1970).
'Renormalization group, inversibility postulates and the finiteness of Z_3', *Rev. Bras. Fis.* **3**, 345 (1973).
'Renormalization Group and composite Field in Wilson's Theory of critical phenomena', *Nuovo Cim.* **18A**, 459 (1973).
'Renormalization Group representation and the Gell-Mann–Low and Callan–Symanzik equations for the photon propagator', *Nuovo Cim.* **23A**, 621 (1974).
'The renormalization group and its relativity principle', *Nuovo Cim.* **23A**, 521 (1974).
'Asymptotic freedom and discontinuities in field theory models for

$g \to 0$', *Phys. Rev.* **D15**, 1565 (1977) (in collaboration with E. Tirapegui).
'Quantum aspects of Classical and Statistical fields', *Phys. Rev.* **A19**, 1350 (1979). (in collaboration with R. Phytian).

3. Phenomenology

'Electrons from μ capture and the radius of Fermi interaction', *Nuovo Cimento* **15**, 31 (1960), (in collaboration with J. C. Houard).
'Limite sur les interactions anormales des muons et leur manifestation dans la diffusion nucléaire à haute énergie', *Nuovo Cimento* **18**, 702 (1960) (in collaboration with L. Goldzahl).
'Non localité et asymétrie de l'interaction électron-neutrino dans le mode K e3', *Journ. de Phys.* **24**, 732 (1963), (in collaboration with Ch. Chahine).
'The ρ-meson in the electro and K-production of N_{33}^{+}', *Physics Letters* **6**, 307 (1963) (in collaboration with J. M. Abillon et G. Bordes).
'Formule de factorisation dans le calcul des sections efficaces inélastiques au voisinage des résonances', *C.R.A.S.* **257**, 1007 (1963) (in collaboration with G. Bordes).
'Modification of the fine structure constant in strong fields or very dense matter', *Nuovo Cimento* **46B**, 122 (1966).
'On the stimulated photon–graviton conversion by an electromagnetic field', *Ann. Inst. H. Poincaré* **8**, 25 (1968) (in collaboration with L. Halpern).
'Are the weak neutral currents an electromagnetic effect?', *Lett. Nuovo Cim.* **12**, 81 (1975).
'Electrodynamic interpretation of neutral interactions of the neutrinos', *Ann. Phys.* **95**, 424 (1975).

4. Phase transitions

'Renormalization group approach to structural phase transition', *J. Phys. C* **7**, 1449 (1974) (in collaboration with A. Holz).
'Collective mode mediated phase transition in two-field system with $\lambda S^2(x) Q^2(x)$-coupling', *Phys. Lett.* **49A**, 177 (1974) (in collaboration with A. Holz).
'Temperature dependence of soft mode frequency above phase transition', *Rev. Bras. Fis.* **4**, 425 (1974) (in collaboration with A. Holz).
'Bound-state-induced phase transition in a two field thermodynamic system with $\lambda S^2(x) Q^2(x)$-coupling', *Nuovo Clim.* **28B**, 105 (1975) (in collaboration with A. Holz).

5. Didactic or review papers in books

'La théorie quantique des champs. La théorie dynamique de la matière'. In *Relativité et Quanta* (*Les grandes théories de la Physique Moderne*), Chap 2, pp. 83–136, Masson Edit., 1968.

Lectures on 'Field theoretical approach to composite particles' and Seminar on the 'Gauge invariance with non zero photon bare mass', IXth Latin American School of Physics, I. Saavedra (ed.), W. A. Benjamin, 1968.

PART ONE

FIELD THEORY AND QUANTIZATION

C. BECCHI, A. ROUET and R. STORA

RENORMALIZABLE THEORIES WITH SYMMETRY BREAKING

ABSTRACT. The description of symmetry breaking proposed by K. Symanzik within the framework of renormalizable theories is generalized from the geometrical point of view. For an arbitrary compact Lie group, a soft breaking of arbitrary covariance, and an arbitrary field multiplet, the expected integrated Ward identity are shown to hold to all orders of renormalized perturbation theory provided the Lagrangian is suitably chosen. The corresponding local Ward identity which provides the Lagrangian version of current algebra through the coupling to an external classical Yang–Mills field, is then proved to hold up to the classical Adler–Bardeen anomaly whose general form is written down. The BPHZ renormalization scheme is used throughout in such a way that the algebraic structure analyzed in the present context may serve as an introduction to the study of fully quantized gauge theories.

1. Introduction

Besides the well-known relevance of broken symmetries to elementary particle physics, further investigations of renormalizable models exhibiting broken symmetries are amply justified by the present understanding of gauge and even supergauge theories. The present status of the subject is well represented by K. Symanzik's 1970 Cargèse lectures where the fundamental phenomena are discussed [1]. Since much of the structure analyzed there can also be found in the study of gauge theories, it is of interest to complete Symanzik's analysis from the algebraic point of view, both in considering an arbitrary compact Lie group as describing the symmetry to be broken, and in analyzing the perturbative analog of current algebra, namely the coupling with a classical Yang–Mills field, which, as is well-known, leads to the definition of the celebrated Schwinger–Bell–Jackiw–Adler–Bardeen anomaly [2]. In fact, although this program had been announced by K. Symanzik, in collaboration with one of us (R.S.) [1], it has not been carried out until now following the stimulation [3] provided by the advent of gauge theories. Also, it seems that recent progress in renormalization theory has allowed a more tractable general treatment than the techniques known in 1970 would have permitted. Most of the present analysis relies on general properties of the perturbative series which stem from locality

E. Tirapegui (Ed.), Field Theory, Quantization and Statistical Physics, 3–32.

and power counting, as summarized by the renormalized action principle of Lowenstein and Lam [4] whose detailed form is one of the highlights of the Bogoliubov–Parasiuk–Hepp–Zimmermann [5] renormalization scheme. Within this framework, explicit bases of local operators of given dimensions are constructed [6], together with the linear relations connecting operators with different dimensions (the so-called Zimmermann identities). To avoid inessential technical complications, we shall only treat cases in which no massless field is involved. Our analysis can, however, be extended without essential modifications to a wide class of models involving massless fields exploiting Lowenstein's and Zimmermann's extension of BPHZ renormalization scheme [7]. Whereas these elementary tools, which are best exploited by means of a repeated application of the implicit function theorem for formal power series [3], suffice to solve most of the algebraic problems at hand, the elimination of some possible anomalies is occasionally performed by looking more deeply into the behaviour of the theory under scaling transformations, which provides some new non-renormalization type statements similar to that which leads to the non-renormalization of the Adler–Bardeen anomaly coefficient [8].

This article is divided into two main parts: Section 2 is devoted to the proof of the integrated Ward identity which expresses symmetry breaking for an arbitrary compact Lie group, with an arbitrary dimension (< 4) and covariance. Section 3 is devoted to a discussion of 'current algebra' which, in the present framework amounts to the proof of a local Ward identity, in the presence of a classical Yang–Mills field, and leads to the definition of the Adler–Bardeen anomaly.

A number of appendices are devoted to the treatment of some technical questions, among which the elimination from the integrated Ward identities of algebraically allowed anomalies consistent with power counting, and details about the cohomology [9] of the gauge Lie algebra associated with the symmetry group (i.e. the Wess–Zumino [10] consistency conditions).

2. Global symmetries broken

2.1. *The Classical Theory*

The general situation is as follows: G is a compact Lie group, $\mathcal{G}$ its Lie algebra: $\mathcal{G} = \mathcal{S} + \mathcal{A}$, $\mathcal{S}$ semi-simple, $\mathcal{A}$ abelian. φ is a field multiplet belonging to a fully reduced finite-dimensional unitary representation D of G, d_φ the canonical dimension of φ. The corresponding infinitesimal transformation of φ is:

$$\delta_X \varphi = - \mathring{t}(X)\varphi, \qquad X \in \mathscr{G}, \tag{1}$$

where $X \rightarrow \mathring{t}(X)$ is the representation of $\mathscr{G}$ induced by D.

Let $\boldsymbol{\beta}$ be a classical field to which is assigned dimension d_β, belonging to a multiplet characterized by another representation $\mathscr{D}$ of G (finite dimensional, fully reduced, unitary, with no identity component)

$$X \rightarrow \mathring{\theta}(X), \qquad X \in \mathscr{G}, \tag{2}$$

the corresponding representation of $\mathscr{G}$.

The symmetry G will be said to be broken with dimension $d_B = 4 - d_\beta$, covariance $\mathring{\boldsymbol{b}}$, belonging to multiplet $\mathscr{D}$, if there exists a Lagrangian $\mathscr{L}(\varphi, \boldsymbol{\beta})$, of maximum dimension four invariant under the simultaneous transformation

$$\varphi \rightarrow {}^g\varphi = D(g^{-1})\varphi, \tag{3}$$
$$\boldsymbol{\beta} + \mathring{\boldsymbol{b}} \rightarrow {}^g(\boldsymbol{\beta} + \mathring{\boldsymbol{b}}) = \mathscr{D}(g^{-1})(\boldsymbol{\beta} + \mathring{\boldsymbol{b}}). \tag{4}$$

The classical field $\boldsymbol{\beta}$ is introduced as an auxiliary item, in order to characterize the breaking described by the space-time independent $\mathring{\boldsymbol{b}}$ according to its dimension, a concept which is meaningful in the renormalizable framework we have in mind. The theory will be truly renormalizable if

$$d_B < 4, \tag{5}$$

i.e. $\mathscr{L}(\varphi, \boldsymbol{\beta})$ will be a polynomial. This criterion, introduced by Symanzik [1] leaves the broken theory with an asymptotic memory of the initial symmetry group. However, the limiting case

$$d_B = 4$$

can also be considered since $\mathscr{L}(\varphi, 0)$ is invariant under simultaneous transformation of φ and $\mathring{\boldsymbol{b}}$ and is not, in general, the most general Lagrangian which is invariant under the residual symmetry group, namely the stability group H_0 of $\mathring{\boldsymbol{b}}$. Clearly, the notion we have introduced only depends on the equivalence classes of D, $\mathscr{D}$ and the orbit of $\mathring{\boldsymbol{b}}$. We shall assume that for some values of the parameters characterizing $\mathscr{L}$, a particle interpretation is possible and that there is an invertible change of parameters between the coefficients of $\mathscr{L}$ and those occurring in normalization conditions on the tree approximations of the corresponding

Green's functions through which masses, coupling constants, etc., are defined. We shall furthermore assume that no vanishing mass parameter occurs in the theory.

When the Lagrangian has a term linear in the quantized field, the particle interpretation requires to perform a field translation

$$\varphi \to \varphi + \mathring{F}$$

through which the linear term is eliminated. $\mathring{F}$ is then defined by:

$$\partial_\varphi \mathscr{L}(\varphi,0)\big|_{\varphi=\mathring{F}} = 0$$

which certainly has a solution which is continuous in the parameters of $\mathscr{L}$ if the mass matrix

$$\mathscr{M} = \partial^2_{\varphi\varphi}\mathscr{L}(\varphi,0)\big|_{\varphi=\mathring{F}}$$

is non-degenerate. As shown in Appendix A, $\mathring{F}$ is then a covariant function of $\mathring{b}$, and consequently, the coefficients of the Lagrangian expressed in terms of the translated field are also covariant. From now on, we shall still denote by φ the translated field and by $\mathscr{L}$ the new Lagrangian from which the linear term is absent.

At this point, the action

$$\mathring{\Gamma}(\varphi,\beta) = \int \mathrm{d}x\, \mathscr{L}(\varphi,\beta)(x) \tag{6}$$

fulfills the 'integrated' Ward identity

$$W(X)\mathring{\Gamma}(\varphi,\beta) \equiv -\int \mathrm{d}x \left\{ \frac{\delta\mathring{\Gamma}}{\delta\varphi}\mathring{t}(X)(\varphi + \mathring{F}) + \frac{\delta\mathring{\Gamma}}{\delta\beta}\mathring{\theta}(X)(\beta + \mathring{b}) \right\}(x) = 0, \qquad X \in \mathscr{G}, \tag{7}$$

which expresses its invariance under the infinitesimal transformation

$$\begin{aligned} \delta_X\varphi &= -\mathring{t}(X)(\varphi + \mathring{F}), \\ \delta_X\beta &= -\mathring{\theta}(X)(\beta + \mathring{b}). \end{aligned} \tag{8}$$

We shall see in the sequel that if one wants to construct a perturbation theory for which a particle interpretation exists, it is necessary to assume the following: Let the Lagrangian be written in the form

$$\mathscr{L} = C_{\#}\mathscr{L}_{\#} = \sum_i C^i_{\#}\mathscr{L}_{\#,i} \tag{9}$$

where the $C^i_{\#}$'s are numerical coefficients and the $\mathscr{L}_{\#,i}$'s all possible local monomials invariant under Equation (8), consistent with the renormalizability requirement. Then there exists an invertible change of variables between the $C^i_{\#}$'s and $\mathring{\boldsymbol{b}}$ on the one hand, and a set of physical parameters (masses, wave function normalizations, coupling constants) occurring in normalization conditions imposed on $\mathring{\Gamma}$, on the other hand, consistent with the symmetry expressed by the Ward identity, but not constrained by power counting. This implies in particular that power counting does not restrict Equation (9) compared to the most general solution of Equation (7), as far as these normalization conditions are concerned (e.g. does not enforce mass rules). Of course, the fulfillment of normalization conditions is only necessary if a particle interpretation is required, the fulfillment of the Ward identity being sufficient if only a theory of Green's functions is aimed at.

One might object that the prescription of the Ward identity Equation (7) does not seem to define the theory in a natural way from the point of view of power counting: a more general scheme would be to have a Ward identity with the following structure:

$$\begin{aligned} W(X)\mathring{\Gamma} \equiv - \int \mathrm{d}x \Bigg\{ & \frac{\delta \mathring{\Gamma}}{\delta \boldsymbol{\varphi}}(t(X)\boldsymbol{\varphi} + \boldsymbol{F}(X)) \\ & + \frac{\delta \mathring{\Gamma}}{\delta \boldsymbol{\beta}}(\theta(X)\boldsymbol{\beta} + b(X)) \Bigg\}(x), \qquad X \in \mathscr{G}, \end{aligned} \tag{10}$$

subject to the algebraic constraint

$$[W(X), W(Y)] = W([X, Y]), \qquad X, Y \in \mathscr{G}, \tag{11}$$

thus expressing the invariance of $\mathscr{L}$ under the transformation

$$\begin{aligned} \delta_X \boldsymbol{\varphi} &= -[t(X)\boldsymbol{\varphi} + \boldsymbol{F}(X)], \\ \delta_X \boldsymbol{\beta} &= -[\theta(X)\boldsymbol{\beta} + \boldsymbol{b}(X)], \qquad X \in \mathscr{G}, \end{aligned} \tag{12}$$

with coefficients constrained by:

$$[t(X), t(Y)] = t([X, Y]), \tag{13a}$$

$$[\theta(X), \theta(Y)] = \theta([X, Y]), \tag{13b}$$

$$t(X)\boldsymbol{F}(Y) - t(Y)\boldsymbol{F}(X) - \boldsymbol{F}([X, Y]) = 0, \tag{13c}$$

$$\theta(X)\boldsymbol{b}(Y) - \theta(Y)\boldsymbol{b}(X) - \boldsymbol{b}([X, Y]) = 0, \tag{13d}$$

according to which

$$X \to t(X),$$
$$X \to \theta(X),$$

are representations of $\mathscr{G}$, and $\boldsymbol{F}(X)$, $\boldsymbol{b}(X)$ are cocycles [9] of $\mathscr{G}$ with values in the representation spaces E and $\mathscr{E}$ of t and θ. It is shown in Appendix B that the requirements which allow a particle interpretation and take into account our definition of symmetry breaking put quite severe restrictions on $t(X)$, $\theta(X)$, namely they can be lifted to the group G, and thus, in particular they are fully reducible. They can thus be obtained from representatives of their equivalence classes which we shall call $\mathring{t}, \mathring{\theta}$ through suitable field renormalizations:

$$\begin{aligned} t(X) &= Z^{-1}\mathring{t}(X)Z, \\ \theta(X) &= z^{-1}\mathring{\theta}(X)z. \end{aligned} \tag{14}$$

It is then shown in Appendix B that $\boldsymbol{F}(X)$, is a coboundary [9]

$$\boldsymbol{F}(X) = t(X)\boldsymbol{F} \tag{15}$$

for some fixed $\boldsymbol{F}$, up to invariant components

$$F_{\#}(X)$$

which vanish for $X \in \mathscr{S}$, the semi-simple part of $\mathscr{G}$. It is finally shown in Appendix B that $\boldsymbol{F}_{\#}(X) \neq 0$ contradicts the assumption that the mass matrix is non-degenerate. Thus the Ward identity (Equation (7)) is actually the most general in the present context, since the coboundary structure of $\boldsymbol{b}(X)$:

$$\boldsymbol{b}(X) = \theta(X)\boldsymbol{b},$$

is implied by our picture of symmetry breaking.

2.2. *Radiative Corrections*

The description of radiative corrections proceeds via the construction of an effective dimension four Lagrangian

$$\mathscr{L}_{\text{eff}} = \mathscr{L} + \hbar\Delta\mathscr{L} \tag{16}$$

without a term linear in the quantized fields, with coefficients formal power series in $\hbar$ such that a renormalized Ward identity holds:

$$W(X)\Gamma \equiv -\int \mathrm{d}x \left\{ \frac{\delta\Gamma}{\delta\varphi}(t(X)\varphi + F(X)) + \frac{\delta\Gamma}{\delta\beta}(\theta(X)\beta + b(X)) \right\}(x) = 0, \qquad X \in \mathscr{G}, \tag{17}$$

subject to the algebraic constraint

$$[W(X), W(Y)] = W([X, Y)] \tag{18}$$

thus expressing the invariance of $\mathscr{L}_{\text{eff}}$ in the sense of the renormalized action principle [4] under the renormalized transformation

$$\begin{aligned} \delta_X\varphi &= -(t(X)\varphi + F(X)), \\ \delta_X\beta &= -(\theta(X)\beta + b(X)), \end{aligned} \tag{19}$$

which coincides with Equation (8) in the lowest order in $\hbar$.

The analysis performed in this section will actually lead to the conclusion that there exists a quantum extension in which the almost naive Ward identity holds:

$$W(X)\Gamma \equiv -\int \mathrm{d}x \left\{ \frac{\delta\Gamma}{\delta\varphi}\mathring{t}(X)(\varphi + F) + \frac{\delta\Gamma}{\delta\beta}\mathring{\theta}(X)(\beta + b) \right\}(x) = 0, \qquad X \in \mathscr{G}, \tag{20}$$

there only F which is determined by the requirement that $\mathscr{L}_{\text{eff}}$ has no term linear in φ and b which has the same stability group H_0 as $\mathring{b}$ pick up radiative corrections. This is to say that one can fulfill quantum invariance under the renormalized transformation

$$\begin{aligned} \delta_X\varphi &= -\mathring{t}(X)(\varphi + F), \\ \delta_X\beta &= -\mathring{\theta}(X)(\beta + b), \end{aligned} \tag{21}$$

as a consequence of the algebraic constraints, Equation (18): as shown in Appendix B the consistency conditions on $t(X)$ and $\theta(X)$ can be used to replace them by $\mathring{t}(X)$ and $\mathring{\theta}(X)$ up to terms which can be interpreted as invariant anomalies to the Abelian Ward identities, Equation (17), where X is restricted to $\mathscr{A}$, the Abelian part of $\mathscr{G}$. Similarly the consistency condition on $b(X)$ leads to

$$b(X) = \mathring{\theta}(X)b$$

where b is arbitrary, and can trivially be chosen to have the same stability group as $\mathring{\boldsymbol{b}}$, (the residual symmetry group). Finally $\boldsymbol{F}(X)$ is found to be of the form $t(X)\boldsymbol{F}$ for some fixed $\boldsymbol{F}$, up to terms which again can be interpreted as invariant anomalies to the Abelian Ward identities. Since it will be shown, c.f., in particular, Appendix C, that the Ward identity, Equation (20), cannot be broken exclusively by invariant Abelian anomalies, one concludes that, modulo a field renormalization, all solutions of Equation (17) are solutions of Equation (20). We thus proceed to try and fulfill Equation (20). Applying the action principle to the variation Equation (21), we find

$$W(X)\Gamma = -\Delta(X)\Gamma \tag{22}$$

where $\Delta(X)$ denotes a dimension four insertion:

$$\begin{aligned} -\Delta(X) &= \int \mathrm{d}x'' \delta_X \mathscr{L}_{eff}''(x) \\ &= \int \mathrm{d}x \{N_4 \delta_X \mathscr{L}_{eff} + \hbar N_4 Q(X)\}(x) \end{aligned} \tag{23}$$

where $\hbar Q(X)$ lumps together radiative corrections.

Since $\mathscr{L}_{\text{eff}}$ and Q belong to finite-dimensional representation spaces for G, they can be reduced into irreducible components. Let $\boldsymbol{\Omega}$, $\partial^\mu \boldsymbol{\Omega}_\mu$, $\partial^\mu \partial^\nu \boldsymbol{\Omega}_{\mu\nu}$ be a basis of covariant local polynomials of dimension smaller than or equal to four which realizes this reduction, which we explicitly split into an invariant part $\boldsymbol{\Omega}_{\#}, \ldots$, and a non-invariant part $\boldsymbol{\Omega}_{b}, \ldots$. We shall denote by $-T(X)\boldsymbol{\Omega}$ the classical variation of $\boldsymbol{\Omega}$ obtained by varying $\boldsymbol{\varphi}$ and $\boldsymbol{\beta}$ according to Equation (21):

$$\delta_X \boldsymbol{\Omega} = -T(X)\boldsymbol{\Omega}. \tag{24}$$

The first step of our analysis will consist in deriving consistency conditions on $\Delta(X)$ which stem from the algebraic properties of $W(X)$ (equation (18)). The most convenient way to do so is to explicitly introduce into the Lagrangian sources linearly coupled to $N_4 \boldsymbol{\Omega}$ thus defining

$$\mathscr{L}_{\text{eff}}(\boldsymbol{\varphi}, \boldsymbol{\beta}, \boldsymbol{\omega}) = \mathscr{L}_{\text{eff}}(\boldsymbol{\varphi}, \boldsymbol{\beta}) + \boldsymbol{\omega}\boldsymbol{\Omega} + \partial^\mu \boldsymbol{\omega}_1 \boldsymbol{\Omega}_\mu + \ldots. \tag{25}$$

In terms of the corresponding $\Gamma(\boldsymbol{\varphi}, \boldsymbol{\beta}, \boldsymbol{\omega})$ the action principle reads:

$$\mathfrak{W}(X)\Gamma(\boldsymbol{\varphi}, \boldsymbol{\beta}, \boldsymbol{\omega}) = \hbar C_X(\boldsymbol{\omega}) \tag{26}$$

where $C_X(\boldsymbol{\omega})$ is a local functional, formal power series in $\boldsymbol{\omega}, \hbar$ and the coefficients of $\mathscr{L}_{\text{eff}}(\boldsymbol{\varphi}, \boldsymbol{\beta})$, lumping together the quantum corrections to

$\delta_X \mathscr{L}_{\text{eff}}(\boldsymbol{\varphi},\boldsymbol{\beta},\boldsymbol{\omega})$, which only depend on $\boldsymbol{\omega}$, at least $O(\omega^2)$ and

$$\mathfrak{W}(X) \equiv W(X) - \int \mathrm{d}x\, \boldsymbol{\omega}_x T(X)\delta_{\omega_x} + \hbar \int \mathrm{d}x [\boldsymbol{B}_X(\omega)\delta_\omega](x). \tag{27}$$

Here $\boldsymbol{B}_X(\omega)$ is a local functional of $\boldsymbol{\omega}$ and its derivatives, formal power series in $\boldsymbol{\omega},\hbar$, etc., which takes into account the $\boldsymbol{\varphi},\boldsymbol{\beta}$ dependent radiative corrections to $\delta_X \mathscr{L}_{\text{eff}}(\boldsymbol{\varphi},\boldsymbol{\beta},\boldsymbol{\omega})$. In particular

$$\hbar \int \mathrm{d}x [\boldsymbol{B}_X(0)\delta_\omega](x)\Gamma = \Delta(X)\Gamma. \tag{28}$$

The second term in $\mathfrak{W}(X)$ comes from the naïve variation of the $\boldsymbol{\omega}$-dependent terms in $\mathscr{L}_{\text{eff}}(\boldsymbol{\varphi},\boldsymbol{\beta},\boldsymbol{\omega})$.

Now, we can iterate Equation (27):

$$[\mathfrak{W}(X),\mathfrak{W}(Y)]\Gamma(\boldsymbol{\varphi},\boldsymbol{\beta},\boldsymbol{\omega}) = \hbar[\mathfrak{W}(X)C_Y(\boldsymbol{\omega}) - \mathfrak{W}(Y)C_X(\boldsymbol{\omega})] \tag{29}$$

since the right-hand side exists. Evaluating both sides at $\omega = 0$ yields

$$\begin{aligned}
&\{-B_0([X,Y]) + B_0(X)[-T(Y) + \hbar B_1(Y)] \\
&\quad - B_0(Y)[-T(X) + \hbar B_1(X)]\}\delta_\omega \Gamma|_{\omega=0} \\
&\equiv -A([X,Y]) \int \mathrm{d}x \delta_{\omega_x} \Gamma|_{\omega=0} = 0,
\end{aligned} \tag{30}$$

where

$$\begin{aligned}
&\boldsymbol{B}_0(X) = \boldsymbol{B}_X(0), \\
&B_1(X) = \int \mathrm{d}x \left[\frac{\delta \boldsymbol{B}_X(\boldsymbol{\omega})}{\delta \boldsymbol{\omega}(0)}\right]_{\omega=0}(x).
\end{aligned} \tag{31}$$

The only difficulty is that

$$\int \mathrm{d}x \delta_{\omega_x} \Gamma \tag{32}$$

does not a priori exist, because $\delta_{\omega_x}\Gamma$ may have a field independent term. However, we have

$$A([X,Y])\left\{\int \mathrm{d}x \delta_{\omega_x}\Gamma - \int \mathrm{d}x \delta_{\omega_x}\Gamma\bigg|_{\varphi=\beta=0}\right\} = 0 \tag{33}$$

i.e.

$$A([X,Y])\left\{\int \mathrm{d}x [\boldsymbol{\Omega}_x - \Omega_x(\boldsymbol{\varphi} = \boldsymbol{\beta} = 0)] + O(\hbar)\right\} = 0. \tag{34}$$

Since the $\mathbf{\Omega}$'s form a complete basis except for constants, we deduce

$$A([X,Y])=0. \tag{35}$$

We shall now split $\mathscr{G}$ into its Abelian part and its semi-simple part:

$$\mathscr{G}=\mathscr{A}+\mathscr{S} \tag{36}$$

and $\boldsymbol{B}_0(X)$ into its invariant and non-invariant parts:

$$\boldsymbol{B}_0(X)=\boldsymbol{B}_0^{\#}(X)+\boldsymbol{B}_0^{b}(X). \tag{37}$$

Equation (35) is a perturbed cocycle condition:

$$\mathscr{T}(X)\boldsymbol{B}_0(Y)-\mathscr{T}(Y)\boldsymbol{B}_0(X)-\boldsymbol{B}_0([X,Y])=\hbar\mathscr{M}(X,Y,\boldsymbol{B}_0) \tag{38}$$

where

$$\mathscr{T}(X)=-T^{t}(X) \tag{39}$$

and $\mathscr{M}$ is linear in its arguments, antisymmetric in X and Y. Let $\{X,X\}$ be a symmetric positive definite invariant form on $\mathscr{G}$ (e.g. Tr $\mathscr{T}(X)\mathscr{T}(X)$) which may be used to raise and lower indices. Let

$$X=X^{\alpha}e_{\alpha},\qquad \mathscr{T}(e_{\alpha})=\mathscr{T}_{\alpha}; \tag{40}$$

e_α being a basis in $\mathscr{G}$.

Equation (38) may be rewritten

$$\mathscr{T}_{\alpha}\mathrm{B}_{0\beta}-\mathscr{T}_{\beta}B_{0\alpha}-f^{\gamma}_{\alpha\beta}B_{0\gamma}=\hbar\mathscr{M}_{\alpha\beta}(B_0). \tag{41}$$

Let

$$\{\mathscr{T},\mathscr{T}\}=\mathscr{T}^{\alpha}\mathscr{T}_{\alpha} \tag{42}$$

$$\{\mathscr{T},\mathscr{T}\}B_{0\beta}-\mathscr{T}_{\beta}\mathscr{T}^{\alpha}B_{0\alpha}=\hbar\mathscr{T}^{\alpha}\mathscr{M}_{\alpha\beta}(B_0) \tag{43}$$

where commutation relations have been used together with the antisymmetry of $f_{\alpha\beta\gamma}$ which is due to the invariance of $\{X,X\}$. Positive definiteness of $\{X,X\}$ insures that $\{\mathscr{T},\mathscr{T}\}$ is strictly positive on the b part, so that using invariance once again which insures that

$$[\mathscr{T}_{\beta},\{\mathscr{T},\mathscr{T}\}]=0 \tag{44}$$

we get

$$B^{b}_{0\beta}=\mathscr{T}_{\beta}\frac{\mathscr{T}^{\alpha}}{\{\mathscr{T},\mathscr{T}\}^{b}}B^{b}_{0\alpha}+\hbar\mathscr{T}^{\alpha}\frac{1}{\{\mathscr{T},\mathscr{T}\}^{b}}\mathscr{M}^{b}_{\alpha\beta}(\boldsymbol{B}_0) \tag{45}$$

i.e.

$$\boldsymbol{B}_0^b(X) = \mathscr{T}(X)\hat{\boldsymbol{B}}_0 + \mathrm{O}(\hbar \boldsymbol{B}_0) \tag{46}$$

where $\hat{\boldsymbol{B}}_0$ is linear in $\boldsymbol{B}_0^b$.

Furthermore, for $e_\alpha, e_\beta \in \mathscr{S}$

$$-f_{\alpha\beta}^{\gamma} B_{0\gamma}^{\#} = \hbar \mathscr{M}_{\alpha\beta}^{\#}(B_0) \tag{47}$$

so that, using the non-degeneracy and invariance of the Killing form for $\mathscr{S}$,

$$B_{0\alpha}^{\#} = \mathrm{O}(\hbar \boldsymbol{B}_0), \qquad e_\alpha \in \mathscr{S}. \tag{48}$$

Assuming for a while that also, as we shall demonstrate in a moment

$$B_{0\alpha}^{\#} = \mathrm{O}(\hbar \boldsymbol{B}_0), \qquad e_\alpha \in \mathscr{A}, \tag{49}$$

we have

$$\boldsymbol{B}_0(X) = \mathscr{T}(X)\hat{\boldsymbol{B}}_0 + \mathrm{O}(\hbar \boldsymbol{B}_0). \tag{50}$$

We shall first show that it is possible to choose $\mathscr{L}_{\mathrm{eff}}$ and $\boldsymbol{F}$ in such a way that

$$\hat{\boldsymbol{B}}_0 = \mathscr{T}^\alpha B_{0\alpha}^b = 0. \tag{51}$$

Indeed, expanding both $\mathscr{L}_{\mathrm{eff}}$ and Q in the Ω bases:

$$\begin{aligned} \mathscr{L}_{eff} &= C_{\#}\boldsymbol{\Omega}^{\#} + C_b \boldsymbol{\Omega}^b, \\ Q(X) &= q_{\#}(X)\boldsymbol{\Omega}^{\#} + q_b(X)\boldsymbol{\Omega}^b, \end{aligned} \tag{52}$$

Equation (51) reads

$$\boldsymbol{C}_b + \hbar \frac{T^\alpha}{\{T,T\}_b} q_{b\alpha} = 0, \tag{53}$$

(which one may check to be independent from the choice of $\{X, X\}$), and is soluble for $\boldsymbol{C}_b$ in terms of $\boldsymbol{C}_{\#}$, $\boldsymbol{F}$, $\boldsymbol{b}$. $\boldsymbol{F}$ is now fixed by the requirement that $\mathscr{L}_{\mathrm{eff}}$ has no term linear in $\boldsymbol{\varphi}$, i.e. the linear part of $\boldsymbol{C}_{\#}\boldsymbol{\Omega}^{\#}$ cancels that of $\boldsymbol{C}_b \boldsymbol{\Omega}^b$, because the corresponding condition is of the type

$$\mathring{\mathscr{M}}(F - \mathring{F}) = \mathrm{O}(\hbar) \tag{54}$$

where $\mathring{\mathscr{M}}$ is the mass matrix of $\mathscr{L}$, which is invertible by hypothesis.

Once $\mathscr{L}_{\mathrm{eff}}$ and $\boldsymbol{F}$ are so adjusted, system (50) is of the form

$$\boldsymbol{B}_0 = \mathrm{O}(\hbar \boldsymbol{B}_0) \tag{55}$$

whose solution is

$$\boldsymbol{B}_0 = 0. \tag{56}$$

$\boldsymbol{b}$ which has been so far left arbitrary will eventually be determined together with $\boldsymbol{C}_{\#}$ in terms of the physical parameters which respect the residual symmetry H_0.

Thus, there remains to prove that

$$\boldsymbol{B}_0^{\#} = 0(\hbar \boldsymbol{B}_0) \tag{57}$$

which requires a more detailed analysis of the way in which the theory behaves under dilatations than is provided by power counting which was used up to now. The idea is to order $\Delta^{\#}$ according to decreasing dimension and analyze the various terms successively [8].

For this purpose, let us associate with each integrated monomial $M^{\alpha_I}_{I,\mu(I)}(x)$ in the components of $\boldsymbol{\varphi}$ and their derivatives:

$$\mathscr{M}^{\alpha}_{I,\mu(I)} = \int \mathrm{d}x\, M^{\alpha_I}_{I\mu(I)}(x) = \int \mathrm{d}x \prod_{i \leqslant I} D^{(\mu(i)}\varphi^{\alpha_i}_i(x) \tag{58}$$

where

$$D^{\mu(i)} = \partial_0^{\mu_0(i)}\partial_1^{\mu_1(i)}\partial_2^{\mu_2(i)}\partial_3^{\mu_3(i)}.$$

I is an integer, α_i is a vector index in the field representation space $E \oplus \mathscr{E}$ a dual test operator

$$X^{\alpha_I}_{I,\mu(I)}(q_I) = \int \mathrm{d}k \prod_{i \leqslant I} D^{\mu(i)}_{q_i}\delta_{\varphi_i^{\alpha}(q_i+k)}. \tag{59}$$

We shall write $M^{\alpha_I}_{I,\mu(I)} \sim M^{\alpha'_I}_{I,\mu'(I)}$ if $M^{\alpha_I}_{I,\mu(I)}$ differs from $M^{\alpha'_I}_{I,\mu'(I)}$ by a divergence, or if $\alpha_I = \alpha_{P(I)}$ and $\mu(I) = \mu'(P(I))$ for some permutation P. One has the following orthogonality property:

$$X^{\alpha_I}_{I,\mu(I)}(0)\mathscr{M}^{\alpha'_{I'}}_{I'\mu'(I')} = N_{I,\mu(I)}\delta_{I,I'}\delta_{\mu(I),\mu'(I')}\delta_{\alpha_I,\alpha_I},$$

where $N_{I,\mu(I)}$ is a non-vanishing normalization factor and

$$\delta_{\mu(I),\mu'(I)}\delta_{\alpha_I,\alpha'_I} = \begin{cases} 1 & \text{if } M^{\alpha_I}_{I,\mu(I)} \sim M^{\alpha'_{I'}}_{I,\mu'(I)}, \\ 0 & \text{otherwise.} \end{cases} \tag{61}$$

Furthermore, it is easy to see that

$$X^{\alpha_I}_{I\mu(I)}(q_I)\mathscr{M}^{\alpha'_{I'}}_{I',\mu'(I')} = N_{I,\mu(I)}\delta_{I,I'}\delta_{\mu(I),\mu'(I')}\delta_{\alpha_I,\alpha'_I} \tag{62}$$

independent of q_I provided that

$$\dim M^{\alpha'_I}_{I'\mu'(I')} \leqslant \dim M^{\alpha_I}_{I,\mu(I)}. \tag{63}$$

Let $\mathscr{G}$ act on $\{X^{\alpha_I}_{I,\mu(I)}(q_I)\}$ according to:

$$X^{\alpha_I}_{I,\mu(I)}(q_I)\, W^h_\alpha \Gamma = (-\,T^h_\alpha X)^{\alpha_I}_{I,\mu(I)}(q_I)\Gamma \tag{64}$$

where

$$W^H_\alpha = -\int \mathrm{d}x\,\{\varphi \mathring{t}^T_\alpha \delta_\varphi + \beta \mathring{\theta}^T_\alpha \delta_\beta\}. \tag{65}$$

T^h_α is the homogeneous part of T_α obtained by putting $\boldsymbol{F} = \boldsymbol{b} = 0$ in Equation (21).
Let

$$X^{i\#}_{i,\mu(I)}(q_i) = \sum_{\alpha_I} C^{i\#\alpha_I}_{I,\mu,(I)} X^{\alpha_I}_{I,\mu(I)}(q_I) \tag{66}$$

be an element of a basis of T^h_a-invariant test operators corresponding to dimension four local polynomials,

$$\begin{aligned} X^{i\#}_{I,\mu(I)}(q_I) W_\alpha \Gamma &= -\,X^{i\#}_{I,\mu(I)}(q_I)\Delta_\alpha \Gamma \\ &= -\,X^{i\#}_{I,\mu(I)}(q_I)\int \mathrm{d}x\,\{\boldsymbol{F}\mathring{t}^T_\alpha \delta_\varphi + \boldsymbol{b}\mathring{\theta}^T_\alpha \delta_\beta\}(x)\Gamma \qquad (67) \\ &= -\,X^{i\#}_{I,\mu(I)}(q_I)\Delta_\alpha + \mathrm{O}(\hbar \boldsymbol{B}^\#_0) \\ &= -\,X^{i\#}_{I\mu(I)}(0)\Delta_\alpha + \mathrm{O}(\hbar \boldsymbol{B}^\#_0). \end{aligned}$$

The first line is a consequence of the anomalous Ward identity, the second one makes use of the T^h_α-invariance of $X^{i\#}_{I,\mu(I)}(q_I)$, the last one follows from Equations (62, 63).

Now for q_I large in the Euclidean region power counting insures that the expression

$$X^{i\#}_{I,\mu(I)}(q_I)\int \mathrm{d}x\,\{F\mathring{t}^T_\alpha \delta_\varphi + b\mathring{\theta}^T_\alpha \delta_\beta\}(x)\Gamma$$

is asymptotically negligible, since dim $\beta > 0$. Hence

$$X^{i\#}_{I,\mu(I)}(0)\Delta_\alpha = \mathrm{O}(\hbar \boldsymbol{B}^\#_0), \qquad \forall i. \tag{68}$$

Thus the dimension four part of $\Delta^\#_\alpha$ is O $(\hbar B^\#_0)$. The analysis of lower dimension terms of $\Delta^\#_\alpha$ is slightly more sophisticated and is given in Appendix C.

The analysis completes the proof of Equation (57)

$$\boldsymbol{B}_0^{\#} = 0(\hbar B_0^{\#})$$

At this point, we have completed the construction of a Green functional fulfilling the Ward identity (Equation 20).

The free power series parameters $\boldsymbol{C}_{\#}$ and $\boldsymbol{b}$ can then be used to fulfill the three normalization conditions which allow a particle interpretation of the theory under the assumptions stated in section A, namely the existence of an invertible transformation from $\mathring{\boldsymbol{C}}_{\#}, \mathring{\boldsymbol{b}}$ to the physical parameters involved in the normalization conditions.

3. The Local Ward Identity ('Current Algebra')

3.1. *The Classical Theory* [12]

Given a Lagrangian $\mathscr{L}(\boldsymbol{\varphi},\boldsymbol{\beta})$ invariant under the global transformation Equation (8), it is easy to introduce a classical gauge field $\boldsymbol{a}_\mu$ of dimension 1, and construct a Lagrangian $\mathscr{L}(\boldsymbol{\varphi},\boldsymbol{\beta},\boldsymbol{a}_\mu)$ invariant under the local gauge transformation:

$$\begin{aligned} \delta_\omega \boldsymbol{\varphi}(x) &= -\mathring{t}(\omega_x)(\boldsymbol{\varphi}_x + \mathring{F}), \\ \delta_\omega \boldsymbol{\beta}(x) &= -\mathring{\theta}(\omega_x)(\boldsymbol{\beta}_x + \mathring{b}), \\ \delta_\omega \boldsymbol{a}_\mu(x) &= \partial_\mu \omega(x) - [\omega_x, a_{\mu x}], \end{aligned} \tag{69}$$

where we have considered $a_\mu(x)$ as well as ω_x as elements of $\mathscr{G}$: it is enough to replace the derivatives occurring in $\mathscr{L}(\boldsymbol{\varphi},\boldsymbol{\beta})$ by covariant derivatives:

$$\begin{aligned} \partial_\mu \varphi &\to \mathrm{D}_\mu \boldsymbol{\varphi} = \partial_\mu \boldsymbol{\varphi} + \mathring{t}(a_\mu)(\varphi + \mathring{F}), \\ \partial_\mu \beta &\to \Delta_\mu \boldsymbol{\beta} = \partial_\mu \boldsymbol{\beta} + \mathring{\theta}(a_\mu)(\boldsymbol{\beta} + \mathring{b}), \end{aligned} \tag{70}$$

and to include gauge invariant terms constructed with a_μ through the antisymmetric covariant tensor

$$G_{\mu\nu} = \partial_\mu a_\nu - \partial_\nu a_\mu - [a_\mu, a_\nu]. \tag{71}$$

The local Ward identity which expresses the invariance of $\mathscr{L}(\boldsymbol{\varphi},\boldsymbol{\beta},\boldsymbol{a}_\mu)$ under the local gauge transformation Equation (69) is:

$$\begin{aligned} \mathscr{W}(\omega)\mathring{\Gamma}(\boldsymbol{\varphi},\boldsymbol{\beta},\boldsymbol{a}_\mu) = \int \mathrm{d}x \{ & \delta_{a_\mu}\mathring{\Gamma}(\partial_\mu\omega + [\omega, a_\mu]) \\ & - \delta_\varphi \mathring{\Gamma}\mathring{t}(\omega)(\varphi + \mathring{F}) - \delta_\beta \mathring{\Gamma}\mathring{\theta}(\omega)(\boldsymbol{\beta} + \mathring{\boldsymbol{b}})\}(x). \end{aligned} \tag{72}$$

The relation of the integrated Ward identity for $\mathscr{G}$ to the local Ward identity for the associated gauge group is:

$$W(\omega) = \mathscr{W}(\omega) \qquad \text{for } \omega \text{ space-time independent.} \tag{73}$$

Note that we have kept as an external field the gauge field a_μ which globally transforms under the adjoint representation $\{f_\alpha\}$ of $\mathscr{G}$; this does not spoil any of the conclusions of the previous section since nowhere the type of Lorentz covariance of the fields was used.

3.2. *Radiative Corrections*

We are going to show in general an anomalous local Ward identity:

$$\mathscr{W}_\alpha(x)\Gamma = G_\alpha(x) \tag{74}$$

where G_α is a dimension four polynomial in the classical gauge field a_μ and its derivatives.

Let us first assume one has proved an integrated Ward identity in the presence of the gauge field:

$$\int \mathrm{d}x \{ \boldsymbol{a}_\mu f_\alpha^T \delta_{a_\mu} + (\varphi + F)\mathring{t}_\alpha^T \delta_\varphi + (\beta + \mathring{b})\mathring{\theta}_\alpha^T \delta_\beta \}(x)\Gamma = 0. \tag{75}$$

Performing a local gauge transformation yields

$$\mathscr{W}_\alpha(x)\Gamma = \mathscr{K}_\alpha(x)\Gamma \tag{76}$$

where $\mathscr{K}_\alpha(x)$ is a dimension four insertion.

It follows from the validity of the integrated Ward identity that

$$\int \mathscr{K}_\alpha(x)\,\mathrm{d}x = 0 \tag{77}$$

hence

$$\mathscr{K}_\alpha(x) = \partial_\mu \mathscr{K}_\alpha^\mu(x) \tag{78}$$

where $\mathscr{K}_\alpha^\mu$ is a dimension three insertion.

Now $\mathscr{K}_\alpha$ fulfills the compatibility condition (10)

$$\begin{aligned}[\delta_{\omega_\alpha(x)}\mathscr{K}_\beta(y) - \delta_{\omega_\beta(y)}\mathscr{K}_\alpha(x) - f_{\alpha\beta}^\gamma \delta(x-y)\mathscr{K}_\gamma(y)]\Gamma \\ = \mathrm{O}(\hbar\mathscr{K})\Gamma\end{aligned} \tag{79}$$

where O $(\hbar\mathscr{K})$ lumps together the radiative corrections in the compatibility condition

$$\{\mathscr{W}_\alpha(x)\mathscr{K}_\beta(y) - \mathscr{W}_\beta(y)\mathscr{K}_\alpha(x) - f_{\alpha\beta}^{\gamma}\delta(x-y)\mathscr{K}_\gamma(y)\}\Gamma = 0. \tag{80}$$

Clearly, the solution of Equation (79) is provided by a system

$$\mathscr{K}_\alpha(x) = K_\alpha(x) + \mathrm{O}(\hbar\mathscr{K}). \tag{81}$$

where $K_\alpha(x)$ is the most general solution of the classical equation [10]

$$\delta_{\omega_\alpha(x)} K_\beta(y) - \delta_{\omega_\beta(y)} K_\alpha(x) - f_{\alpha\beta}^{\gamma}\delta(x-y)K_\gamma(y) = 0. \tag{82}$$

As usual this writing summarizes the result of testing Equation (80) by a basis of field monomials at zero field, zero momentum. Furthermore, since $\mathscr{K}_\alpha$ is a divergence, so is K_α. It is shown in Appendix D that K_α is of the form

$$K_\alpha(x) = \delta_{\omega_\alpha(x)} \int \mathrm{d}x\, K(x) + g_\alpha(x) \tag{83}$$

where $g_\alpha(x)$ does not depend on the quantized field and is not the gauge variation of any local functional of dimension less than or equal to four. This last term leads to the anomaly

$$g_\alpha(x)\Gamma = G_\alpha(x) \tag{84}$$

and does not contribute any radiative correction to the right-hand side of Equation (80) [12]. Now, recall that

$$\mathscr{K}_\alpha(x) = \delta_{\omega_\alpha(x)} \int \mathrm{d}y\, \mathscr{L}_{eff}(y) + \hbar\mathscr{Q}_\alpha(x) = K_\alpha(x) + \mathrm{O}(\hbar(\mathscr{K} - g)). \tag{85}$$

Let us choose a suitable basis:

$$\mathscr{L}_{\mathrm{eff}} = C_{\#}\mathscr{L}^{\#} + C_b^{(1)}\mathscr{L}_1^b + C_b^{(2)}\mathscr{L}_2^b \tag{86}$$

where

$$\delta_{\omega^\alpha(x)} \int \mathrm{d}x\, \mathscr{L}^{\#}_{(x)} = 0, \qquad \delta_{\omega^\alpha(x)} \int \mathrm{d}x\, \mathscr{L}^b_{1,2}(x) \neq 0,$$

$$\int \mathrm{d}x\, \delta_{\omega^\alpha(x)} \int \mathrm{d}y\, \mathscr{L}_1^b(y) = 0, \qquad \int \mathrm{d}x\, \delta_{\omega^\alpha(x)} \int \mathrm{d}y\, \mathscr{L}_2^b(y) \neq 0. \tag{87}$$

since K_α is a linear combination of terms from $\mathscr{L}_1^b$ one may write:

$$\hbar \mathscr{Q}_\alpha(x) = \delta_{\omega^\alpha(x)} \int \mathrm{d}y \, \boldsymbol{q}_1^b \mathscr{L}_1^b(y) +$$

$$+ g_\alpha(x) - \delta_{\omega_\alpha(x)} \int C_2^b \mathscr{L}_2^b(y) \mathrm{d}y + \mathrm{O}(\hbar(\mathscr{K} - g)) \qquad (88)$$

for some $\boldsymbol{q}_1^b$. Now, testing $\mathscr{K}_\alpha$ with monomials from $\delta_{\omega_\alpha(x)} \int \mathrm{d}y \mathscr{L}_2^b(y)$ gives zero by hypothesis. Choosing $\boldsymbol{C}_1^b$ such that

$$\boldsymbol{C}_1^b + \boldsymbol{q}_1^b = 0 \qquad (89)$$

which is possible [3] since $\boldsymbol{q}_1^b$ is $\hbar$ times a formal power series in the C's and $\hbar$, Equation (85) reduces to

$$\mathscr{K}_\alpha(x) - g_\alpha(x) = \mathrm{O}(\hbar(\mathscr{K} - g)) \qquad (90)$$

whose unique solution is

$$\mathscr{K}_\alpha(x) = g_\alpha(x). \qquad (91)$$

At this point, then,

$$\mathscr{K}_\alpha(x)\Gamma = g_\alpha(x)\Gamma = G_\alpha(x) \qquad (92)$$

namely, we have proved the anomalous local Ward identity

$$\mathscr{W}_\alpha(x)\Gamma = G_\alpha(x). \qquad (93)$$

As shown in Appendix C, G_α can always be chosen of the form:

$$G_\alpha(x) = \partial^\mu K_{\mu\alpha}(x) \qquad (94)$$

with

$$K_\mu^\alpha(x) = \varepsilon_{\mu\nu\rho\sigma} [D^{\alpha\beta\gamma}(\partial^\nu a_\beta^\rho) a_\gamma^\sigma + F^{\alpha\beta\gamma\delta} a_\beta^\gamma a_\gamma^\rho a_\delta^\sigma](x) \qquad (95)$$

where

$$F^{\alpha\beta\gamma\delta} = \tfrac{1}{12}[D^{\alpha\beta\lambda} f_\lambda^{\gamma\delta} + D^{\alpha\gamma\lambda} f_\lambda^{\delta\beta} + D^{\alpha\delta\lambda} f_\lambda^{\beta\gamma}] \qquad (96)$$

and $D^{\alpha\beta\gamma}$ is a symmetric covariant tensor with three indices in the adjoint representation of $\mathscr{G}$, which parametrizes the general form of the Adler–Bardeen anomaly.

Conclusion

We have completed a number of points of Symanzik's program on the renormalization of renormalizable theories with symmetry breaking.

For models without massless particles, we have been able to deal with an arbitrary compact internal symmetry Lie group, and prove the integrated Ward identities characteristic of a superrenormalizable breaking with given covariance. The corresponding anomalous local Ward identity – the perturbative expression of current algebra – is then proved in all generality and a compact formula exhibited for the corresponding Adler–Bardeen anomaly.

Our perturbative treatment fails if power counting mixes with geometry to produce e.g. mass rules, since in this case a particle interpretation of the theory is no longer possible. The breakdown of our treatment generated by this phenomenon is quite more dramatic in models involving massless particles. This happens in particular if, due to tree approximation mass rules there are more massless scalar fields than Goldstone bosons (pseudo-Goldstone bosons [13]). In this case, even the construction of a Green function theory needs a deep modification of the perturbative scheme [14].

Another limiting case which is worth mentioning occurs when the breaking has dimension four if the most general invariant Lagrangian formed with the quantized field and the direction $\boldsymbol{b}$ which characterizes the breaking is not the most general Lagrangian invariant under the residual symmetry group H_b.

Acknowledgments

One of us (R.S.) wishes to thank K. Symanzik for the intense and informative exchange of correspondence on the subject of this paper, as well as enlightening discussions during the year 1970.

We wish to thank our colleagues of the Service de Physique Théorique, Saclay, where most of this work was performed, for their kind hospitality.

Appendix A

A classical action is viewed as a local functional whose argument is indefinitely differentiable with fast decrease. In the present case

$$\mathring{\Gamma}(\varphi,\boldsymbol{\beta}) = \int \mathrm{d}x\, \mathscr{L}(\varphi,\boldsymbol{\beta})(x) \tag{A.1}$$

where $\mathscr{L}$ is a classical Lagrangian density, without a constant term, defined up to a divergence. We shall limit ourselves to renormalizable Lagrangians, according to the conventional power counting theory through which fields $\boldsymbol{\varphi}$ are assigned dimensions connected with the structure of the quadratic part of $\mathscr{L}$, the dimension of $\boldsymbol{\beta}$ being a priori given, namely Lagrangians of dimension smaller than or equal to four. If dim $\boldsymbol{\beta} < 0$, renormalizable Lagrangians are polynomials. Assuming that $\mathscr{L}$ has no term linear in $\boldsymbol{\varphi}$, which can be achieved by performing a field translation, we see that the integrated Ward identity Equation (7) is only meaningful if

$$\left.\frac{\delta\mathring{\Gamma}}{\delta\beta}\right|_{\varphi=\beta=0}\mathring{\theta}(x)\mathring{\boldsymbol{b}}=0 \tag{A.2}$$

which we shall assume. The field translation parameter appropriate to get rid of the term linear in $\boldsymbol{\varphi}$ from a Lagrangian which is an invariant formed with $\boldsymbol{\varphi}$ and $\boldsymbol{\beta}+\mathring{\boldsymbol{b}}$ of course depends on $\mathring{\boldsymbol{b}}$. In terms of untranslated variables, $\mathring{\Gamma}$ $(\boldsymbol{\varphi},\mathbf{0})$ is an invariant polynomial formed with $\boldsymbol{\varphi}$ and $\mathring{\boldsymbol{b}}$: $\mathscr{T}(\boldsymbol{\varphi},\mathring{\boldsymbol{b}})$. Hence $\mathring{\boldsymbol{F}}$ is defined by

$$\frac{\partial\mathscr{T}}{\partial\boldsymbol{\varphi}}(\mathring{\boldsymbol{F}},\mathring{\boldsymbol{b}})=0 \tag{A.3}$$

and the Ward identity implies

$$\frac{\partial\mathscr{T}}{\partial\boldsymbol{\varphi}}\mathring{t}(X)\boldsymbol{\varphi}+\frac{\partial\mathscr{T}}{\partial\mathring{\boldsymbol{b}}}\mathring{\theta}(X)\mathring{b}=0. \tag{A.4}$$

Differentiating Equation (A.3) with respect to $\mathring{\boldsymbol{b}}$ and Equation (A.4) with respect to $\boldsymbol{\varphi}$ at $\boldsymbol{\varphi}=\mathring{F}$ yields

$$\frac{\partial^2\mathscr{T}}{\partial\boldsymbol{\varphi}\,\partial\boldsymbol{\varphi}}\left[\mathring{t}(X)\mathring{F}-\frac{\partial\mathring{F}}{\partial\mathring{\boldsymbol{b}}}\mathring{\theta}(X)\mathring{\boldsymbol{b}}\right]=0 \tag{A.5}$$

which, under the assumption that the mass matrix

$$\mathring{\mathscr{M}}=\left.\frac{\partial^2\mathscr{T}}{\partial\varphi\,\partial\varphi}\right|_{F,B} \tag{A.6}$$

is non-degenerate implies that $\mathring{\boldsymbol{F}}$ is a covariant function of $\mathring{\boldsymbol{b}}$:

$$\mathring{t}(X)\mathring{\boldsymbol{F}}=\frac{\partial\mathring{F}}{\partial\mathring{b}}\mathring{\theta}(X)\mathring{\boldsymbol{b}}. \tag{A.7}$$

Similarly, the other coefficients of $\mathscr{L}$ are covariant functions of $\mathring{b}$.

The second part of Equation (A.2) is then a consequence of these covariance properties.

Appendix B

This appendix is devoted to the reduction of the Ward identity to canonical form.

B.1. *The Tree Approximation*

We first show that the representations

$$\begin{matrix} X \to t(X), \\ X \to \theta(X), \end{matrix} \qquad X \in \mathscr{G} = \mathscr{A} + \mathscr{S}, \tag{B.1}$$

are fully reducible. This is automatic for $X \in \mathscr{S}$, *the semi-simple part of* $\mathscr{G}$. For $X \in \mathscr{A}$, the Abelian part of $\mathscr{G}$, this is a consequence of the assumption that the kinetic part of $\mathscr{L}$ is Hermitian non-degenerate, which insures that $X \to t(X)$ is fully reducible. Then, the cocycle condition

$$t(X)\boldsymbol{F}(Y) - t(Y)\boldsymbol{F}(X) - F([X,Y]) = 0 \tag{B.2}$$

can be solved as follows (cf. Equations (40)–25)): let $\{X,X\}$ be a non-degenerate invariant quadratic form on $\mathscr{G}$, used to raise and lower indices. Reducing Equation (B.2) to components:

$$t_\alpha \boldsymbol{F}_\beta - t_\beta \boldsymbol{F}_\alpha - f^\gamma_{\alpha\beta} \boldsymbol{F}_\gamma = 0 \tag{B.3}$$

yields

$$\{t,t\}\boldsymbol{F}_\beta \equiv (t^\alpha t_\alpha)\boldsymbol{F}_\beta = t_\beta(t^\alpha \boldsymbol{F}_\alpha). \tag{B.4}$$

Thus restricting $\boldsymbol{F}_\alpha$ to its non invariant part F^b_α:

$$\boldsymbol{F}^b_\alpha = \frac{1}{\{t,t\}} t_\alpha(t^\beta \boldsymbol{F}_\beta) = t_\alpha \boldsymbol{F}^b. \tag{B.5}$$

Similarly, using the non-degeneracy of the Killing form, of $\mathscr{S}$, the semi-simple part of $\mathscr{G}$, we get also

$$\boldsymbol{F}^\#_\alpha = t_\alpha \boldsymbol{F}^\# \qquad e_\alpha \in \mathscr{S} \tag{B.6}$$

so that only

$F^\#_\alpha, e_\alpha \in \mathscr{A}$ (the Abelian part of $\mathscr{G}$) is left undetermined.

Performing the field translation $\varphi \to \varphi - \boldsymbol{F}$ one has thus to find a polynomial Lagrangian $\mathscr{L}$ invariant under

$$\begin{aligned} \delta_X \boldsymbol{\varphi}^b &= -t(X)\varphi^b, \\ \delta_X \boldsymbol{\varphi}^\#(X) &= F^\#(X) \\ \delta_X \boldsymbol{\beta} &= -\theta(X)(\boldsymbol{\beta} + \boldsymbol{b}). \end{aligned} \tag{B.7}$$

In the transformation law for β, we have assumed that

$$\boldsymbol{b}(X) = \theta(X)\boldsymbol{b}$$

which conforms with the idea of symmetry breaking. If $\boldsymbol{\varphi}^b$ is to couple polynomially with $\boldsymbol{\beta}$, since the tensorial powers of $X \to t(X)$ are fully reducible, so is $X \to \theta(X)$. Furthermore θ is assumed not to contain any component of the identity. It is then easy to see that, due to the polynomial character of $\mathscr{L}$, those components of $\boldsymbol{\varphi}^{\#}$ for which $F_{\#}(X) \neq 0$ *do not couple.*

B.2. Radiative Corrections

We shall first show that the representation property

$$[t(X), t(Y)] = t([X, Y]) \tag{B.8}$$

with

$$t(X) = \mathring{t}(X) + \mathrm{O}(\hbar) \tag{B.9}$$

where $t(X)$ is a formal power series in $\hbar$, implies that

$$t(X) = Z^{-1}\mathring{t}(X)Z \tag{B.10}$$

for some formal power series Z:

$$Z = \mathbb{1} + \mathrm{O}(\hbar) \tag{B.11}$$

Let first $X \in \mathscr{S}$, the semi-simple part of $\mathscr{G}$, let

$$t(X) = \sum_0^\infty t_n(X) \tag{B.12}$$

$$Z = \sum_0^\infty Z_n \tag{B.13}$$

be the formal power series for $t(X)$, Z, respectively. First we may choose

$$Z_0 = \mathbb{1} \tag{B.14}$$

thanks to a symmetric wave function renormalization. The first term in the expansion of Equation (B.8) reads

$$[\mathring{t}(X), t_1(X)] - [\mathring{t}(Y), t_1(X)] - t_1([X, Y]) = 0 \tag{B.15}$$

which is a cohomology equation; hence, due to the semi-simplicity of $\mathscr{G}$,

$$t_1(X) = [\mathring{t}(X), Z_1] \tag{B.16}$$

for some Z_1.

Let us now assume that

$$t(X) = (Z^{(n-1)})^{-1}\mathring{t}(X)Z^{(n-1)} + \tau_n(X) \tag{B.17}$$

with

$$\tau_n(X) = O(\hbar^n) \tag{B.18}$$

which is true for $n = 1$ with

$$Z^{(1)} = \mathbb{1} + Z_1 \tag{B.19}$$

The term of Equation (B.8) of lowest non-vanishing order reads:

$$\begin{aligned}&[\mathring{t}(X), \{Z^{(n-1)}\tau_n(Y)(Z^{(n-1)})^{-1}\}_n]\\ &\quad - [\mathring{t}(Y), \{Z^{(n-1)}\tau_n(X)(Z^{(n-1)})^{-1}\}_n]\\ &\quad - \{Z^{(n-1)}\tau_n([X, Y])(Z^{(n-1)})^{-1}\}_n = 0.\end{aligned} \tag{B.20}$$

Hence, using the same cohomological result:

$$(Z^{(n-1)}\tau_n(X)(Z^{(n-1)})^{-1})_n = [\mathring{t}(X), Z_n] \tag{B.21}$$

for some Z_n, so that

$$\tau_n(X) = ((Z^{(n-1)})^{-1}[\mathring{t}(X), Z_n]Z^{(n-1)})_n \tag{B.22}$$

Thus

$$t_n(X) = ((Z^{(n)})^{-1}\mathring{t}(X)Z^{(n)})_n \tag{B.23}$$

with

$$Z^{(n)} = Z^{(n-1)} + Z_n. \tag{B.24}$$

As a conclusion, we may choose $t(X) = \mathring{t}(X), X \in \mathscr{S}$ up to a field renormalization.

Now, for $X \in \mathscr{A}$, the Abelian part of $\mathscr{G}$, any anomaly in the Ward identity can be considered as a breaking of the canonical Ward identity through an insertion which, up to $O(\Delta^2)$, is Abelian invariant, and thus, cannot occur as a consequence of the argument at the end of Section 2 for the anomaly in $t(X)$, and of the argument in Appendix C for the anomaly in $F(X)$.

Appendix C

Elimination of Soft Invariant Anomalies from the Integrated Ward Identity

Once the dimension four anomalies have been eliminated as indicated in the text, one has a Ward identity of the form:

$$W(X)\Gamma = \boldsymbol{C}_{\#}(X)\boldsymbol{\Delta}_{\#}\Gamma + \mathrm{O}(\hbar C_{\#}) \tag{C.1}$$

where the components of $\boldsymbol{\Delta}_{\#}$ are invariant soft. Introducing a set of invariant external fields $\boldsymbol{\eta}$ with dimensions strictly between 0 and 4, repeating the argument in the text leads to a Ward identity

$$W(X)\Gamma^{(\eta)} = \tilde{\boldsymbol{C}}_{\#}(X)\boldsymbol{\Delta}^{(\eta)}_{\#}\Gamma^{(\eta)} + \mathrm{O}(\hbar\tilde{C}_{\#}) \tag{C.2}$$

where $\Delta^{(\eta)}_{\#}$ are soft, η-dependent invariant insertions and $\Gamma^{(\eta)}$, the new η-dependent vertex functional.

Let now λ be any parameter of the theory, and let

$$D_\lambda = \partial_\lambda - \int \mathrm{d}x(\partial_\lambda\mathscr{F}, \delta_{\phi(x)}) \tag{C.3}$$

where

$$\mathscr{F} = (\boldsymbol{F};\boldsymbol{b}), \qquad \phi = (\boldsymbol{\varphi};\boldsymbol{\beta}). \tag{C.4}$$

Then

$$[D_\lambda, W(X)] = 0 \tag{C.5}$$

In particular, let ∂_μ be the operator which scales all the parameters of the theory according to their dimensions, (the first term in the Callan–Symanzik equation), and D_μ the associated invariant operator (cf. Equation (C.3)). Since in the tree approximation $D_\mu\mathscr{L}^{(\eta)}$ is invariant, soft, namely of the form

$$D_\mu\mathscr{L}^{(\eta)} = \boldsymbol{m}\int \mathrm{d}x\,\delta_{\eta(x)}\mathscr{L}^{(\eta)} \tag{C.6}$$

one deduces, using the Zimmermann identities:

$$\begin{aligned} D_\mu\Gamma^{(\eta)} &= m\int \mathrm{d}x\delta_{\eta(x)}\Gamma^{(\eta)} + \hbar\{\boldsymbol{d}_{\#}\boldsymbol{\Delta}^{(h)}_{\#}\Gamma^{(\eta)} + \boldsymbol{d}_b\boldsymbol{\Delta}_b\Gamma^{(\eta)}\} \\ &= m\int \mathrm{d}x\,\delta_{\eta(x)}\Gamma^{(\eta)} + \hbar\sum_\lambda e_\lambda D_\lambda\Gamma^{(\eta)} + \hbar\,\boldsymbol{d}_b\boldsymbol{\Delta}^{(h)}_b\Gamma^{(\eta)} \end{aligned} \tag{C.7}$$

where $\Delta_{\#}^{(h)}$, respectively $\Delta_b^{(h)}$ are invariant, respectively non-invariant, hard insertions.

Hence, combining Equations (C.2) and (C.7),

$$W(X)\hbar\, \boldsymbol{d}_b \Delta_b^{(h)} \Gamma^{(\eta)} = \left(-\boldsymbol{m} \int \mathrm{d}x\, \delta_{\eta(x)} + D_\mu \right) \tilde{C}_{\#}(X) \Delta^{(\eta)} + \mathrm{O}(\hbar \tilde{C}_{\#})$$
$$= \hbar\, \boldsymbol{d}_b\, T(X) \Delta_b^{(h)} + \mathrm{O}(\hbar^2 d_b). \tag{C.8}$$

Looking at the dimension four terms yields

$$\hbar d_b = \mathrm{O}(\hbar \tilde{C}_{\#}). \tag{C.9}$$

Looking at dimension three terms yields

$$\hbar\, \boldsymbol{d}_b\, T(X) \Delta_b^{(h)}|_3 = D_\mu \tilde{C}_{\#}(X) \Delta_{\#}^{(\eta)}|_3 + \mathrm{O}(\hbar \tilde{C}_{\#})$$
$$= \tilde{C}_{\#}(X) \Delta_{\#}^{(\eta)}|_3 + \mathrm{O}(\hbar \tilde{C}_{\#}). \tag{C.10}$$

Thus,

$$\hbar d_{b,3} = \mathrm{O}(\hbar \tilde{C}_{\#}). \tag{C.11}$$

Hence

$$\tilde{C}_{\#,3}(X) = \mathrm{O}(\hbar \tilde{C}_{\#}). \tag{C.12}$$

Iterating this procedure down to lower dimensions yields the desired result

$$\tilde{C}_{\#}(X) = \mathrm{O}(\hbar \tilde{C}_{\#}) \tag{C.13}$$

Hence

$$\tilde{C}_{\#}(X) = 0. \tag{C.14}$$

Appendix D

Cohomology of the Gauge Lie Algebra

We shall analyze the structure of

$$K_\alpha(x) = \partial_\mu K_\alpha^\mu(x) \tag{D.1}$$

solution of the cocycle condition (Cf. Equation (82))

$$\delta_{\omega^\alpha(x)} K_\beta(y) - \delta_{\omega^\beta(y)} K_\alpha(x) - f_{\alpha\beta}^\gamma \delta(x-y) K_\gamma(y) = 0 \tag{D.2}$$

(the Wess–Zumino consistency condition [10]).

Integrating first Equation (D.2) over x shows that $\partial_\mu K^\mu_\alpha(x)$ transforms like the regular representation under global transformations. Indeed, upon x-integration Equation (D.2) reduces to

$$T_\alpha K_\beta(y) - f^\gamma_{\alpha\beta} K_\gamma(y) = 0 \tag{D.3}$$

where

$$T_\alpha = \int \mathrm{d}x \, \delta_{\omega^\alpha(x)} \tag{D.4}$$

is the infinitesimal generator of the global transformation.

Now, K^μ_α can be decomposed in a unique way along the null space N_∂ of ∂:

$$\partial: K^\mu_\alpha \to \partial_\mu K^\mu_\alpha \tag{D.5}$$

and its complement N^c_∂, both of which are invariant under G since G is compact and its action commutes with ∂_μ:

$$K^\mu_\alpha = K^\mu_\alpha|_{N_\partial} + K^\mu_\alpha|_{N^c_\partial}. \tag{D.6}$$

Since Equation (D.3) implies

$$T_\alpha K^\mu_\beta|_{N_\partial} - f^\gamma_{\alpha\beta} K^\mu_\gamma|_{N^c_\partial} \in N \tag{D.7}$$

the stability of N^c_∂ implies

$$T_\alpha K^\mu_\beta|_{N^c_\partial} - f^\gamma_{\alpha\beta} K^\mu_\gamma|_{N^c_\partial} = 0. \tag{D.8}$$

Thus, one may choose K^α_μ to transform according to the regular representation.

We shall now expand K^α_μ in increasing powers of $\{a^\alpha_\mu\}$. Let $\overset{\circ}{K}{}^\mu_\alpha$ be the term of K^μ_α independent of $\{a^\alpha_\mu\}$. We may write

$$\partial_\mu \overset{\circ}{K}{}^\mu_\alpha = \delta_{\omega^\alpha(x)} \int a_{\beta\mu}(y) \overset{\circ}{K}{}^{\beta\mu}(y) \mathrm{d}y + \partial_\mu L^\mu_\alpha \tag{D.9}$$

where L^μ_α is linear in $\{a^\alpha_\mu\}$.

Let $\overset{1}{K}{}^\mu_\alpha$ be the term of $K^\mu_\alpha + L^\mu_\alpha$ linear in $\{a^\alpha_\mu\}$, one may similarly write

$$\partial_\mu \overset{1}{K}{}^\mu_\alpha = \delta_{\omega^\alpha(x)} \tfrac{1}{2} \int a_{\beta\mu}(y) \overset{1}{K}{}^{\beta\mu}(y) \mathrm{d}y + \partial_\mu Q^\mu_\alpha(x) \tag{D.10}$$

where $Q^\alpha_\mu(x)$ is now quadratic in $\{a^\alpha_\mu\}$, provided that

$$\overset{1}{K}{}^{\mu\nu}_{\alpha\beta}(x, y) \equiv \delta_{a^\alpha_\mu(x)} \overset{1}{K}{}^\nu_\beta(y) - \delta_{a^\beta_\nu(y)} \overset{1}{K}{}^\mu_\alpha(x) = 0. \tag{D.11}$$

Now, from the $\{a^\alpha_\mu\}$ independent part of Equation (D.2), we get

$$\partial^x_\mu \partial^y_\nu \overset{1}{K}{}^{\mu\nu}_{\alpha\beta}(x,y) = 0. \tag{D.12}$$

The only possible form of $\overset{1}{K}{}^{\mu\nu}_{\alpha\beta}(x,y)$ consistent with power counting, symmetry and condition (D.12) is:

$$\overset{1}{K}{}^{\mu\nu}_{\alpha\beta}(x,y) = (\Box g^{\mu\nu} - \partial^\mu\partial^\nu)\delta(x-y)\Gamma_{\underline{\alpha\beta}} \tag{D.13}$$

where $\Gamma_{\underline{\alpha\beta}}$ is antisymmetric invariant. Then.

$$\overset{1}{\tilde{K}}{}^\mu_\alpha(x) = \overset{1}{K}{}^\mu_\alpha(x) + \tfrac{1}{2}[\partial^\mu(\partial a)^\beta - \Box a^{\beta\mu}]\Gamma_{\underline{\alpha\beta}} \tag{D.14}$$

does fulfill Equation (D.11) and

$$\partial_\mu \tilde{K}^\mu_\alpha(x) = \partial_\mu \overset{1}{K}{}^\mu_\alpha(x). \tag{D.15}$$

Thus,

$$\partial_\mu \overset{1}{K}{}^\mu_\alpha(x) = \delta_{\omega\alpha(x)} \tfrac{1}{2} \int dy\, a_{\beta\mu}(y) \overset{1}{\tilde{K}}{}^{\beta\mu}(y) + \partial_\mu Q^\mu_\alpha \tag{D.16}$$

for some Q^μ_α quadratic in $\{a^\mu_\alpha\}$.

Similarly we proceed considering the terms $\overset{2}{K}{}^\mu_\alpha$ of $K^\mu_\alpha + Q^\mu_\alpha$ quadratic in $\{a_{\beta\mu}\}$. The most general form of $\overset{2}{K}{}^\mu_\alpha$ such that

$$\partial_\mu K^\mu_\alpha \neq 0$$

is:

$$\begin{aligned}\overset{2}{K}{}^\mu_\alpha = {} & \partial_\nu (a^{\beta\mu} a^{\gamma\nu})\varepsilon^{(1)}_{\alpha\beta\gamma} + (a^{\beta\mu} \overleftrightarrow{\partial}_\nu a^{\gamma\nu})\varepsilon^{(2)}_{\alpha\beta\gamma} \\ & + a^{\beta\nu}\partial^\mu a^{\gamma\nu}\varepsilon^{(3)}_{\alpha\beta\gamma} + D_{\alpha\beta\gamma}\varepsilon^{\mu\nu\rho\sigma}\partial_\nu a^\beta_\rho a^\gamma_\rho\end{aligned} \tag{D.17}$$

in which

$$\begin{aligned} & D_{\alpha\beta\gamma} = D_{\alpha\gamma\beta}, \\ & \varepsilon^{(1)}_{\alpha\beta\gamma} = \varepsilon^{(1)}_{\alpha\gamma\beta}. \end{aligned} \tag{D.18}$$

The part of the cocycle condition on $\partial_\mu \overset{2}{K}{}^\mu_\alpha$ (Equation (D.2)), which is linear in $\{a_{\alpha\mu}\}$:

$$\partial_\mu \delta_{a^\alpha_\mu(x)} \partial_\nu \overset{2}{K}{}^\gamma_\beta(y) - \begin{pmatrix} \mu \leftrightarrow \nu \\ \alpha \leftrightarrow \beta \\ x \leftrightarrow y \end{pmatrix} = 0 \tag{D.19}$$

yields after a lengthy but straightforward calculation:

$$\begin{aligned} 2\varepsilon^{(2)}_{\beta\gamma\alpha} &= -\varepsilon^{(3)}_{\alpha\beta\gamma}, \\ \varepsilon^{(1)}_{\alpha\beta\gamma} &= \tfrac{1}{4}(\varepsilon^{(3)}_{\underbrace{\gamma\alpha\beta}} - 2\varepsilon^{(3)}_{\underbrace{\beta\gamma\alpha}}), \end{aligned} \tag{D.20}$$

with

$$\varepsilon^{(3)}_{\underbrace{\gamma\,\alpha\beta}} = (\varepsilon^{(3)}_{\gamma\alpha\beta} + \varepsilon^{(3)}_{\alpha\alpha\gamma}). \tag{D.21}$$

Then one finds that

$$\begin{aligned} \partial_\mu \overset{2}{K}{}^\mu_\alpha(x) &= \delta_{\omega^\alpha(x)} \int \mathrm{d}y\,[a^\mu_\alpha a^{\beta\nu}\partial_\mu a^{\gamma\nu}\varepsilon^{(3)}_{\alpha\beta\gamma}](y) \\ &\quad + \partial_\mu D_{\alpha\beta\gamma}\,\varepsilon^{\mu\nu\rho\sigma}\partial_\nu a^\beta_\rho a^\gamma_\sigma. \end{aligned} \tag{D.22}$$

The structure of the D-term can be further reduced: let

$$Q = \tfrac{1}{6}\int \mathrm{d}x\,\varepsilon^{\mu\nu\rho\sigma}(\partial_\mu a^\alpha_\nu)a^\beta_\rho a^\gamma_\rho(D_{\underbrace{\beta\gamma}\alpha} + D_{\underbrace{\beta\alpha}\gamma}) \tag{D.23}$$

where

$$\begin{aligned} D_{\underbrace{\beta\gamma}\alpha} &= (D_{\beta\gamma\alpha} - D_{\gamma\beta\alpha}), \\ D_{\underbrace{\beta\alpha}\gamma} &= (D_{\beta\alpha\gamma} - D_{\gamma\alpha\beta}). \end{aligned} \tag{D.24}$$

Then,

$$\partial_\mu D_{\alpha\beta\gamma}\,\varepsilon^{\mu\nu\rho\sigma}(\partial_\nu a^\beta_\rho)a^\gamma_\sigma = (\delta_{\omega^\alpha(x)}Q)^{(2)} + \partial_\mu \overset{2}{\tilde{K}}{}^\mu_\alpha, \tag{D.25}$$

where now

$$\overset{2}{\tilde{K}}{}^\mu_\alpha = \tilde{D}_{\alpha\beta\gamma}\,\varepsilon^{\mu\nu\rho\sigma}(\partial_\nu a^\beta_\rho)a^\gamma_\sigma \tag{D.26}$$

with $\tilde{D}_{\alpha\beta\gamma}$ totally symmetric:

$$\tilde{D}_{\alpha\beta\gamma} = \tfrac{1}{6}[D_{\alpha\beta\gamma} + D_{\beta\gamma\alpha} + D_{\gamma\alpha\beta} + D_{\alpha\gamma\beta} + D_{\gamma\beta\alpha}].$$

In order to perform the last step we put together Equations (D.9), (D.16) and (D.25) and omitting the ~ sign obtain:

$$\begin{aligned} \partial_\mu K^\mu_\alpha &= \partial_\mu \varepsilon^{\mu\nu\rho\sigma}[D_{\alpha\beta\gamma}(\partial_\nu a^\beta_\rho)a^\gamma_\sigma + F_{\alpha\beta\gamma\delta}a^\beta_\nu a^\gamma_\rho a^\delta_\sigma] \\ &\quad + \partial_\mu G_{\alpha\beta\gamma\delta}a^{\beta\mu}a^\gamma_\nu a^{\delta\nu} + \delta_{\omega^\alpha(x)}K \end{aligned} \tag{D.27}$$

where

$$K = \int \mathrm{d}x\,(a_{\beta\mu}\mathring{K}^{\beta\mu})(x) + \tfrac{1}{2}\int \mathrm{d}x\,(a_{\beta\mu}\overset{1}{\tilde{K}}{}^{\beta\mu})(x) + Q \tag{D.28}$$

and $F_{\alpha\beta\gamma\delta}$ is antisymmetric under a permutation of $\beta\gamma\delta$ whereas $G_{\alpha\beta\gamma\delta}$ is symmetric with respect to $\gamma\delta$.

The cocycle condition Equation (D.2) first yields

$$\begin{aligned} &f^{\beta}_{\tau\alpha} D_{\beta\gamma\delta} + f^{\beta}_{\tau\gamma} D_{\beta\alpha\delta} + f^{\beta}_{\tau\delta} D_{\beta\alpha\gamma} = 0 \\ &f^{\beta}_{\tau\alpha} F_{\beta\lambda\gamma\delta} + f^{\beta}_{\tau\lambda} F_{\alpha\beta\gamma\delta} + f^{\beta}_{\tau\gamma} F_{\alpha\lambda\beta\delta} + f^{\beta}_{\tau\delta} F_{\alpha\lambda\gamma\beta} = 0 \end{aligned} \tag{D.29}$$

which ascertain that both D and F are invariant tensors, and, besides:

$$6F_{\underbrace{\alpha\gamma\tau\delta}} = 2f^{\beta}_{\underbrace{\alpha\delta} } D_{\underbrace{\tau\beta\gamma}} \tag{D.30}$$

where the symmetrization with respect to $\alpha\tau$ is as usual indicated by the $\underbrace{\quad}$ sign. A particular solution of Equation (D.30) is given by

$$F_{\alpha\beta\gamma\delta} = \tfrac{1}{12}\left[f^{\lambda}_{\gamma\delta} D_{\alpha\beta\lambda} + f^{\lambda}_{\alpha\beta} D_{\alpha\gamma\lambda} + f^{\lambda}_{\beta\gamma} D_{\alpha\delta\lambda}\right] \tag{D.31}$$

And the most general solution differs from it by a totally antisymmetric F. This can be seen as follows: Equation (D.30) can be rewritten, using the covariance of D (Equation (D.29)):

$$F_{\alpha\beta\gamma\delta} + 6F_{\gamma\beta\alpha\delta} = 2f^{\delta}_{\beta\tau} D_{\tau\alpha\gamma}. \tag{D.32}$$

It suffices then to write

$$\begin{aligned} 24F_{\alpha\beta\gamma\delta} = {} & 6[F_{\alpha\beta\gamma\delta} - F_{\beta\alpha\gamma\delta} + F_{\gamma\alpha\beta\delta} - F_{\delta\alpha\beta\gamma}] \\ & + 6[F_{\alpha\beta\gamma\delta} + F_{\beta\alpha\gamma\delta}] + 6[F_{\alpha\beta\gamma\delta} + F_{\gamma\beta\alpha\delta}] + 6[F_{\alpha\beta\gamma\delta} + F_{\delta\beta\gamma\delta}] \end{aligned} \tag{D.33}$$

use Equation (D.32) three times and notice that the first term in the right-hand side of (D.33) is totally antisymmetric, i.e. is actually a solution of the homogenous version of Equation (D.30):

$$F_{\underbrace{\alpha\beta}\,\gamma\delta} = 0 \tag{D.34}$$

which, taking into account the antisymmetry of F with respect to its last three indices defines totally antisymmetric F's. Totally antisymmetric F's do not really contribute to the anomaly since their contributions to $\partial_{\mu} K^{\mu}_{\alpha}$ is the gauge variation of a local functional

$$\partial_{\mu} \varepsilon^{\mu\nu\rho\sigma} F_{\alpha\beta\gamma\delta} a^{\beta}_{\nu} a^{\gamma}_{\rho} a^{\delta}_{\sigma} = \delta_{\omega^{\alpha}(x)} \frac{1}{4} \int \mathrm{d}x\, \varepsilon^{\mu\nu\rho\sigma} F_{\tau\beta\gamma\delta} \cdots (a^{\tau}_{\mu} a^{\beta}_{\nu} a^{\gamma}_{\rho} a^{\delta}_{\sigma})(x) \tag{D.35}$$

if F is totally antisymmetric.

Finally, the consistency condition, Equation (D.2) implies the symmetry properties on G:

$$G^{\alpha\delta\beta\gamma} = G^{\delta\alpha\beta\gamma} = G^{\beta\gamma\alpha\delta}$$

together with the covariance of G. The contribution of G to $\partial_\mu K^\mu_\alpha$ is again the gauge variation of a local functional

$$\partial_\mu G_{\alpha\delta\beta\gamma} a^{\delta\mu} a^{\beta\nu} a^\gamma_\nu = \tfrac{1}{4} \delta_{\omega^\alpha(x)} \int \mathrm{d}x\, G^{\beta\gamma\delta\eta} \ldots (a^\mu_\beta a_{\gamma\mu} a^\nu_\delta a_{\eta\nu})(x)$$

The Adler–Bardeen anomaly is thus reduced to its canonical form (Equations (95) and (96)), up to compensable terms:

$$K^\mu_\alpha = \varepsilon_{\mu\nu\rho\sigma} [D_{\alpha\beta\gamma} (\partial^\nu a^\rho_\beta) a^\sigma_\gamma + F_{\alpha\beta\gamma\delta} a^{\beta\nu} a^{\gamma\rho} a^{\delta\sigma}]$$

with

$$F_{\alpha\beta\gamma\delta} = \tfrac{1}{12} [D_{\alpha\beta\lambda} f^\lambda_{\gamma\delta} + D_{\alpha\gamma\lambda} f^\lambda_{\delta\beta} + D_{\alpha\delta\lambda} f^\lambda_{\beta\gamma}].$$

University of Geneva
Max-Planck Institut für Physik und Astrophysik, München
Centre de Physique Theorique, C.N.R.S. Marseille

NOTES AND REFERENCES

[1] Symanzik, K., in D. Bessis (Ed.), *Cargèse Lectures in Physics* (1970), Vol. 5, Gordon and Breach, 1972.
[2] Adler, S., in S. Deser, M. Grisaru and H. Pendleton (eds.) *Lectures on Elementary Particles and Quantum Field Theory*, 1970 Brandeis University Summer Institute of Theoretical Physics, M.I.T. Press Cambridge, Mass., 1970.
[3] Becchi, C., Rouet, A., and Stora, R. (a) Commun. Math. Phys., to be published; (b) *Ann. Phys.*, to be published.
[4] Lowenstein, J. H., *Commun. Math. Phys.* **24**, 1 (1974).
Lam, Y. M. P., *Phys. Rev.* **D.6**, 2145 (1972); *Phys. Rev.* **D.7**, 2943 (1973).
[5] Zimmermann, W., *Ann. Phys.* **77**, 536 (1973); *Ann. Phys.* **77**, 570 (1973).
[6] Gomes, M., Lowenstein, J. H., *Phys. Rev.* **D.7**, 550 (1973).
[7] Lowenstein, J. H., and Zimmermann, W., *Nucl. Phys.* **B86**, 77 (1975).
Lowenstein, J. H., *Commun. Math. Phys.* **47**, 53 (1976).
Clark, T., and Lowenstein, J. H., *Nucl. Phys.* **113B**, 109 (1976).
[8] Lowenstein, J. H., and Schroer, B., *Phys. Rev.* **D.7**, 1929 (1973).
Becchi, C., *Commun. Math. Phys.* **33**, 97 (1973).
[9] For a brief summary see [3b], Appendix A.

[10] Wess, J., and Zumino, B., *Phys. Lett.* **37B**, 95 (1971).
[11] Bardeen, W., *Phys. Rev.* **184**, 1848 (1969), eqn. 44.
[12] Becchi, C., *Commun. Math. Phys.* **39**, 329 (1975).
[13] Weinberg, S., *Phys. Rev. Lett.* **29**, 1698 (1972).
[14] Bandelloni, G., Becchi, C., Blasi, A., and Collina, R., 'Renormalization of Models with Radiative Mass Generation', to be published in *Commun. Math. Phys.*

J. CALMET and A. VISCONTI

COMPUTING METHODS IN QUANTUM ELECTRODYNAMICS

ABSTRACT. Algebraic and numerical computing methods used for calculations in quantum electrodynamics are reviewed. Computer algebra systems suitable for high energy physics computations are presented. The impact of these methods on the evaluation of the Lamb shift in hydrogen and of the anomalous magnetic moment of leptons is shown.

1. Introduction

One of the most important applications of computing methods in physics is the evaluation of radiative corrections at high order (i.e. sixth or higher order) in Quantum Electrodynamics (QED) [1]. QED is the only field theory for which theoretical and experimental results have always been in agreement. This is true for very different domains such as astronomy, atomic, nuclear and subnuclear physics. This agreement has been checked with a precision of a few parts per million because experimental techniques have constantly been improved and because of the possibility to calculate Feynman diagrams at high orders. These calculations are in most cases impossible to perform by hand due to the length and complexity of the expressions handled. The analysis of the diagrams is made possible by the availability of powerful computer algebra systems, while the n-tuple integrals obtained are integrated over by means of either numerical or symbolic methods.

We will review two physical effects. The Lamb shift and the anomalous magnetic moment of leptons. The Lamb shift in hydrogen is the measure of the splitting between the $2P_{1/2}$ and $2S_{1/2}$ energy levels. The lepton anomaly is the measure of the deviation from the magnetic moment as predicted by the Dirac theory. Both effects can be studied in the same framework by introducing the matrix element of vertex diagrams

$$u(p_2)\Gamma_\mu(p_2,p_1)u(p_1). \tag{1.1}$$

* On leave of absence from Centre de Physique Theorique, Section II, CNRS, Marseille.

E. Tirapegui (Ed.), Field Theory, Quantization and Statistical Physics, 33–57.

Current conservation implies that the tensorial decomposition of (1.1) is

$$u(p_2)\Gamma_\mu(p_2,p_1)u(p_1) = u(p_2)[\gamma_\mu F_1(q_2) - (i/2m)\sigma_{\mu\nu}q_\nu F_2(q_2)]u(p_1) \quad (1.2)$$

where q is the momentum transfer, γ_μ a Dirac matrix and $\sigma_{\mu\nu}$ is proportional to the commutator of γ_μ and γ_ν. F_1 and F_2 are the Dirac and Pauli form factors.

F_1, which in the nonrelativistic case gives the charge density by Fourier transformation, can be expressed as

$$F_1(p^2) = 1 + \sigma(p^2/m^2) + \dots \quad (1.3)$$

σ is called the slope of F_1 and gives the Lamb shift (ΔE) through the equation

$$\Delta E = (Ze^2/m^2)\sigma|\psi(0)|^2 \quad (1.4)$$

where ψ is the electron wave function. The lepton anomaly is calculated from the F_2 Pauli form factor by putting the outgoing photon on the mass-shell:

$$a_l \equiv \tfrac{1}{2}(g_l - 2) = F_2(0). \quad (1.5)$$

The question we want to answer is the following: what part of a QED calculation can be successfully performed with the help of computing techniques presently available? A more ambitious goal would be to ask whether completely automated computations are possible in QED. This would require several steps.

(i) One has first of all to select the Feynman diagrams which are associated with the physical process under consideration. This is indeed a trivial matter for low perturbative orders but not so for higher orders.

(ii) We need then to find out which ones of the selected graphs are convergent or divergent. This step requires the enumeration of all the divergences included in a diagram.

(iii) For any diagram we have to perform the following operations:

(a) Dirac matrix algebra.

(b) Integration over internal momenta of the graph which has been parametrized through different parametrization techniques.

(c) Numerical – or when possible analytical – integration over the Feynman parameters.

(iv) When a diagram is divergent, steps (b) and (c) of (iii) can only be carried on after the renormalization procedure: one then gets convergent results.

In the following sections we will emphasize the use of computer algebra systems for QED calculations. Presently only the step iii (c) cannot be completely handled by these techniques. The next section is devoted to this step.

2. Integration methods

The calculation of the integrals over the Feynman parameters is straightforward in principle but very tedious in practice. These integrals are always convergent once the renormalization procedure has been carried on, but have a very complicated structure. They are n-tuples integrals of rational functions having poles or branch points in the integration domain. The location of these integrable singularities is generally difficult and makes the usual Monte Carlo methods ineffective. More sophisticated programs have been set up. The basic idea due to Sheppey, consists in combining the Riemann sum definition of an integral with an adaptive and iterative Monte Carlo method.

The values obtained have always been confirmed when an analytic integration has then be possible. There is no mathematical proof that the iteration converges toward the exact value of the integral. For some special cases such as powers of logarithms for instance, it is possible to show that the result is one standard derivation too low [2].

It must be emphasized that during the process of calculating a diagram the most time consuming task is the numerical evaluation of the integrals over the Feynman parameters. A typical diagram contributing to the sixth-order radiative correction to the electron anomaly requires a few minutes of computing time to perform the algebraic part of the calculation while the numerical integration may take several hours to get a precise value.

In this section, we will review briefly the different methods used to evaluate the Feynman integrals. We also discuss the present possibilities of symbolic integrators.

2.1. *The RIWIAD Program*

This program [3] is an illustration of the original idea of Sheppey previously quoted. It is used for the numerical evaluation of the n-dimensional integrals ($1 \leqslant n \leqslant 10$) encountered in high energy physics. It is known under different acronyms which correspond to different versions. The first one, SHEP (1963), is due to Sheppey. It was then modified by Dufner at SLAC and became SPCINT (1967). Finally RIWIAD (1970) is due to Lautrup, it is the version available in the CERN program library for instance. For obvious reasons this program is sometimes called LSD.

To sketch the main features of RIWIAD let us consider then n-dimensional integral.*

$$I = \int_0^1 \mathrm{d}\chi_n \cdots \int_0^1 \mathrm{d}\chi_1 f(\chi_1, \dots, \chi_n). \tag{2.1}$$

The limits of integration are 0 and 1 on each axis. The integrand f must behave in the integration domain like a function with integrable singularities.

We divide the pth axis into $p_{\max}$ intervals. (i,j) represents the jth interval on the ith axis. To a set of infinitesimal intervals.

$$\{(1,j),(2,k),(3,l)\ldots\ldots\}$$

corresponds a point $\chi \in \mathbb{R}^n$ such that

$$\chi = \chi(j,k,l\ldots)$$

and an elementary volume around this point

$$\mathrm{d}(1,j)\,\mathrm{d}(2,k)\,\mathrm{d}(3,l)\ldots \tag{2.2}$$

where $\mathrm{d}(i,j)$ is the length of the jth internal on the ith axis. An elementary contribution to the integral is thus given by

$$v = f(\chi(j,k,1,\ldots))\,\mathrm{d}(1,j)\,\mathrm{d}(2,k)\,\mathrm{d}(3,l)\ldots \tag{2.3}$$

v is called a subvolume. An approximation to the integral I is given by the sum V over all subvolumes

$$V = \sum_{\substack{1\leqslant j\leqslant j_{\max}\\ 1\leqslant k\leqslant k_{\max}}} f(\chi(j,k,l\ldots))\,\mathrm{d}(1,j)\,\mathrm{d}(2,k)\,\mathrm{d}(3,l)\ldots \tag{2.4}$$

*It may be easier to understand the different steps of the method by considering the simpler integral

$$\int_0^1 \mathrm{d}\chi_2 \int_0^1 \mathrm{d}\chi_1 f(\chi_1,\chi_2)$$

The total number of subvolumes is given by

$$J_{\max} = j_{\max}\, k_{\max}\, l_{\max} \cdots$$

and V is divided into subvolumes v_J such that

$$V = \sum_{1}^{J_{\max}} v_J. \tag{2.5}$$

J means a particular choice of intervals such as the one given in (2.2) for instance. One of the main step of the method consists in evaluating v_J. The intervals are not infinitesimal but of finite length and (2.4) is not therefore a good approximation of the integral. We generate in each subvolume v_J, p random points where the integrand is calculated. An average value $\langle v_J \rangle$ is defined by

$$\langle v_J \rangle = \frac{1}{p} \sum_{k=1}^{p} v_J^k \tag{2.6}$$

where v_J^k is one of the value of the integrand calculated at a random point in v_J. The estimated error on the calculation of v_J is given by the variance:

$$\sigma^2(v_J) = \frac{1}{p-1} \sum_{k=1}^{p} (v_J^k - \langle v_J \rangle)^2 \tag{2.7}$$

and the total error is taken as

$$\sigma^2(V) = \sum_{J=1}^{J_{\max}} \sigma^2(v_J). \tag{2.8}$$

In practice the number of random points in each subvolume is set equal to 2. These are the basic operations which will be repeated in the iterative process until the integral is known with the accuracy requested by the user. If the iteration requests s steps and if V_q is the value of V for the qth step, the average value $\langle V \rangle$ is defined by the formula of weighted means

$$\langle V \rangle = \frac{\sum_{q=1}^{s} V_q w(q)}{\sum_{q=1}^{s} w(q)} \tag{2.9}$$

where $w(q)$ is the relative weight assigned to the measure V_q. How to choose the weight $w(q)$? It seems reasonable enough to take it proportional to the inverse of the variance. We thus write

$$w(q) = \frac{V_q^2}{\sigma^2(V_q)} \tag{2.10}$$

and the accumulated standard derivation is then

$$\sigma_{\text{tot}} = \frac{\langle V \rangle}{\sqrt{\sum_{q=1}^{s} w(q)}}. \tag{2.11}$$

The iterative process obviously aims at improving the accuracy of the calculation. This implies that the method must also be adaptive. It must find the regions where the estimation is poor and put more subvolumes in these regions, without an undue sacrifice in the rest of the integration space. This is realized either by changing the length of the intervals used in the last step of the iteration or even by increasing the number of intervals on some axis (and therefore decrease this number for the other axis).

To change the interval length $\mathrm{d}(i,j)$ one uses a formula of the type

$$\mathrm{d}(i,j) \to \mathrm{d}'(i,j) = \mathrm{d}(i,j)\left[\log \frac{\sigma^2(V)}{\sigma^2(i,j)}(1 - \mathrm{d}(i,j))\right]^{\delta} \tag{2.12}$$

where $\sigma^2(i,j)$ is the sum of the variances found in all subvolumes enclosing $\mathrm{d}(i,j)$. δ is an adjustable constant, $0 \leqslant \delta \leqslant 1$. A similar formula decides whether or not the number of intervals on a given axis must be changed. The adaptive part of this integration method is rather empirical. This is not surprising since to find the regions where the integrand is not smooth is the most difficult task for this kind of method. Quite often it is advisable to perform a change of variables in order to get a better behaving integrand.

Despite this empirical adaptive part this program gives very good results in the evaluation of Feynman diagrams. Whenever an analytic calculation of an integral previously calculated by RIWIAD has been possible both results have been in agreement.

In practice the input consists of the dimension of the integral, the accuracy requested, the maximum number of subvolumes to be used and the maximum number of iterations to be performed. The program stops either when the desired accuracy is reached or when all iterations have been done. Several output options allow to have very detailed informations on the calculation at each step of the iteration for all axis: values of the variance and the integral for each interval. When calculating a contribution to the electron anomaly in sixth order an accuracy of 0.01 is usually required. On a CDC 7600 the simplest integrals are calculated in 10 minutes while the most difficult diagrams have required almost 3 hours of computing time [4]. To improve the accuracy of a whole set of diagrams contributing at that order, Cvitanovic and Kinoshita [5] had to use almost 60 hours on the same computer. A feature of the program allows to punch intermediate results on cards (or store them on disk) and to restart a new series of iterations.

2.2. *Recent Progress in Numerical Integration*

For several years RIWIAD was the only program available to evaluate the multidimensional integrals appearing in QED calculations. Although, as previously mentioned, it never fails to give the right answer it would be safer to have different possibilities to check this kind of calculations. A drawback of RIWIAD is the use of stratified sampling. This implies that for a given number of calculations of the integrand (M), for high dimensions (n) the adaptation of the axis intervals (N) is not very efficient since

$$M = 2N^n. \tag{2.13}$$

It is then difficult to reduce the variance σ^2. This remark led Lepage [6] to use a slightly different method in his program VEGAS. Its principle is easily explained when considering the simplest case

$$I = \int_0^1 d\chi f(\chi). \tag{2.14}$$

The variance can be reduced through a change of variable $\chi = g(y)$, with a uniform distribution of the points y_i. One gets:

$$I = \int_0^1 dy g'(y) f(g(y)) \approx \frac{1}{M} \sum_{i=1}^{M} g'(y_i) f(g(y_i)). \tag{2.15}$$

It is well known that σ^2 is minimal when

$$\frac{1}{g'(y)} = \frac{|f(\chi)|}{\int_0^1 d\chi |f(\chi)|}. \tag{2.16}$$

In VEGAS, $g'(y)$ is set as follows: $g'(y) = N\Delta\chi_i$. The increment $\Delta\chi_i$ is such that: $\chi_i - \Delta\chi_i \leqslant \chi \leqslant \chi_i$. It follows that:

$$\chi = g(y) = \Delta\chi_i(Ny - (i-1)) + \sum_{j=1}^{i-1} \Delta\chi_j. \tag{2.17}$$

During the successive iterations the increments $\Delta\chi_i$ are adjusted such that the variance minimizing Equation (2.16) holds. The calculation of the integrand is concentrated in the regions where it is large, whether or not there is a peak. The number of intervals on each axis is independent of the number of computations of the integrand. A consequence is that the adaptativity is more efficient than in RIWIAD for high dimensions ($n > 4$). For low dimensions RIWIAD appears to work better. The present trend is to calculate integrals with both programs and to take then the weighted average of the results.

A program written by Sasaki [7] uses as one of the variance reducing methods a technique very similar to the one implemented in VEGAS. DIVONNE, written by Friedman [8] is based on a Monte Carlo method with stratified sampling to reduce the variance. One of its main features is that the axis splitting and the error estimation do not depend on a variance calculation a la Monte Carlo. It localizes very efficiently the peaks of the integrand, even sharp ones, and concentrates the evaluation in these regions.

2.3. *Other Integration Methods*

Before reviewing in the next paragraph the capabilities of symbolic integrators we briefly sketch some methods which, although only used by their authors, have given very important results.

The first one is due to Levine and Wright [9]. They noticed that the most troublesome integrable singularities in the integrals over Feynman parameters lie on the faces ([0, 1]) of the integration hypercube. Indeed when these singularities are inside the hypercube it is possible to concentrate the evaluation of the integrand in their neighbourhood in order to reduce the error of the calculation in these regions. This is not as efficient when the singularity is on the hypercube. The method consists in finding a change of variables which translates the singularities inside the integration domain. Then a classical Gauss method is used for the integration.

We now consider the analytic calculation of these integrals. This was the method first used for this kind of problems (Karplus and Kroll,

Peterman, Soto ...) [10] at low orders. Calculations at high orders require dedication, patience, ingenuity and resourcefulness but they are very valuable because this is the only way to check the numerical evaluations. For several years two groups are pursuing the analytical calculation of the electron anomaly in sixth order. Remiddi and collaborators [11] consider a 'dispersion relations' type method such as the original one of Källen or Terentiev [10]. Introducing Spence and Nielsen's functions they succeeded in calculating many diagrams. The other method, by Levine and Roskies [12], makes use of some invariance properties of the vertex diagrams to express their contributions in terms of Gegenbauer polynomials. These two approaches lead to the same kind of integrals and can be combined [13]. These methods can be in a way considered as being seminumerical: if the analytic calculation is not completely possible the last remaining integrations can be performed numerically.

The contributions of order 4 can be calculated analytically in a few minutes by the SINAC program of Peterman [14]. The integrals are expressed in terms of Nielsen functions or generalized polylogarithms and handled by the program. Originally written in SCHOONSCHIP [15], a REDUCE [16] version exists under the name of RSIN [17].

2.4. *Symbolic Integration*

Symbolic integration means analytic integration performed completely by computer. The methods quoted in the previous section, although they are a splendid piece of work, are only particular versions of a tool that every physicist hopes for: a symbolic integrator. Indeed as we shall see in the following section computer algebra systems offer a wide range of capabilities to solve several classes of problems but fall well behind the users expectations in the domain of integration.

Considering the importance of this problem we think that it is useful to review the state of the art in this area. We will first describe the 'best' symbolic integrator presently available to users. Second we review the recent progress in this domain at the research level. The most recent review article on this topic is by Davenport and Norman in [18]. We consider only indefinite integrals. Although very interesting work has been done on finite integrals and limit evaluations we do not review it.

MACSYMA [19] is the computer algebra system which provide users with the most complete symbolic integrator. It is known as SIN and comes from the original work of Moses [20]. The strategy of SIN includes three stages. During the first stage one tries to find the solution by using a table of derivatives of known functions. If this process fails, a decision algorithm begins the second stage by directing the computation to one of the eleven following specialized methods:

(i) Elementary functions of exponentials.
(ii) Substitution for an integral power.
(iii) Substitution for a rational root of a linear fraction.
(iv) Binomial conversion to Chebyschev form.
(v) Arctrigonometric substitutions.
(vi) Elementary function of trigonometric functions.
(vii) Rational function times an exponential.
(viii) Rational functions.
(ix) Rational function times a log or arctrigonometric function with a rational argument.
(x) Rational function times an elementary function of log $(a+bx)$.
(xi) Expansion of the integrand.

In the last stage of SIN, either integration by parts or the Risch algorithms [21] are applied to the integral being calculated. The strategy of SIN aims to efficiency. Indeed the first two stages solve many simple integrations by using methods which are very efficient when considering both computing times and core used. The last stage is much more costly. One of the major drawbacks of symbolic integrators is that they request a large size memory, usually 100 to 120 K of 36 bit words. This is also true for SIN. Apart from elementary functions it can integrate Spence, Gauss and error functions. It must be added that some other computer algebra systems possess adapted copies of SIN but do not distribute them to users.

As mentioned by Moses the calculation of indefinite integrals has been studied through three different approaches: methods of artificial intelligence, technological resources of computers, mathematical study. Nowadays the third approach, illustrated by the work of Risch on the theory of integration, gives way to a spectacular breakthrough. Let us review the present state of the art of this discipline.

Since Liouville and Hermite a theory exists for the integration of rational functions. Such an integral can be written as:

$$\int R(\chi)\mathrm{d}\chi = U(X) + \sum_i \log V_i(\chi) \tag{2.18}$$

where U and V_i are also rational functions. Therefore for this class of integrals only technical difficulties arise for their evaluation. They are the following ones: squarefree decomposition, partial fraction decomposition, solving sets of linear equations, g.c.d. and factorization of polynomials and arithmetic over algebraic number fields. In all these domains significant progress has been made making it possible set up an efficient integrator for rational functions.

For transcendental integrands of the logarithmic and exponential types the theoretical difficulty has been solved by Risch who has extended the theory of Liouville to this case. He found a constructive algorithm which gives the integral in a form similar to (2.18). The algorithm of Risch has been implemented for this class of functions, but some implementation problems still remain and this package is not ready for distribution yet.

A third class of integrands is actively studied: the algebraic integrands. They are functions f which can be defined by a polynomial identity of the form $P(\chi, f(\chi)) = 0$, with integer coefficients. A simple example is radicals, such as $f(\chi) = (1 + \chi^2)^{1/2}$. For a long time it was only possible to calculate this class of integrals by using heuristic methods. Now, by combining the Risch algorithm with some techniques of algebraic geometry, a general solution seems to be possible [22]. Some work is also in progress on the class of higher transcendental functions. It is known how to treat error functions or dilogarithms but the general solution is not yet found. It is closely related to the concept of integration is closed forms.

The conclusion of this brief review is that if during the recent years users had the impression that no progress was being made in the domain of symbolic integration, in fact an impressive amount of work has been successfully carried out.

3. Computer algebra systems

The choice of a language to perform numerical calculations is obvious and its name is FORTRAN. The situation is very different for symbolic computations. In this domain there is no single dominant system.

If the priority is transportability between different computers, a possible choice could be LISP [23] (acronym for List Processor). Indeed a LISP system is available on almost all computers, including mini ones (PDP 11, SOLAR ...). Although they may have quite different syntaxes, programs are easily adapted by automatic translation. Another reason for choosing LISP is that for many years it was the only language available on some computers to do high energy physics calculations (Univac 1100 for instance until 1975). These are some of the reasons which led some physicists to use it in the past [24, 25]. Actually the choice of LISP presents many drawbacks. The most important is the lack of a program library, except for list processing all other operations must be programmed. The consequence is that as soon as a computer algebra system – which is constantly improved by its designer – is available, the choice of LISP is not advisable.

Presently two systems are mainly used by elementary particle physicists: REDUCE and SCHOONSCHIP. Levine got very nice results with his ASHMEDAI system, MACSYMA is now used in this field and a new efficient algorithm for trace calculation makes SAC 2 potentially interesting. These are the systems we will briefly describe. It must be emphasized that some other systems have important applications in different domains of physics [26, 27].

3.1. *REDUCE*

This language for algebraic and symbolic manipulations has been designed by A.C. Hearn [16, 28]. It has the following capabilities:
- expansion and ordering of polynomials and rational functions.
- symbolic differentiation.
- substitution and pattern matching in a wide variety of forms.
- calculation of the g.c.d. of two polynomials.
- automatic and user controlled simplification of expressions.
- calculation with symbolic matrices (inverse, determinant...).
- calculation for high energy physics including spin 1/2 and spin 1 algebra.
- tensor operations.
- extensibility of the system allowing user defined procedures.

This list is taken from the user's manual and show the capabilities of the version presently distributed. REDUCE has evolved from a package specialized to high energy physics calculations to one of the most general computer algebra systems. A whole group is now working on this language. The version presently available has not changed, apart some improvements, much from the 1973 one. It gives a very incomplete idea of the research conducted by the Utah group. It includes very different topics such as compiler study, LISP, algorithm improvement to name a few

[29]. A very different version is announced for a not too soon future. It is presently known under the name of MODE REDUCE. It will present brand new features such as modularity and the definition of a mode for each type of expressions, among others.

REDUCE is written in LISP. In fact it needs only a subpart, called 'Standard LISP', of this language in order to improve the transportability to many different computers. The programming aspect is ALGOL-like. Versions are presently available for IBM 360 and 370, PDP 10 and 20, Univac 1100, Burroughs 6700. It has recently been implemented on a Honeywell-Bull [30]. Although mainly suitable for interactive use, it can also be run in batch mode.

3.2. *SCHOONSCHIP*

This system [15] written in COMPASS (assembly code) by Veltman for the CDC 6000/7000 is much appreciated by all its users. Indeed its programming style is very simple to learn and it is very efficient. It makes use of the technology of the CDC mainly the rapid access to magnetic storage of information. It is also very well debugged, a feature users always enjoy. This program is specialized to high energy physics calculation and therefore can handle vectors, operators, indices, scalar products and functions. It operates mainly by definitions and substitutions. The choice of input and output devices is very flexible. Recently Strubbe [15] has adapted SCHOONSCHIP to the IBM 360/370 series by translating it in PL1 and using for the implementation macros in assembly code. This system has been heavily used at CERN.

3.3. *ASHMEDAI*

Written by Levine [31] in FORTRAN this system is quite similar to SCHOONSCHIP when considering its field of application, structure and capabilities. The original Univac 1100 version does use some of the specific features of FORTRAN V (bit manipulation) but FORTRAN IV versions are available for IBM 360 or PDP 10 for instance. The choice of FORTRAN insures transportability and efficiency. It implies also that the size of the core needed for a given calculation is much less than with LISP based systems. A drawback is that the extensibility of this program is non-trivial. As for SCHOONSCHIP, ASHMEDAI operates through definitions and substitutions, it handles also the algebra of Dirac matrices.

It allows one to work with polynomials but not with rational functions. It is not a general purpose computer algebra system. This system is mainly used by Levine and his collaborators although it is available in several computing centers.

These three systems share a common feature: their distributed version is very efficient to perform calculations in the field of theoretical physics. This is no longer true for the two systems now mentioned.

3.4. *MACSYMA*

This system [19, 32] deserves a very special attention in the domain of computer algebra. The reasons are its technical features, the wide range of its capabilities and the number of users. Written in MACLISP (LISP version of the project MAC at M.I.T.) for a PDP 10 of 512 K-words located at M.I.T., it requests at least 135 K to be run. It is not distributed and accessible only via a network. It is probably the most general computer algebra system available today. From the beginning of its design it aims at attracting users from all fields and at being as efficient as possible. For example it offers four different types of internal representation of expressions: general one (default one), rational functions, power series and Poisson series (for general relativity calculations). Each of these representations comes with its own algorithms. Specialized packages, such as the symbolic integrator previously described, are also available. As for REDUCE an important group is working on the improvement of the system. Furthermore users contribute in some way to the extension of its capabilities. In this respect some work in progress on the calculation of Feynman diagrams in QED and QCD [33, 34] will most probably make MACSYMA very competitive for high energy physics calculations. Although it is not easily accessible several hundred scientists have used this system and successful user's conferences are regularly organized.

3.5. *ALDES/SAC* 2

The SAC 2 system [35] of Collins is a new system consisting of the previous SAC 1 [36] algorithms written in the ALDES (Algorithm Descriptor) language of Loos [37]. SAC is often considered as the ideal tool for measuring and developing algorithms. ALDES requests as input algorithms written in Knuth's representation [38]. The programming style is therefore rather unusual for most physicists. A new algorithm for trace calculation of expressions including Dirac matrices by Woll [18] can be the starting point of a high energy physics module. Physicists may have therefore the possibility to use this system in the future.

Apart from the short-term prospects mentioned in this section two technological developments will most probably have a great impact on computer algebra systems. The first one is the spreading use of networks. The second one will be the availability of computer algebra machines.

4. Feynman Diagrams Analysis

The calculation of a given physical process at a given order of the perturbative expansion consists, roughly speaking, of three different steps. During the first one the contributing diagrams are collected. Then one must look for divergent ones in order to apply the renormalization procedure during the last step. This third step deals with the actual calculation which finally gives multi-dimensional integrals over the Feynman parameters. An important part of this calculation is the Dirac matrix algebra.

4.1. *Diagrams Generation*

Diagram generation may seem completely irrelevant for present-day calculations. Indeed it is quite simple to collect all the diagrams contributing to an effect of order 4, 5 or 6. But this is no longer obvious when dealing with renormalizable theories where the number of graphs increases strongly with the order [38]. Several programs for the enumeration of diagrams have been written for scalar theories, QED or non-abelian gauge theories. Very different approaches have been adapted when designing them: Wick's theorem for ordered products and REDUCE [39], combinatorial method and FORTRAN [25], generating functional and LISP [25], topological analysis and LISP [40]. Functional techniques of field theory unable to get the number and weight of each diagram for different classes of theories [41].

4.2. *Divergence Recognition*

When a diagram is ultra violet divergent, renormalization theory sets the number and the type of the counter-terms which have to be subtracted in order to get a finite contribution. In fact this is an obvious stage of the calculation which does not presently need any computer assistance. But if one has in mind some kind of completely automated diagram calculator this must be programmed. A REDUCE routine exists which by topological inspection of a diagram lists its proper and internal divergences (of vacuum polarization, electron self-energy, vertex and overlapping types) [42].

There is no standard method for the elimination of infrared divergences. Among different ad hoc methods the most efficient one seems to be the intermediate renormalization approach used in [43].

4.3. *Dirac Matrices Algebra*

As previously emphasized the only computer algebra systems used for QED calculations are those which incorporate algorithms for computing expressions including Dirac matrices. Two operations must be performed: reduction (sum over repeated matrix indices) and traces.

(i) *Reduction.* Dirac matrices are defined by the equation.

$$\gamma_\mu \gamma_\nu + \gamma_\nu \gamma_\mu = 2g_{\mu\nu}. \tag{4.1}$$

This anticommutation relation does not lead to an efficient algorithm for summing over repeated indices because of the large amount of intermediate terms generated. Two methods are available: the Chisholm identities or the Kahane algorithm.

If S denotes a string of γ-matrices and S_R the reversed string:

$$S = \gamma_{\alpha_1} \cdots \gamma_{\alpha_n}, S_R = \gamma_{\alpha_n} \cdots \gamma_{\alpha_1} \tag{4.2}$$

the Chisholm identities [44] can be written as:

$$\begin{aligned} &\gamma_\mu S \gamma_\mu = 2S_R, \qquad \text{(n odd)} \\ &\gamma_\mu S \gamma_\mu = (\mathrm{Tr}\, S_R) I - \gamma_5 \mathrm{Tr}(\gamma_5 S_R), \qquad \text{(n even)} \\ &\not{p} S \not{p} = -p^2 S_R + \tfrac{1}{2}\not{p}\,\mathrm{Tr}(\not{p} S_R) + \tfrac{1}{2}\not{p}\gamma_5 \,\mathrm{Tr}(\gamma_5 \not{p} S_R), \qquad (n\ \text{odd}) \\ &\not{p} S \not{p} = -p^2 S_R + \tfrac{1}{2}\gamma_\nu \not{p}\,\mathrm{Tr}\,(\not{p}\gamma_\nu S_R), \qquad \text{(n even)} \end{aligned} \tag{4.3}$$

where Tr is the trace operator and $\not{p} = \Sigma_1^4 \gamma_\alpha p^\alpha$. These identities can be used to reduce products of Dirac matrices. But this reduction is not 'minimal' with respect to the number of intermediate terms created. This is achieved by an algorithm due to Kahane [45]. It is a constructive algorithm which directly gives the reduced form. It is very easy to implement and has been adopted by most computer algebra system. A very clear description with simple examples can be found in [46]. This algorithm is only valid for a space of dimension 4; it is not valid when the matrices to be summed over belong to two different trace operators. In this case the following Chisholm identities must be applied:

$$\begin{aligned} &\gamma_\mu S \,\mathrm{Tr}\,(\gamma_\mu S') = 2\{S'\}S, \\ &\mathrm{Tr}(\gamma_\mu S)\mathrm{Tr}(\gamma_\mu S') = 2\,\mathrm{Tr}(\{S\}S'), \end{aligned} \tag{4.4}$$

where the brace symbol { } stands for the sum of the string and of its reversed: $\{S\} = S + S_R$. Hearn and Chisholm have generalized the Kahane algorithm to this special case [46].

(ii) *Traces.* It is usually satisfactory enough to take the trace of an expression by applying the well known equation:

$$\begin{aligned} &\mathrm{Tr}(\gamma_{\mu_1} \cdots \gamma_{\mu_{2n}}) = g_{\mu_1\mu_2}\mathrm{Tr}(\gamma_{\mu_3} \cdots \gamma_{\mu_{2n}}) + \ldots + g_{\mu_1\mu_{2n}}\mathrm{Tr}(\gamma_{\mu_2} \cdots \gamma_{\mu_{2n-1}}), \\ &\mathrm{Tr}(\gamma_{\mu_1} \cdots \gamma_{\mu_{2n+1}}) = 0, \\ &\mathrm{Tr}(\gamma_\mu \gamma_\nu) = 4g_{\mu\nu}. \end{aligned} \tag{4.5}$$

From an algorithm complexity study point of view (4.5) is a very poor algorithm. Indeed, it requires asymptotically exponential computing times. This remark led Woll [18] to set up an algorithm based on the idea that Dirac matrices are generating elements of a Clifford algebra of dimension 16. A string of matrices is thus decomposed on the basis of this algebra after reduction by Kahane algorithm. To take the trace is then straightforward. The nice feature of this method is that few intermediate terms are generated and the computing time is a polynomial function of the computing times for the arithmetic of multivariate polynomials. This method is not covariant. This can, in fact, be useful if a numerical integration follows the trace calculation. In any case another algorithm is available to transform the result into a covariant expression.

4.4. *Calculation of Diagrams*

Once the algebra of Dirac matrices part of the calculation is done the remaining part of the computation of the contribution of Feynman diagrams is extremely simple. Several methods are available to get the parametric form. We do not want to review them and will consider only the simpler classical one which can be found in any textbook on QED. A

Feynman integral can be written as:

$$\int \frac{\mathrm{d}^4 k_1 \dots \mathrm{d}^4 k_n}{a_1 a_2 \dots a_n} \tag{4.6}$$

where the a_i are of the form:

$$a_i = \left(\sum_r (\xi_r p_r + \eta_r k_r) \right)^2 + m^2 \tag{4.7}$$

ξ_r and η_r can take the value 0, $\pm$ 1. p_i and k_i are, respectively, the external and internal momenta. m is the lepton mass.

Feynman parameters are introduced through the identity:

$$\frac{1}{a_1 \dots a_n} = (n-1)! \int_0^1 \mathrm{d}\alpha_1 \dots \int_0^1 \mathrm{d}\alpha_n \frac{\delta\left(\sum_1^n \alpha_i - 1\right)}{\left(\sum_1^n \alpha_i a_i\right)^n}. \tag{4.8}$$

Similar identities can be established when some of the factors a_i are equal. When the integral at hand is divergent, counter-terms are subtracted by means of the following equation:

$$\frac{1}{a^n} - \frac{1}{b^n} = n \int_0^1 \mathrm{d}u \frac{a-b}{[u(a-b)+b]^{n+1}}. \tag{4.9}$$

Finally the integration over the four-dimensional internal momenta is readily performed by the identity:

$$\int \mathrm{d}^4 k \frac{(k^2)^{m-2}}{(k^2+a^2)^n} = \frac{i\pi^2}{(a^2)^{n-m}} \frac{\Gamma(m)\Gamma(n-m)}{\Gamma(n)} \tag{4.10}$$

after elimination of the odd powers of k in the numerator. The common features of these operations is that they can be carried on by algorithms which cover all possible cases. This elementary method of calculation is not possible when working with non-Abelian gauge theories. Dimensional regularization must then be used and (4.10) is replaced by:

$$\int \frac{d^{2n}k}{(k^2 - L + i\epsilon)^m} = (-1)^m i\pi^n \frac{1}{L^{n-m}} \frac{\Gamma(m-n)}{\Gamma(m)} \tag{4.11}$$

When the residues of the poles at $n = 4$ add up to zero the limit $n \to 4$ is taken and the final result obtained. Also one of the basic equations of the Dirac algebra must be slightly changed. With the same notations:

$$\gamma_\mu \gamma^\mu = \delta^\mu_\mu = 4$$

becomes

$$\gamma_\mu \gamma^\mu = \delta^\mu_\mu = 2n.$$

Gastmans *et al.* [47] have described which changes have to be made in REDUCE in order to have a valid Dirac algebra whatever the value of n is.

Such Feynman diagrams manipulation programs have been written in LISP [48], REDUCE [39, 47, 49], SCHOONSCHIP [41], ASHMEDAI [9, 12] and MACSYMA [33, 34]. This list is a very limited one of all the work completed during the last ten years with the aid of computer algebra systems. A very interesting approach has been given by Cvitanovic when he was calculating the electron anomaly with Kinoshita: he used the text-editor of the PDP-10, TECO, to generate the Feynman integrands [50].

More elaborate methods to get the parametric form of Feynman integrals are described by Brenner [33] and Kinoshita and Cvitanovic [41]. The calculation of a physical effect requires either to square the amplitudes in order to get the differential cross-sections, or to project the amplitude on a given basis to get the contributions to the lepton anomaly or to the Lamb shift. These are straightforward calculations for algebra systems.

5. Theoretical and Experimental Results

We now review some of the most important theoretical results which have been obtained by using the methods described in the preceding sections. We consider only two physical effects: the Lamb–Retherford effect for hydrogen and the anomalous magnetic moment of the electron and of the muon. It is indeed for these calculations that the impact of computing methods has been the most important. Nowadays they are used in the study of many other processes, the jet production in e^+–e^- scattering for example.

5.1. *The Lamb–Retherford Effect*

Equation (1.4) shows that to calculate the Lamb shift in hydrogen consists in computing the slope σ of the Dirac form factor $F_1(q^2)$. This in turn amounts in getting the terms proportional to γ_μ in the tensorial decomposition (1.2) of the vertex diagrams $\Gamma_\mu(p_1, p_2)$. The present experimental accuracy requests the evaluation of the 4th-order radiative corrections to σ.

$$\sigma^{(4)} = m^2 \frac{\mathrm{d}}{\mathrm{d}p^2} F_1^{(4)}(p^2)\Big|_{p^2=0} \tag{5.1}$$

which includes seven diagrams. Following the pioneering work of Bethe, the first calculations were completed by Weneser, Bersohn and Kroll in 1953, then by Soto and later on by different groups in 1970 [10]. The analytic results obtained are in complete agreement. Another contribution to the Lamb shift comes from the binding effects (2nd-order electron self energy) for high orders of $Z\alpha$. The two existing calculations of this contribution by Erickson [51] and Mohr [52] are in disagreement and lead to different theoretical results.

$$\Delta E_E^{\mathrm{th}}(2S_{1/2} \to 2P_{1/2}) = \begin{matrix} 1057.916(10)\,\mathrm{MHz} & [51], \\ 1057.864(14)\,\mathrm{MHz} & [52], \end{matrix}$$

The most recent experimental results have been obtained by Lundeen and Lipkin [53] and by Andrews and Newton [54]:

$$\Delta E_E^{\mathrm{exp}}(2S_{1/2} \to 2P_{1/2}) = \begin{matrix} 1057.893(20)\,\mathrm{MHz} & [53], \\ 1057.862(20)\,\mathrm{MHz} & [54]. \end{matrix}$$

Improved experimental results are expected and the theoretical discrepancy must therefore be resolved.

5.2. *The Lepton Anomalous Magnetic Moments*

We first consider the electron anomaly. It is well known that one diagram contributes to the second order. It gives the famous Schwinger contribution: $\alpha/2\pi$.

In fourth order, seven diagrams must be calculated. They give the following contribution: $-0.328478\ (\alpha/\pi)^2$.

This number was first calculated by hand by several groups [10]. It is a very long and tedious work and the first published result had some wrong partial contributions. Inspection of the results show that the only special functions appearing are di- and tri-logarithmic functions which can now be computed by SINAC [14] in less than 4 minutes of computing time on a CDC 7600.

The calculation of the 72 diagrams contributing to the sixth order is an extremely difficult task although symmetry considerations reduce to 40 the number of diagrams which must be actually computed. They are usually divided into six classes by gauge invariance considerations.

(1) 6 diagrams including the light by light scattering diagrams.
(2) 12 diagrams including second-order vacuum polarizations.
(3) 4 diagrams including fourth-order vacuum polarizations.
(4) 6 diagrams with 3 photon exchange.
(5) 20 diagrams with 2 photon exchange.
(6) 24 diagrams with 1 photon exchange.

Without reviewing in detail all the different works on their evaluation it is important to mention that each of these diagrams has been calculated numerically at least twice and that 51 are now known analytically. The analytic calculations have been performed by two groups: Remiddi *et al.* [11 and references therein] and Levine *et al.* [12, 55], they are still in progress. Presently all the diagrams of classes 2, 3, 6 and most of those of class 5 are analytically known. These results have confirmed the numerical evaluations of Cvitanovic and Kinoshita [5, 43] and Levine–Wright [9] for classes 4, 5, 6, Brodsky–Kinoshita [56] and Calmet–Perrottet [57] for class 2. For class 1, only numerical calculations have yet been made. The possibility of a complete analytic solution is not presently resolved. The first evaluation is by Aldins *et al.* [58] and their result was later on improved by Calmet–Peterman [4]. A few other calculations, not always in agreement, have then been made (see reference 7 in [11]) the very last one by Levine and Engelmann (quoted by Kinoshita in [59]) is both a confirmation and a big improvement of the original ones. They get: $0.370986(20)\ (\alpha/\pi)^3$. Their semi-numerical method is a good check of the results obtained with the RIWIAD program described in Section 2.

The sixth-order contribution to the electron anomaly is:

$$1.1835(61)(\alpha/\pi)^3$$

This gives the following theoretical value for the electron anomaly

$$a_e^{th} = 1\,159\,652\,569\,(150) \times 10^{-12}$$

This value is in agreement with the last published experimental value of Dehmelt [60]:

$$a_e^{exp} = 1\,159\,652\,410\,(200) \times 10^{-12}$$

but the last measurement would give (Kinoshita in [59]):

$$a_e^{exp} = 1\,159\,652\,200\,(40) \times 10^{-12},$$

which breaks the agreement between theory and experiments. In order to restore it one must take into account more subtle contributions such as a virtual muon loop (2.8×10^{-12}), hadronic effects (1.6×10^{-12}) and weak interactions (0.05×10^{-12} in the Weinberg–Salam model). This also emphasizes the importance of the calculation of the 8th-order contribution presently carried on by Kinoshita and Lindquist [61]. In this order 891 diagrams contribute; 430 of them are topologically distinct and must be calculated. To give an idea of the difficulty of this work it is enough to mention that a single diagram, computed by SCHOONSCHIP can generate as many as 10 000 terms in its final expression, that renormalization requires to subtract up to 15 counter-terms and that the final parametric integrals which have to be integrated over are 10-tuple ones. Presently the whole symbolic calculation is finished and the contribution of some classes of diagrams are already known. They are those consisting of the 2nd-order vertices including vacuum polarizations of order 2, 4 and 6 and 4th-order vertices including vacuum polarizations of order 2 and 4. The respective results are:

$$0.0766\,(6)(\alpha/\pi)^4 \text{ and } -0.5238\,(10)\,(\alpha/\pi)^4.$$

We now consider the muon anomaly. We have noticed that in the electron case the effect of a virtual muon loop is small. The situation is completely different for the muon. Indeed the anomaly of a lepton

$$a_l = \tfrac{1}{2}(g_l - 2) \tag{5.2}$$

can be expressed as

$$a_l = a_1 + a_2(m_l/m_{l'}) \tag{5.3}$$

because the anomaly is a dimensionless number. It follows that

$$\begin{aligned} a_e &= a_1 + a_2(m_e/m_\mu) \\ a_\mu &= a_1 + a_2(m_\mu/m_e) \end{aligned} \tag{5.4}$$

thus:

$$a_\mu - [a_e - a_2(m_e/m_\mu)] = a_2(m_\mu/m_e). \tag{5.5}$$

Once the electron anomaly is known, to determine the muon anomaly it is only necessary to calculate the diagrams with virtual electron insertions.

Once again the contributions are calculated order by order. The mass of the muon being much greater than the electron mass, not only do QED effects contribute but also hadronic and weak interaction contributions have to be taken into account. The diagrams have been evaluated up to eighth order. At that order, the calculation is partly numerical and partly an estimation precise enough to consider this contribution as known. A detailed review of the muon anomaly can be found in [62]. Since this review was published no significant new results have been obtained. We therefore quote the theoretical value given in [62].

$$a_\mu^{\text{th}} = 1\,165\,920.6\,(12.9) \times 10^{-9}$$

which is in perfect agreement with the experimental value of Bailey *et al.* [63]

$$a_\mu^{\text{exp}} = 1\,165\,924\,(8.5) \times 10^{-9}.$$

The main conclusion which can be drawn from these results is that the computing methods used in QED are very efficient. Without these methods none of the calculations reviewed would have been possible. Also all effects have been evaluated by different methods which have given results in complete agreement.

A class of problem has not been mentioned in this section: bound state problems. Indeed it is possible to find in the literature some examples such as the study of a possible stable bound state of a positron and a helium atom [63] or the lower order vacuum polarization around a point charge [64] but the methods used are rather hybrid ones where symbolic manipulations do not play a very important role. For bound state problems, calculations are no longer covariant and require different approximations. This forbids an algorithmic approach to this class of problems and therefore the systematic use of computer algebra systems.

Institute für Informatik I, Universität Karlsruhe
University of Aix-Marseille I and CPT II,
Marseille

References

[1] Jouvet, B., 'La théorie quantique des champs. Les théories dynamiques de la matière', in *Relativité et Quanta. Les grandes théories de la Physique Moderne*, pp. 83–136, Masson and Cie., Paris, 1968. Or any other book on Quantum Field Theory.

[2] Calmet, J., and Peterman, A., unpublished.

[3] Lautrup, B., in A. Visconti (ed.), *Proc. 3rd Int. Colloq. on Adv. Comput. Meth. in Theor. Phys.*, Marseille, 1971.
[4] Calmet, J., and Peterman, A., *Phys. Lett.* **47B**, 369 (1973).
[5] Cvitanovic, P., Kinoshita, T., *Phys. Rev.* **D10**, 4007 (1974).
[6] Lepage, G. P., *J. Comp. Phys.* **27**, 192 (1978).
[7] Sasaki, T., in A. Visconti (ed.), *Proc. 4th Coll. Adv. Comput. Meth. in Theor. Phys.*, St. Maximin, 1977.
[8] Friedman, J., SLAC-PUB-2006 (Rev.), July 1978.
[9] Levine, M. J., and Wright, J., *Phys. Rev.* **D8**, 3171 (1973).
[10] A complete list of references can be found in: Lautrup, B. E., Peterman, A., and de Rafael, E., *Phys. Rep.* **3**, 193 (1972).
[11] Barbieri, R., *et al.*, *Nucl. Phys.* **B144**, 329 (1978).
[12] Levine, M. J., and Roskies, R., *Phys. Rev.* **D9**, 421 (1974).
[13] Levine, M. J., Remiddi, E., and Roskies, R., *Phys. Rev.* **D20**, 2068 (1979).
[14] Maison, D., and Peterman, A., *Comp. Phys. Comm.* **7**, 121 (1974).
[15] Strubbe, H., *Comp. Phys. Comm.* **18**, 1 (1979).
[16] Hearn, A. C., 'REDUCE 2 User's Manual', Univ. of Utah (1973), unpublished.
[17] Fox, J. A., and Hearn, A. C., *J. Comp. Phys.* **14**, 301 (1974).
[18] 'Symbolic and Algebraic Computation'. In E. W. Ng (ed.) Proc. of EUROSAM 79, Marseille, June 1979, *Lecture Notes in Computer Science*, **72** (1979), Springer-Verlag.
[19] MATHLAB Group, MACSYMA Reference Manual, Version 9, M.I.T., (1977).
[20] Moses, J., *Communications ACM*, **14**, 548 (1971).
[21] Risch, R. H., *Bulletin A.M.S.* **76**, 605 (1970).
[22] Reports of J. H. Davenport and of B. M. Trager in ref. 18.
[23] See for instance the report of M. King and P. Hayes, in E. Charniak and Y. Wilks (eds.), *Computational Semantics*, North-Holland, 1976.
[24] Campbell, J. A., in *Computing as a Language of Physics*, International Atomic Energy Agency, Vienna, 1972, p. 391.
[25] Calmet, J., and Perrottet, M., *J. Comp. Phys.* **7**, 191 (1971).
[26] J. Fitch, in ref. [18].
[27] Calmet, J., in H. Bacry (ed.), *Proc. of the Coll. 'Regards sur la Phys. Théor. d'Aujourd' hui'*, Marseille, July 1979, Editions du CNRS (to be published in 1980).
[28] Hearn, A. C., in A. Visconti (ed.), *Proc. 3rd Int. Coll. on Adv. Comput. Meth. in Theor. Phys.*, Marseille, June 1973.
[29] REDUCE Newsletters, **1–6**, A. C. Hearn (ed.), Univ. of Utah.
[30] Siret, Y., IMAG Grenoble, Personal Communication.
[31] Levine, M. J., and Roskies, R., in A. Visconti (ed.), *Proc. 4th Coll. on Adv. Comput. Meth. in Theor. Phys.*, St. Maximin, March 1977.
[32] Moses, J., ACM-SIGSAM Bulletin **8**, 105 (1974).
[33] Brenner, R. L., in V. E. Lewis (ed.), *Proceedings '1979 MACSYMA User's Conference'*, Washington, July 1979, M.I.T., 1979.
[34] Wolfram, S., *ibid.*
[35] Collins, G. E., Abstract in A. Visconti (ed.), *Proc. 4th Int. Coll. on Adv. Comput. Meth. in Theor. Phys.*, St. Maximin, March 1977.
[36] Collins, G. E., in S. R. Petrick (ed.), *Proc. Second Symp. on Symbolic and Algebraic Manipulations*, Los Angeles, 1971, A.C.M., 1971.

[37] Loos, R., ACM-SIGSAM Bulletin **10**, 15 (1976).
[38] Lautrup, B. E., *Phys. Lett.* **69B**, 109 (1977).
[39] Campbell, J. A., and Hearn, A. C., *J. Comp. Phys.* **5**, 280 (1970).
[40] Sasaki, T., *J. Comp. Phys.* **22**, 189 (1976).
[41] Cvitanovic, P., Lautrup B. E., and Pearson, R. B., Princeton preprint, April 1978.
[42] Calmet, J., *ACM-SIGSAM Bulletin* **8**, 74 (1974).
[43] Cvitanovic, P., and Kinoshita, T., *Phys. Rev.* **D10**, 3978, 3991 (1974).
[44] Chisholm, J. S. R., *Nuovo Cimento*, **30**, 426 (1963); and *Proc. Camb. Phil. Soc.* **48**, 2, 300 (1952).
[45] Kahane, J., *J. Math. Phys.* **9**, 1732 (1968).
[46] Hearn, A. C., in A. Visconti (ed.), *Proc. 2nd Int. Coll. on Adv. Comp. Meth. in Theor. Phys.*, Marseille, June 1971.
[47] Gastmans, R., Van Proyen, A., and Verbaeten, P., to appear in *Comp. Phys. Comm.*
[48] Calmet, J., PhD. Thesis, Aix-Marseille University, 1970, unpublished.
[49] Calmet, J., *Comp. Phys. Comm.* **4**, 199 (1972).
[50] Cvitanovic, P., in A. Visconti (ed.), *Proc. 3rd Int. Coll. on Adv. Comp. Meth. in Theor. Phys.*, Marseille, June 1973.
[51] Erickson, G. W., *Phys. Rev. Lett.* **27**, 780 (1971).
[52] Mohr, P. J., *Phys. Rev. Lett.* **34**, 1050 (1975).
[53] Lundeen, S. R., and Lipkin, F. M., *Phys. Rev. Lett.* **34**, 1368 (1975).
[54] Andrews, D. S., and Newton, G., *Phys. Rev. Lett.* **37**, 1254 (1976).
[55] Levine, M. J., and Roskies, R., *Phys. Rev.* **D14**, 2191 (1976).
[56] Brodsky, S. J., and Kinoshita, T., *Phys. Rev.* **D3**, 356 (1971).
[57] Calmet, J., and Perrottet, M., *Phys. Rev.* **D3**, 2101 (1971).
[58] Aldins, J., *et al.*, *Phys. Rev.* **D1**, 2378 (1970).
[59] Kinoshita, T., in H. Bacry (ed.), *Proc. of the Coll. 'Regards sur la Phys. Théor d'-Aujourd'hui'*, Marseille, July 1979. Editions du CNRS (to be published in 1980).
[60] Van Dyck, Jr., R. S., Schwinberg, P. B., and Dehmelt, H. G., *Phys. Rev. Lett.* **38**, 310 (1977).
[61] Kinoshita, T., and Lindquist, W. B., preprints CLNS-423, 424, 426. Cornell Univ., 1979.
[62] Calmet, J., Narison, S., Perrottet, M., and de Rafael, E; *Rev. Mod. Phys.* **49**, 21 (1977).
[63] Gentleman, W. M., *Proc. of the ACM National Conference*, Boston 1972, p. 840, ACM, 1972.
[64] Fullerton, L. W., and Rinker Jr., G. A., *Phys. Rev.*, **A13**, 1283 (1976).

GERARD CLEMENT

CLASSICAL MECHANICS OF AUTOCOMPOSITE PARTICLES

ABSTRACT. A class of theories of fields on $(n+1)$-dimensional space-time valued on a compact n-manifold admit classical localized solutions, which are interpreted as systems of autocomposite particles. A covariant method for extracting the classical mechanics of these autocomposite particles from the underlying field theory is presented.

1. Introduction

The term 'autocomposite particle' was coined by Bernard Jouvet in 1967 [1]. Its meaning is best explained on an example. Consider the model quantum field theory defined by the unrenormalized Lagrangian density

$$\mathscr{L} = \tfrac{1}{2}(\partial_\mu \varphi_0 \partial^\mu \varphi_0 - \mu_0^2 \varphi_0^2) + \tfrac{1}{2}(\partial_\mu \psi_0 \partial^\mu \psi_0 - m_0^2 \psi_0^2) + \tfrac{1}{2} g_0 \varphi_0 \psi_0^2. \tag{1.1}$$

In the limit $\mu_0^2 \to \infty, g_0/\mu_0^2$ finite (which corresponds, in particular, to $Z_3 = 0$), the Euler equation for the field $\varphi_0(x)$ reduces to the constraint

$$\varphi_0 = \frac{g_0}{2\mu_0^2}\psi_0^2, \tag{1.2}$$

which may be interpreted as meaning that the φ-particle is a bound state of the ψ-particles (allocomposite particle). Similarly, in the limit $m_0^2 \to \infty$, g_0/m_0^2 finite (corresponding to $Z_2 = 0$), we obtain the constraint

$$\psi_0 = \frac{g_0}{m_0^2}\varphi_0 \psi_0, \tag{1.3}$$

meaning that the ψ-particle is composed of the φ-particle and of itself (oedipian autocomposite particle). If both renormalization constants Z_3 and Z_2 are zero, then there are no longer any elementary particles in the theory, only (gigogne) autocomposite particles.

The foregoing example is not really satisfactory, in that there is no

E. Tirapegui (Ed.), Field Theory, Quantization and Statistical Physics, 59–76.

classical solution to the constraint (1.3), while a particle, whether elementary or composite, can always be described in quasi-classical terms. However, examples were subsequently found [2,3] of autocomposite field models for which the compositeness constraints have a classical solution. Such examples shall be exhibited below.

Assuming then that an autocomposite field theory has a classical meaning, what is that meaning? In a classical field theory, elementary particles can be described as point singularities of the fields. If there are no elementary particles ($Z_3 = Z_2 = 0$), then there must be no singularities at the classical level; however, there may appear local concentrations of energy-momentum, which may be interpreted as extended particles. Thus, autocomposite particles, if they exist, are the same thing as localized solutions of a classical field theory, or solitons.

As will be recalled in Section 2, such solutions do indeed exist for specific autocomposite field models. The well-known properties of solitons vindicate, from the phenomenological point of view, the use of the term 'autocomposite particles' to describe these solutions. However, one would like to go a step further and set up a theory of the motion of autocomposite particles similar to that of elementary (point) particles. The purpose of the present contribution is to show how, for a specific class of autocomposite field models (in which the number of independent scalar fields is equal to the number of space dimensions), one can extract from the underlying field theory such a (relativistic) classical mechanics of autocomposite particles.

In Section 3 of this paper, it will be seen that one can associate to a multi-soliton solution of a field theory of the type considered a multiplicity of families of world-lines (one family for each soliton), thus providing a kinematical description of the multi-soliton system. Dynamics are discussed in the next sections, first from a local point of view, then from a global point of view. Local equations of motion, giving the local acceleration along each world-line in terms of effective potentials, are derived from the field equations in Section 4. The energy-momenta and the angular momenta of the various solitons of the system are obtained in Section 5 as integrals on space-like surfaces orthogonal to the corresponding families of world-lines, and the resulting global equations of motion are discussed. The case of internal symmetries of the autocomposite field theory is investigated in Section 6. In the final section, three directions for further investigation are outlined.

2. AUTOCOMPOSITE SCALAR FIELDS AND SOLITONS

Let us consider the model field theory defined by the classical Lagrangian density

$$\mathscr{L} = \tfrac{1}{2}(\partial_\mu \vec{\phi} . \partial^\mu \vec{\phi} - m_0^2 \vec{\phi}^2) + \frac{g_0}{4}(\vec{\phi}^2)^2, \tag{2.1}$$

where $\vec{\phi}$ has $(p+1)$ components. The corresponding Euler equations reduce, in the limit $m_0^2 \to \infty$, $m_0^2/g_0 = v^2$ finite, to the constraint

$$m_0^2 \vec{\phi} = g_0 \vec{\phi}(\vec{\phi}^2), \tag{2.2}$$

which means classically that the vector field $\vec{\phi}$ is constrained to vary on a p-sphere of radius v. Carrying back this constraint in the original Lagrangian, we obtain the modified Lagrangian density

$$\mathscr{L} = \tfrac{1}{2}\partial_\mu \vec{\phi} . \partial^\mu \vec{\phi} + K(\vec{\phi}^2 - v^2), \tag{2.3}$$

where K is a Lagrange multiplier. It may be recalled that Lagrangian densities such as (2.3) are closely related [3–5] to Sugawara–Sommerfeld current-algebra models [6].

From now on, we shall choose the number p of independent scalar fields equal to the number n of space dimensions. It will be seen in the next section that this choice ensures the possibility of a satisfactory kinematical description of autocomposite particles. That it allows dynamically the existence of such solutions to the field equations will be recalled now, on specific examples.

(a) $n = 1$. Putting $\phi^1 = v \cos\theta$, $\phi^2 = v \sin\theta$, the theory defined by (2.3) is simply that of a free field θ subject to a periodicity condition; this does not admit localized solutions. However we can allow, in our heuristic derivation, for the possibility of dynamical symmetry breaking by modifying the final Lagrangian density to

$$\mathscr{L} = \tfrac{1}{2}\partial_\mu \vec{\phi} . \partial^\mu \vec{\phi} + \mu^2 v \phi^1 + K(\vec{\phi}^2 - v^2). \tag{2.4}$$

This defines the well-known sine-Gordon model, for which one has analytical multi-soliton solutions. It will be convenient for our purpose to rewrite the Lagrangian density (2.4) in the 'normal' form:

$$\mathscr{L} = \tfrac{1}{2}F(\varphi)[\partial_\mu \varphi \partial^\mu \varphi - \lambda^2], \tag{2.5}$$

where

$$F(\varphi) = (\mathrm{ch}(\mu/\lambda)\varphi)^{-2}, \tag{2.6}$$

and $\lambda = 2\mu\nu$ (the usual form of the sine-Gordon model may be recovered by putting $\varphi = (\lambda/\mu)$ Log tg $\theta/4$).

(b) $n = 2$. In this case, the model defined by Lagrangian density (2.3) has analytical static k-soliton (or k-antisoliton) solutions for any k. These solutions were first obtained by Belavin and Polyakov [7] in the context of two-dimensional ferro-magnets, and were rediscovered independently by the author [8] and Matsuda [9]. Moreover, such solutions were extended to the case of fields taking their values in any surface Σ homeomorph to the sphere, for any geometry of space [8], and the resulting static general-relativistic gravitational fields were also computed [8]. These results have again recently been extended by Peremolov [10], who has shown the existence of static multi-soliton solutions for free fields in two space dimensions taking their values in any simply-connected compact homogeneous Kähler manifold.

(c) $n = 3$. A scaling argument due to Derrick [11] shows that there can be no localized solution for scalar field theories in more than two space dimensions if the Lagrangian is quadratic in the derivatives. To evade Derrick's argument, the Lagrangian density (2.3) can be modified to allow for a quartic derivative coupling. This was first done by Skyrme [12], who proposed a model given by a Lagrangian density of the type

$$\mathscr{L} = \tfrac{1}{2}\partial_\mu\vec{\phi}.\partial^\mu\vec{\phi} + \tfrac{1}{4}\gamma\,[(\partial_\mu\vec{\phi}.\partial^\mu\vec{\phi})^2 - (\partial_\mu\vec{\phi}.\partial_\nu\vec{\phi})(\partial^\mu\vec{\phi}.\partial^\nu\vec{\phi})] + K(\vec{\phi}^2 - v^2). \tag{2.7}$$

The quartic term in (2.7) can be viewed as a one-loop counterterm in the superpropagator regularization of (2.3) [13]. Skyrme has proved the existence of a static localized solution to this model, but the analytical form of this solution is not known. There are strong indications [12, 13] that such a model might provide the basis for a realistic hadrodynamics.

3. Kinematics

The heuristic line of reasoning we have followed at the beginning of the preceding section may be extended to other Lagrangian densities, such as

$$\mathscr{L} = \tfrac{1}{2}\partial_\mu\vec{\phi}.\partial^\mu\vec{\phi} - F[A(\vec{\phi})], \tag{3.1}$$

to which correspond the autocompositeness constraint

$$A(\vec{\phi}) = A_0, \tag{3.2}$$

where A_0 is an extremum of the function $F(A)$. Thus we are led to study, as autocomposite field models, theories of fields defined on $(n+1)$-dimensional curved space-time M and valued on an n-manifold Σ, which shall be assumed to be homeomorph to the n-sphere, and differentiable.

For $n = 2$ or 3, Lagrangian densities such as (2.3) or (2.7) depend only on the derivatives of the field; a trivial solution of the corresponding Euler equations is the vacuum solution which maps all points of M on an arbitrary fixed point Ω of Σ. For $n = 1$, the Lagrangian density is of the type

$$\mathscr{L} = \tfrac{1}{2}\partial_\mu\vec{\phi}.\partial^\mu\vec{\phi} - V(\vec{\phi}) + K[A(\vec{\phi}) - A_0], \tag{3.3}$$

and the vacuum solution maps all points of M on a point Ω of Σ for which the potential V is minimum. Localized solutions of the field theory are then (for any n) defined as non-constant smooth functions which map the world-line at spatial infinity ($\vec{x} = \infty$) on Ω (this definition is easily seen to be relativistically invariant). The existence of such solutions shall be assumed in this section.

Let us choose on $\Sigma - \{\Omega\}$ a local system of coordinates X^a ($a = 1, \ldots, n$) such that $\vec{X}_\Omega = \infty$, and let the distance on this manifold be

$$\mathrm{d}S^2 = \gamma_{ab}(\vec{X})\mathrm{d}X^a\mathrm{d}X^b. \tag{3.4}$$

The n-component field $\vec{\varphi}$ is then defined as the mapping $\boldsymbol{x} \to \vec{X} = \vec{\varphi}(\boldsymbol{x})$. The associated topological current is

$$k^\mu = \frac{1}{n!}\frac{\sqrt{\gamma}}{\sqrt{|g|}}\varepsilon^{\mu\nu\ldots\sigma}\varepsilon_{a\ldots c}\partial_\nu\varphi^a \ldots \partial_\sigma\varphi^c, \tag{3.5}$$

where $\gamma = \det\gamma_{ab}$, $g = \det g_{\mu\nu}$ ($g_{\mu\nu}(\boldsymbol{x})$ being a suitable metric on M, with signature $+ - \ldots -$), and

$$A = \int \mathrm{d}^n X\sqrt{\gamma}. \tag{3.6}$$

The current k^μ being divergenceless, the topological charge

$$K = \int \mathrm{d}^n x\sqrt{|g|}\,k^0 = A^{-1}\int \mathrm{d}^n x\sqrt{\gamma}\det(\partial\vec{\varphi}/\partial\vec{x}) \tag{3.7}$$

is a constant of the motion.

At any fixed time, the mapping $\vec{\varphi}$ may be locally inverted so that, to each regular point $\vec{X}$ of Σ (such that $\det(\partial\vec{\varphi}/\partial\vec{x}) \neq 0$), there correspond l

points $\vec{x}_{(i)}(\vec{X},t)$ of the n-surface $t=$ constant. Thus,

$$K = A^{-1}\sum_{i=1}^{l}\int \mathrm{d}^n X\,|\det(\partial\vec{x}/\partial\vec{\varphi})|\sqrt{\gamma}\,\det(\partial\vec{\varphi}/\partial\vec{x}) = k \qquad (3.8)$$

where k is the degree of the mapping:

$$k = \sum_{i=1}^{l} \operatorname{sign}\det(\partial\vec{\varphi}/\partial\vec{x}) \qquad (3.9)$$

for an arbitrary regular point $\vec{X}$. We can also write

$$l = l_+ + l_-, \qquad k = l_+ - l_-, \qquad (3.10)$$

where l_+ and l_- are the number of antecedents of $\vec{X}$ with positive and negative signature respectively.

The coordinates X^a can always be chosen such that γ_{ab} is maximum and equal to δ_{ab} for $\vec{X}=0$. In this case, the topological charge density is maximum in the vicinity of the points $\vec{x}_{(i)}(0,t)$, which may be considered as the "centres" of the various solitons. The solution $\vec{\varphi}(\vec{x},t)$ may then be interpreted as representing a system of l_+ solitons and l_- antisolitons.

Letting now t vary, to each regular point $\vec{X}$ of Σ there correspond l world-lines $\Gamma_{(i)}(\vec{X})$ of equations

$$\varphi^a(x_{(i)}) = X^a. \qquad (3.11)$$

In particular, one may view the 'zero-world-lines' $\Gamma_{(i)}(0)$ as the world-lines of a fictitious system of l point particles 'equivalent' to the real system of extended particles. In fact, this definition is somewhat arbitrary, as the $\vec{X}$-world-lines for any given $\vec{X}$ contain qualitatively the same information and may be used to define another 'equivalent' system. Moreover, a full kinematical description of the system obviously necessitates the use of all the $\vec{X}$-world-lines.

It should be noted that neither l_+, l_-, or l needs be a constant of the motion. However, the 'soliton number' k (the topological charge) being constant, the number l of $\vec{X}$-world-lines is constant modulo 2. Thus we have at the classical level the possibility of pair creation or annihilation of world-lines (as may be observed for instance in the sine-Gordon model, in the case of soliton-antisoliton scattering). In this first paper, which is mainly concerned with the similarities between classical mechanics for autocomposite particles and for point particles, it shall be assumed that these phenomena do not occur, so that the number l of world-lines is a constant of the motion. Noting that the world-lines are the lines of current for the topological current k^μ defined by (3.5), a sufficient condition for this is that the vector $\boldsymbol{k}$ is everywhere time-like. We shall assume this stronger condition.

We now proceed to set up a relativistic kinematical description in terms of co-moving coordinates. Putting

$$G^{ab} = g^{\mu\nu}\partial_\mu\varphi^a\partial_\nu\varphi^b = \boldsymbol{\partial}\varphi^a.\boldsymbol{\partial}\varphi^b, \tag{3.12}$$

it is easily seen that the condition that $\boldsymbol{k}$ is time-like is equivalent to the condition that the matrix G^{ab} is negative definite. This enables us to define, at each space-time point $\boldsymbol{x}_{(i)}$ on the world-line $\Gamma_{(i)}(\vec{X})$, n linearly independent basis vectors $\underset{(i)}{\boldsymbol{E}}_a$ orthogonal to the world-line by

$$\underset{(i)}{\boldsymbol{E}}_a = \frac{\partial \boldsymbol{x}_{(i)}}{\partial X^a} = \underset{(i)}{G}_{ab}\,\boldsymbol{\partial}\varphi^b(\boldsymbol{x}_{(i)}) \tag{3.13}$$

where G_{ab} is the matrix inverse of G^{ab}.

Let us now define the local $(n+1)$-velocity $\boldsymbol{u}_{(i)}$ along the world-line $\Gamma_{(i)}(\vec{X})$ by

$$u^\mu = \frac{k^\mu}{\|\boldsymbol{k}\|} = \frac{1}{n!}\frac{\sqrt{|\bar{G}|}}{\sqrt{|g|}}\varepsilon^{\mu\nu\ldots\sigma}\varepsilon_{a\ldots c}\partial_\nu\varphi^a\ldots\partial_\sigma\varphi^c, \tag{3.14}$$

where $\bar{G} = \det G_{ab}$, and the local proper time along the world-line by

$$\mathrm{d}s_\Gamma = \boldsymbol{u}.\mathrm{d}\boldsymbol{x}, \tag{3.15}$$

from which it also follows that

$$\boldsymbol{u} = \frac{\partial \boldsymbol{x}}{\partial s_\Gamma} \tag{3.16}$$

(derivative along the world-line). The time-like coordinate X^0 is then defined by

$$\mathrm{d}X^0 = \sqrt{G^{00}}\,\mathrm{d}s_\Gamma, \tag{3.17}$$

where $G^{00}(\boldsymbol{x})$ is such that $\mathrm{d}X^0$ is a perfect differential[1] (note that, due to the arbitrariness in the definition of X^0, it is always possible to choose this coordinate so that, for instance, $G^{00}(X^0, \vec{X} = 0) = 1$). The corresponding basis vector

$$\boldsymbol{E}_0 = \sqrt{G_{00}}\,\boldsymbol{u}, \tag{3.18}$$

where $G_{00} = (G^{00})^{-1}$, is by construction orthogonal to the n basis vectors $\boldsymbol{E}_a$.

We have thus defined a system of curvilinear coordinates X^α ($\alpha = 0, 1, \ldots, n$), and the associated metric $G_{\alpha\beta}$ (with $G_{0a} = 0$). The components of the $(n+1)$-velocity in the $\boldsymbol{E}$-basis are

$$U^\alpha = \frac{\partial X^\alpha}{\partial s_\Gamma} = \sqrt{G^{00}}\,\delta_0^\alpha \tag{3.19}$$

which means that the X-coordinates are co-moving.

In the case of a non-static solution, it is interesting to compute the covariant derivatives of the local velocity $\boldsymbol{u}_{(i)}$. These are given by

$$D_\beta \boldsymbol{u} = \nabla_\beta U^\alpha \boldsymbol{E}_\alpha = \Gamma^a_{\beta 0} U^0 \boldsymbol{E}_a \tag{3.20}$$

in co-moving coordinates, where $D_\beta = \partial/\partial X^\beta$, and $\Gamma^\alpha_{\beta\gamma}$ is the X-connexion. Thus the local covariant acceleration, which measures the deviation of the world-lines from geodesics, is given in X-coordinates by

$$A^a = \nabla U^a/\nabla s_\Gamma = -\,G^{ab} D_b \operatorname{Log}\sqrt{G_{00}}, \tag{3.21}$$

and the local velocity dispersion is, in X-coordinates,

$$\nabla_b U^a = \frac{1}{2} G^{ac} \frac{\partial G_{bc}}{\partial s_\Gamma}. \tag{3.22}$$

4. Local Equations of Motion

The dynamics of a theory of a p-component field $\vec{\varphi}(\boldsymbol{x})$ may be derived from an action principle, with the action

$$S = \int \mathrm{d}^{n+1} x \sqrt{|g|}\, \mathscr{L}(\vec{\varphi}, \partial_\mu \vec{\varphi}). \tag{4.1}$$

Relativistic invariance of the theory demands that the Lagrangian density $\mathscr{L}$ depends on the field derivatives only through the scalar products $\boldsymbol{\partial}\varphi^a . \boldsymbol{\partial}\varphi^b = G^{ab}$. Specializing now to an autocomposite field theory, with $p = n$, and carrying out the coordinate transformation $x^\mu \to X^\alpha$ introduced in the preceding section, the action may be written as a sum over the families of world-lines:

$$S = \sum_{i=1}^{l} \int \mathrm{d}^{n+1} X \sqrt{|\underset{(i)}{G}(\vec{X})|}\, \mathscr{L}(\vec{X}, \underset{(i)}{G^{ab}}(\vec{X})), \tag{4.2}$$

where $G = \det G_{\alpha\beta} = G_{00}\,\bar{G}$. All that follows shall be derived from this form of the action.

Varying the action (4.2) with respect to the unknown 'fields' $G_{\alpha\beta}$ yields

$$\delta S = -\frac{1}{2}\sum_{i=1}^{l}\int \mathrm{d}^{n+1}X\sqrt{|\underset{(i)}{G}|}\,\underset{(i)}{T}^{\alpha\beta}\,\delta \underset{(i)}{G}_{\alpha\beta}, \tag{4.3}$$

where $T^{\alpha\beta}$ is the canonical energy-momentum tensor in X-coordinates:

$$T^{\alpha\beta} = -\frac{2}{\sqrt{|G|}}\frac{\partial(\sqrt{|G|}\mathscr{L})}{\partial G_{\alpha\beta}}. \tag{4.4}$$

From

$$G_{\alpha\beta} = g_{\mu\nu}D_\alpha x^\mu D_\beta x^\nu, \tag{4.5}$$

it follows that

$$\delta G_{\alpha\beta} = 2g_{\mu\nu}D_\alpha x^\mu D_\beta \delta x^\nu + \partial_\lambda g_{\mu\nu}D_\alpha x^\mu D_\beta x^\nu \delta x^\lambda. \tag{4.6}$$

Carrying (4.6) into (4.3), and integrating by parts, we obtain

$$\begin{aligned}\delta S = &-\sum_{i=1}^{l}\int \mathrm{d}^{n+1}X\,D_\beta[\sqrt{|\underset{(i)}{G}|}\,\underset{(i)}{T}^{\alpha\beta}g_{\mu\nu}(\boldsymbol{x}_{(i)})D_\alpha x^\mu_{(i)}\delta x^\nu_{(i)}]\\ &+\sum_{i=1}^{l}\int \mathrm{d}^{n+1}X\,\{D_\beta[\sqrt{|\underset{(i)}{G}|}\,\underset{(i)}{T}^{\alpha\beta}g_{\mu\nu}(\boldsymbol{x}_{(i)})D_\alpha x^\mu_{(i)}]\\ &-\tfrac{1}{2}\sqrt{|\underset{(i)}{G}|}\,\underset{(i)}{T}^{\alpha\beta}\partial_\nu g_{\lambda\mu}(\boldsymbol{x}_{(i)})D_\alpha x^\mu_{(i)}D_\beta x^\lambda_{(i)}\}\delta x^\nu_{(i)}.\end{aligned} \tag{4.7}$$

The action principle applied to variations of the independent 'fields' $x^\mu_{(i)}(\boldsymbol{X})$ constrained to be zero on a prescribed boundary n-surface thus yields the Euler equations

$$D_\beta[\sqrt{|G|}\,T^{\alpha\beta}g_{\mu\nu}D_\alpha x^\mu] - \tfrac{1}{2}\sqrt{|G|}\,T^{\alpha\beta}\partial_\nu g_{\lambda\mu}D_\alpha x^\mu D_\beta x^\lambda = 0. \tag{4.8}$$

Because of the general covariance of the theory, this may be locally evaluated in the special coordinate system $x^\mu = X^\mu$, and we obtain (lowering indices with the metric $G_{\alpha\beta}$)

$$D_\beta[\sqrt{|G|}\,T^\beta_\alpha] - \tfrac{1}{2}\sqrt{|G|}\,T^{\beta\gamma}D_\alpha G_{\beta\gamma} = 0, \tag{4.9}$$

which means that the covariant divergence [14] of the energy-momentum tensor is (not surprisingly) zero:

$$\nabla_\beta T^\beta_\alpha = 0. \tag{4.10}$$

In fact, these $(n+1)$ equations cannot be independent, as there are only n independent fields. It follows from $G_{0a} = 0$ and from the fact that

$\mathscr{L}$ does not depend on G_{00} that

$$T^{0\beta} = -\mathscr{L}G^{0\beta}; \tag{4.11}$$

hence,

$$D_\beta(\sqrt{|G|}\,T_0^\beta) = -D_0(\mathscr{L}\sqrt{|G|}) = \tfrac{1}{2}\sqrt{|G|}\,T^{\alpha\beta}D_0G_{\alpha\beta}, \tag{4.12}$$

using the definition (4.4) of the energy-momentum tensor. Thus Equation (4.10) for $\alpha = 0$ reduces to an identity. The n independent dynamical equations are

$$\nabla_\beta T_a^\beta = \bar{\nabla}_b T_a^b + \tfrac{1}{2}G^{00}D_bG_{00}(T_a^b - \delta_a^b T_0^0) = 0, \tag{4.13}$$

where the symbol $\bar{\nabla}_b$ stands for the spatial covariant derivative. Putting

$$T^{ab} = \tau^{ab} - \mathscr{L}G^{ab}, \tag{4.14}$$

with $\tau^{ab} = -2\partial\mathscr{L}/\partial G_{ab}$, and using (3.21), Equations (4.13) may be written

$$\bar{\nabla}_b T_a^b = \tau_{ab}A^b. \tag{4.15}$$

These equations can be inverted if $\tau = \det \tau_{ab} \neq 0$, giving the local acceleration in X-coordinates:

$$A^a = \hat{\tau}^{ac}\bar{\nabla}_b T_c^b, \tag{4.16}$$

where $\hat{\tau}_{ac}$ is the matrix inverse of τ_{ac}.

Let us discuss, as an example, the autocomposite field models for $n = 1$ or 2 defined by the Lagrangian density [8]

$$\mathscr{L} = \tfrac{1}{2}\gamma_{ab}(\vec{\varphi})G^{ab} - V(\vec{\varphi}), \tag{4.17}$$

where γ_{ab} is the metric on $\Sigma - \{\Omega\}$ defined in Section 3, and $V = (\lambda^2/2)\gamma_{11}$ for the case $n = 1$ ('normal' form of the Lagrangian), $V = 0$ for $n = 2$. Then,

$$\tau_{ab} = \gamma_{ab}, \tag{4.18}$$

and Equations (4.16) take the form

$$A^a = G^{bc}(\hat{\Gamma}_{bc}^a - \Gamma_{bc}^a) + \hat{\gamma}^{ab}D_bV, \tag{4.19}$$

where $\hat{\Gamma}_{bc}^a$ is the connexion on $\Sigma - \{\Omega\}$.

The corresponding static equations, obtained by putting A^a equal to zero, yield the static solutions of these models [8]. In fact, these solutions were obtained in Reference 8 as solutions of the equations $T_{ab} = 0$ which are indeed solutions of (4.15). The static conditions $T_{ab} = 0$, or the equivalent conditions in x-coordinates $T_{ij} = 0$, have also been shown generally [15] to derive from the metric-independent saturation of the topological lower bound to the action.

We are here interested in the dynamical content of equations (4.19), and more specially in the motion of the 'centres' $\vec{X} = 0$ for which $\gamma_{ab}(0) = \delta_{ab}$, $\hat{\Gamma}^a_{bc}(0) = 0$. The equations of motion of the centres are thus simply

$$A^a = -G^{bc}\Gamma^a_{bc} = \frac{1}{\sqrt{|\bar{G}|}} D_b(\sqrt{|\bar{G}|}\, G^{ab}), \qquad (\vec{X} = 0). \tag{4.20}$$

In the case $n = 1$, assuming now that space-time is flat, these can be written in x-coordinates as

$$\frac{\partial^2 x^\mu}{\partial s^2_\Gamma} = -\frac{\partial x^\mu}{\partial X}\frac{\partial x^\nu}{\partial X}\sqrt{|G^{11}|}\,\partial_\nu(\sqrt{|G^{11}|}), \qquad (X = 0), \tag{4.21}$$

which may be approximated, in the non-relativistic limit, by

$$\frac{\partial^2 x}{\partial t^2} \simeq -\left(\frac{\partial x}{\partial X}\right)^2 \sqrt{|G^{11}|}\,\partial_1(\sqrt{|G^{11}|}), \qquad (X = 0), \tag{4.22}$$

or, noting that in the non-relativistic limit $\partial x/\partial X = \pm\sqrt{|G_{11}|}$,

$$\frac{\partial^2 x_{(i)}}{\partial t^2} \simeq -\frac{\partial}{\partial x}\operatorname{Log}(\lambda^{-1}\sqrt{|G^{11}_{(i)}|}), \qquad (X = 0), \tag{4.23}$$

which are the Newtonian equations of motion for the centres.

These may be checked on the example of the sine-Gordon model, for the soliton-soliton solution in the centre-of-mass referential:

$$\varphi = \frac{\lambda}{\mu}\varepsilon(x)\operatorname{Log}\left|\frac{u \operatorname{sh}\mu\gamma x}{\operatorname{ch}\mu\gamma u t}\right|, \tag{4.24}$$

where $\gamma = (1 - u^2)^{-1/2}$. In this case we find

$$G^{11} = -\lambda^2\gamma^2(\coth^2\mu\gamma x - u^2 \operatorname{th}^2\mu\gamma u t), \tag{4.25}$$

which gives, in the non-relativistic limit ($u \ll 1$, whence $\mu\gamma x \gg 1$ for finite φ):

$$\operatorname{Log}(\lambda^{-1}\sqrt{|G^{11}|}) \simeq 2\mathrm{e}^{-2\mu\gamma|x|} + \frac{u^2}{2\operatorname{ch}^2\mu\gamma u t} + \ldots, \tag{4.26}$$

so that the effective Newtonian potential is

$$V_{\text{eff}}(x) = 2M\mathrm{e}^{-2\mu\gamma|x|}, \tag{4.27}$$

where $M = 2\lambda^2/\mu$ is the soliton mass (such a potential has also been derived by Vinciarelli [16], using a different method). It is easily seen that this potential leads approximately to the correct motion for the centres of the two solitons.

In the case $n = 2$, Equations (4.20) or their non-relativistic counterpart are of a non-Newtonian form. This is not surprising, as the existence of static n-soliton solutions means that the effective static potential is zero (the same happens for a system of gravitating point particles in two space dimensions or, in three space dimensions, for a system of charged particles whose static gravitational attraction and electric repulsion balance exactly). These equations shall be further investigated elsewhere.

5. Energy-Momentum and Angular Momentum

It is assumed in this section that space-time M is flat, and that the metric $g_{\mu\nu}$ is Minkowskian. The total energy-momentum of the field is defined, in terms of the divergenceless energy-momentum tensor ($t^{\mu\nu}$ in x-coordinates) by

$$P^\mu = \int \mathrm{d}\pi_\nu t^{\mu\nu}, \tag{5.1}$$

which does not depend on the space-like n-surface Π.

For a multi-soliton solution, P^μ may be written as the sum of the energy-momenta of the various solitons. There are different ways to do this depending on the choice of Π. A convenient choice is the n-surface $X^0 = \text{constant}$, for which $\mathrm{d}\pi_\nu = \mathrm{d}^n X \sqrt{|\bar{G}|}\, u_\nu$, so that

$$P^\mu = \sum_{i=1}^{l} P^\mu_{(i)}, \tag{5.2}$$

with

$$P^\mu_{(i)}(X^0) = \int \mathrm{d}^n X \sqrt{|\bar{G}_{(i)}|}\, u_{(i)\nu} t^{\mu\nu}_{(i)}. \tag{5.3}$$

Carrying out the coordinate transformation $X^\alpha \to x^\mu$, and using (3.13), (3.16), and (3.17), we obtain

$$t_{\mu\nu} = T^0_0 u_\mu u_\nu + T_{ab}\partial_\mu \varphi^a \partial_\nu \varphi^b. \tag{5.4}$$

Hence, taking into account (4.11) and the orthogonality of $\boldsymbol{u}$ and $\partial\varphi^a$,

$$P^\mu_{(i)}(X^0) = \int \mathrm{d}^n X \sqrt{|\bar{G}_{(i)}|}(-\mathscr{L}_{(i)}) u^\mu_{(i)}. \tag{5.5}$$

This formula for the energy-momentum of the ith soliton may alternatively be derived directly from the action principle. If the action is computed between the two boundary n-surfaces $X^0(1)$ and $X^0(2)$, the variation of the action resulting from an infinitesimal variation of the n-surface $X^0(2)$ is, according to (4.7),

$$\begin{aligned}\delta S &= -\sum_{i=1}^{l} \int \mathrm{d}^{n+1} X D_\beta [\sqrt{|\underset{(i)}{G}|}\, T^{\alpha\beta}_{(i)} g_{\mu\nu} D_\alpha x^\mu_{(i)} \delta x^\nu_{(i)}] \\ &= -\sum_{i=1}^{l} \int_{X^0(2)} \mathrm{d}^n X \sqrt{|\underset{(i)}{\bar{G}}|}\, T^0_{(i)0}\, u_{(i)\nu}\, \delta x^\nu_{(i)},\end{aligned} \tag{5.6}$$

using Gauss's theorem. The variational definition of the energy-momentum

$$P_{(i)\mu} = -\frac{\delta S}{\delta x^\mu_{(i)}}, \tag{5.7}$$

then yields again formula (5.5).

In the case of a scattering solution, we expect that, when $X^0 \to \pm\infty$, the local velocities tend to constants. It follows from the local equations of motion (4.16) that the metric coefficients G_{ab} tend to the static solution, and Equations (3.22) then imply that there can be no dispersion. Asymptotically, the energy-momentum of the ith soliton is therefore

$$P^\mu_{(i)}(\infty) = M u^\mu_{(i)}(\infty), \tag{5.8}$$

where M is the soliton mass (in the case $n = 2$, although there are an infinity of static solutions, the static energy per soliton is still unique [8]).

The total energy-momentum (5.1) is a constant of the motion, but not the individual soliton energy-momenta (except for a static solution). To compute their dependence on X^0 we use the Euler equations (4.8) which, for a Minkowskian metric $g_{\mu\nu}$, reduce to

$$D_0(\sqrt{|G|}(-\mathscr{L})G^{00} D_0 x^\mu) + D_a(\sqrt{|G|}\, T^{ab} D_b x^\mu) = 0. \tag{5.9}$$

It follows that

$$D_0 P^\mu_{(i)} = -\int \mathrm{d}^n X D_a(\sqrt{|\underset{(i)}{G}|}\, T^{ab}_{(i)} D_b x^\mu_{(i)}), \tag{5.10}$$

which gives, using Gauss's theorem,

$$D_0 P^\mu_{(i)} = -\int_{\sigma_\infty(X^0)} \mathrm{d}\sigma_a \sqrt{|\underset{(i)}{G}|}\, T^a_{(i)b}\, \partial^\mu \varphi^b_{(i)}, \tag{5.11}$$

where $\sigma_\infty(X^0)$ is an $(n-1)$-surface at $\vec{X}$-infinity in the n-surface X^0.

Equations (5.11) are the global equations of motion for the various solitons. It is remarkable that the 'forces' (the right-hand sides of these equations) depend only on the behaviour of the fields at $\vec{X}$-infinity, that is in the vicinity of the point Ω of the compact manifold Σ. Note that the convergence of the integrals (5.5) implies $\lim_{\vec{X}\to\infty}|\vec{X}|^n\sqrt{|\bar{G}|}\,\mathscr{L}=0$. We shall assume that this implies also $\lim_{\vec{X}\to\infty}|\vec{X}|^n\sqrt{|\bar{G}|}\,T^b_a=0$. and that G_{00} stays finite everywhere. Then a necessary condition for the existence of non-static multi-soliton solutions is that, at least on portions of $\sigma_\infty(X^0)$, $\partial_\mu\varphi^b\to\infty$ (more quickly than $|\vec{\varphi}|$), which implies $G_{ab}=0$.

In the particular case $n=1$, the point Ω has, at fixed X^0, l antecedents, one of which is $x=\infty$, and the $(l-1)$ others at finite distance 'separate' the various solitons. For a scattering solution, the field φ in the vicinity of the outermost solitons may be approximated, at sufficiently large times, by Lorentz-transforms of the static solution $\varphi(x)=\pm\lambda(x-a)$ [8], so that $\partial_\mu\varphi\to$ constant; thus only the 'separators' contribute to the global acceleration (5.11).

For instance, for soliton-soliton scattering, the separator is, by reason of symmetry, the centre of mass x_G. In the centre-of-mass referential ($x_G=0$) the energies of the two solitons are obviously equal and constant, while their momenta $P^1_\pm$ obey

$$D_0P^1_\pm=\mp\sqrt{G_{00}}\,T^1_1]_{x=0}=\pm\tfrac{1}{2}\sqrt{G_{00}}\,F|G^{11}|]_{x=0}. \tag{5.12}$$

Nothing that, along the centre-of-mass world-line Γ_G, $\mathrm{d}X^0=\sqrt{G^{00}}\,\mathrm{d}t$, this may be integrated to give

$$P^1_\pm=\pm\left[\int_0^t \mathrm{d}t\,\tfrac{1}{2}F|G^{11}|\right]_{x=0}. \tag{5.13}$$

In the example of the sine-Gordon model, this procedure yields

$$P^1_\pm=\pm M\gamma u\,\mathrm{th}\,\mu\gamma ut \tag{5.14}$$

(which may also be obtained by direct integration of the momentum density t^{01} on the half-line $t=$ constant, $x>0$), where t may be re-expressed in terms of X^0 (measured along Γ_G). This may be compared with the purely kinetic contribution

$$Mu^1=\pm M\lambda|G^{11}|^{-1/2}u\,\mathrm{th}\,\mu\gamma ut, \tag{5.15}$$

along the 0-world-line, with G^{11} given by (4.25).

The method used to define the energy-momenta of the various solitons may also be applied to the angular momentum tensor. The total angular momentum relative to the origin of x-coordinates is

$$M^{\mu\nu} = \int d\pi_\lambda (t^{\mu\lambda} x^\nu - t^{\nu\lambda} x^\mu). \tag{5.16}$$

Choosing for Π the n-surface $X^0 = \text{constant}$, this may be written as

$$M^{\mu\nu} = \sum_{i=1}^{l} M^{\mu\nu}_{(i)}, \tag{5.17}$$

with the angular momentum of the ith soliton given by

$$M^{\mu\nu}_{(i)}(X^0) = \int d^n X \sqrt{|\bar{G}_{(i)}|}(-\mathscr{L}_{(i)})(u^\mu_{(i)} x^\nu_{(i)} - u^\nu_{(i)} x^\mu_{(i)}), \tag{5.18}$$

which may also be derived from the action principle via (5.6). The method already used for the energy-momentum gives for the X^0-derivative of this angular momentum:

$$D_0 M^{\mu\nu}_{(i)} = - \int_{\sigma\ (X^0)} d\sigma_a \sqrt{|G_{(i)}|}\, T^{ab}_{(i)} (D_b x^\mu_{(i)} . x^\nu_{(i)} - D_b x^\nu_{(i)} . x^\mu_{(i)}). \tag{5.19}$$

6. Internal symmetries

For the autocomposite field theories under study, an internal symmetry may be defined as a local coordinate transformation on $\Sigma - \{\Omega\}$ which leaves the action (4.1) or (4.2) invariant. For the $n = 2$ models, as the Lagrangian density (4.17) with $V = 0$ depends only on the invariant $\gamma_{ab} G^{ab}$, these transformations are the general coordinate transformations. The same is true of the Skyrme model defined by (2.7).

For any infinitesimal symmetry transformation $\varphi^a \to \varphi^a + \psi^a(\vec{\varphi})$, the current

$$J^\mu = \frac{\partial \mathscr{L}}{\partial(\partial_\mu \varphi^a)} \psi^a = \tau_{ab} \partial^\mu \varphi^b \psi^a \tag{6.1}$$

is divergenceless:

$$\partial_\mu J^\mu = 0. \tag{6.2}$$

It follows that the charge defined by

$$Q = \int \mathrm{d}\pi_\nu J^\nu \tag{6.3}$$

does not depend on the space-like n-surface Π. We may take advantage of this fact to compute the charge Q on the surface $X^0 =$ constant, and obtain the very simple result

$$Q = \sum_{i=1}^{l} \int \mathrm{d}^n X \sqrt{|\underset{(i)}{\bar{G}}|}\, u_{(i)\nu} J^\nu_{(i)} = 0, \tag{6.4}$$

owing to the orthogonality of $\boldsymbol{u}$ and $\partial\varphi^a$. This result may again be derived directly from the action principle using (5.6), and noting that $T^{0a} = 0$.

Thus, the dynamical charge carried by any localized solution of an autocomposite field theory is identically zero, provided (3.17) is integrable.[1] While this does not completely exclude the possibility of localized solutions with non-zero charges, it is doubtful whether these exist for $n = 2$ (in this case, the one-soliton time-dependent charged solution found by Duff and Isham [17] can be shown to be non-localizable.

7. Conclusion

In this paper, we have investigated a formulation of classical mechanics for autocomposite particles, in a class of models of n interacting scalar fields on $(n + 1)$-dimensional space-time. The key to this formulation is that a localized solution of these models is a mapping (of space-time into an n-manifold) of degree k. This mapping may be inverted, so that the original fields may be taken as local space-like coordinates X^a, a local time-like coordinate X^0 being then defined so that the coordinates X^α are co-moving. The l inverse mappings ($l \geqslant |k|$; we have restricted ourselves to the case $l =$ constant) $\boldsymbol{x}_{(i)}(\boldsymbol{X})$ give the original (e.g. Minkowskian) coordinates as a set of l fields depending on the co-moving coordinates X^α. The classical mechanics of the l solitons may then be extracted from the classical theory of these fields.

After this preliminary exploration, further investigations should be conducted along the following three major axes:

(1) *Classical mechanics*. Among the many problems which remain are:
– the extension of the present formalism to the case of non-constant l (implying classical, real or virtual, pair creation or annihilation);

–the definition of the centre of mass of a soliton, which is obvious only in the static case;
–the study of the scattering of autocomposite particles;
–the Hamiltonian formulation;
–the finding of exact or approximate solutions, for the cases $n = 2$ or 3, of Equations (4.16), supplemented by the condition that space-time is flat, i.e. that the X-Riemann tensor vanishes:

$$R_{\alpha\beta\gamma\delta} = 0, \tag{7.1}$$

which for n = 2 reduces to the vanishing of the X-Ricci tensor,

$$R_{\alpha\beta} = 0; \tag{7.2}$$

for the case $n = 2$, a possible tool for this might be a suitable modification of the post-Newtonian approximation (used in general relativity), starting from the known static solutions.

(2) *General relativity*. The coupled Einstein-autocomposite field equations have been solved for the static $n = 2$ case in [8]. The present formalism being generally covariant is appropriate for the treatment of the dynamical case. The dynamical equations are then the Einstein equations in X-coordinates

$$R_{\alpha\beta} - \tfrac{1}{2} R G_{\alpha\beta} = \chi T_{\alpha\beta}, \tag{7.3}$$

from which follow Equations (4.10), hence also Equations (4.16). The problems which should then be investigated are:
–the finding of (approximate) solutions to these equations;
–the general-relativistic definition of total energy-momentum and angular momentum of autocomposite particles;
–in the case $n = 3$, the possibility of solutions with unusual space-time topology.

(3) *Quantum mechanics*. It has been suggested [16] that the knowledge of the classical mechanical properties of solitons might be useful for their approximate quantization, via the correspondence principle. A more radical approach, in the spirit of the classical formalism presented here, would be to quantize the inverse 'fields' $x^{\mu}_{(i)}(X)$ using the canonical quantization procedure (field quantization in curvilinear coordinates on flat space-time), and to extract (just as in the classical case) the quantum mechanics of the solitons from the quantum theory of their coordinate fields. This should provide an interesting alternative to the usual soliton quantization procedure [18] based on collective coordinates. For the

case $n = 3$, this approach should be used to investigate the generation of spin [19], which is known to appear, at the quantum level, in Skyrme's model [12, 20], and of isospin.

Dept. de Physique, Université de Constantine, Algeria

NOTE

[1] dX^0 is a perfect differential if the integrability conditions $\varepsilon^{\mu\nu\rho} u_\mu \partial_\nu u_\rho = 0$ are satisfied; these are of course identically satisfied for $n = 1$, but not necessarily for $n \geq 2$. However, most of the following results may readily be extended to the case of an arbitrary time-like coordinate X^0.

REFERENCES

[1] Jouvet, B., and Le Guillou, J. C., *Nuovo Cimento* **49**, 677 (1967).
[2] Le Guillou, J. C., *Nuovo Cimento* **54A**, 3 (1968).
[3] Clément, G., Rapport interne P.A.M., 70–05 (1970).
[4] Freundlich, Y., and Lurié, D., *Phys. Rev.* **D1**, 1660 (1970).
[5] Clément, G., Thèse de doctorat, Paris-Sud, 1974.
[6] Sugawara, H., *Phys. Rev.* **170**, 1659 (1968).
C. M. Sommerfield, *Phys. Rev.* **176**, 2019 (1968).
[7] Belavin, A. A. and Polyakov, A. M., *JETP Lett.* **22**, 245 (1975).
[8] Clément, G., *Nucl. Phys.* **B114**, 437 (1976).
[9] Matsuda, T., *Prog. Theor. Phys.* **57**, 620 (1977).
[10] Perelomov, A. M., *Comm. Math. Phys.* **63**, 237 (1978).
[11] Derrick, G. H., *J. Math. Phys.* **5**, 1252 (1964).
[12] Skyrme, T. H. R., *Proc. Roy. Soc. (London)* **A260**, 127 (1961); *J. Math. Phys.* **12**, 1735 (1971).
[13] Pak, N. K., and Tze, H. C., *Ann. Phys.* **117**, 164 (1979).
N. K. Pak, *Nuovo Cimento* **46A**, 431 (1978).
[14] Landau, L., and Lifchitz, E., *The Classical Theory of Fields*, Addison-Wesley, Reading, Mass., 1971.
[15] Hoscya, A., *Progr. Theor. Phys.* **59**, 1781 (1978).
[16] Vinciarelli, P., *Phys. Lett.* **61B**, 80 (1976).
[17] Duff, M. J., and Isham, C. J., *Nucl. Phys.* **B108** 130 (1976); *Phys. Rev.* **D16**, 3047 (1977).
[18] Gervais, J. L., Jevicki, A., and Sakita, B., *Phys. Rev.* **D12**, 1038 (1975).
Tomboulis, E., and Woo, G., *Ann. Phys.* **98**, 1 (1976).
[19] Finkelstein, D., and Rubinstein, J., *J. Math. Phys.* **9**, 1762 (1968).
Williams, J. G., *J. Math. Phys.* **13**, 1165 (1972).
[20] Williams, J. G., *J. Math. Phys.* **11**, 2611 (1970).

S. DESER

EXCLUSION OF STATIC SOLUTIONS IN GRAVITY-MATTER COUPLING

ABSTRACT. Coupled Einstein-matter systems of various spins are shown to possess no static non-singular solutions. In particular, this is demonstrated for spin 1/2 with or without torsion by iterating the Dirac operator, and partial results are given for supergravity.

INTRODUCTION

It is well-known that there are no non-singular static solutions for free massless or massive fields of any spin in flat space, basically as a consequence of the energy-momentum relations governing propagation of their excitations. This property persists also (at least in four dimensions) in non-Abelian gauge theories, including general relativity, despite their self-coupling. There is therefore no reason to expect dramatic changes when the matter systems are coupled to gravity. However, the derivation is sufficiently different in the case of spinor fields to warrant discussion. Complications arise here for two reasons: in contrast to spin 0 or 1, covariant derivatives are unavoidable and the necessity of dealing with the second-order form of the Dirac equation brings in curvature dependence explicitly. These difficulties are also of interest as an introduction to the same problem in supergravity, representing a new type of gauge system.

After reviewing pure gravity and its coupling to spins 0 and 1, we will consider spin 1/2 and discuss the associated second-order equation it satisfies in presence of gravity, with or without torsion. The desired result will then be established there. We conclude with some partial results for supergravity based on the iterated Rarita–Schwinger equation.

GRAVITY-INTEGRAL SPIN COUPLING

Static gravitational fields are characterized by the existence of coordinate systems in which both all time dependence ($\partial_0 g_{\mu\nu}$) and the mixed (g_{0i}) components of the metric vanish. Although there are no non-singular solutions in the more general stationary ($g_{0i} \neq 0$) situation [1] either, we shall restrict ourselves to the static case throughout. The Einstein equations in the presence of sources reduce to

$$^3G^{ij} + N^{-1}(g^{ij}\nabla^2 - D^iD^j)N = \tfrac{1}{2}T^{ij} \qquad (1)$$

E. Tirapegui (Ed.), Field Theory, Quantization and Statistical Physics, 77–84.

together with the initial value constraints

$$^{3}R = T^{0}_{0}, \qquad 0 = T^{0}_{i}. \tag{2}$$

Here $N = \sqrt{-g_{00}}$, and all operations are with respect to the three-dimensional metric g_{ij}. Tracing (1), we obtain, using (2),

$$\nabla^2 N = 0 \tag{3}$$

if the sources vanish. This is an elliptic equation whose only solution is $N =$ const., since there are no other harmonic functions for asymptotically flat space. But then it follows that G_{ij} vanishes. Ricci and Riemann tensors being equivalent in three dimensions, the three-space is therefore flat, and so is the full space-time, since we can construct coordinate systems in which $g_{ij} = \delta_{ij}$ while $g_{0\mu} = -\delta_{0\mu}$. Thus, there are no static source-free gravitational solutions.

Consider next coupling to spin 0 or 1 fields. We shall show that they cannot have static excitations in a curved but static geometry. As a consequence, their stress tensor will also vanish; the discussion above will then guarantee flatness as well. For a scalar field ϕ, the field equation

$$\partial_\mu(\sqrt{-g}g^{\mu\nu}\partial_\nu)\phi = 0 \tag{4}$$

reduces to

$$\partial_i(N\sqrt{^3g}\,g^{ij}\partial_j\phi) = 0 \tag{5}$$

in a static space. This is just the Laplace equation, but for the factor N which enters because $\sqrt{-g} = N\sqrt{^3g}$. However, from its definition, the sign of N is everywhere the same in the absence of singularities. This is therefore sufficient to prove that ϕ is constant (and vanishes, with appropriate boundary conditions). We multiply (5) by ϕ and integrate over three-space to obtain

$$\int d^3x[\sqrt{^3g}\,g^{ij}\partial_i\phi\,\partial_j\phi]N = 0. \tag{6}$$

Since the bracket, $(\nabla\phi)^2$, is non-negative, it must vanish, and hence $\partial_i\phi$ and the stress-tensor also vanish. Clearly, the proof could be extended to include a massive ϕ field or any self-interaction $V(\phi)$ which has the same sign as the mass term everywhere. In that case, the slightly stronger statement $\phi = 0$ (required to have vanishing $T_{\mu\nu}$ there) would follow.

For the Maxwell field the proof is again a simple generalization of the flat space one, where the vanishing of E and B follows from the fact that both are simultaneously curl-and divergence-free when there is no time independence. The Maxwell equations are

$$\partial_\mu[\sqrt{-g}g^{\mu\alpha}g^{\nu\beta}F_{\alpha\beta}]=0, \qquad F_{\alpha\beta}\equiv\partial_\alpha A_\beta-\partial_\beta A_\alpha. \tag{7}$$

For static systems, these reduce to

$$\partial_i[N\sqrt{{}^3g}g^{ij}g^{00}F_{j0}]=0, \tag{8a}$$

$$\partial_i[N\sqrt{{}^3g}g^{ij}g^{kl}F_{jl}]=0. \tag{8b}$$

In a static gauge, where $\partial_0 A_i=0$ (which can always be achieved) we multiply (8a) by A_0 and integrate to obtain

$$0=\int \mathrm{d}^3x[\sqrt{{}^3g}g^{ij}\partial_i A_0\partial_j A_0]N^{-1} \tag{9}$$

from which it follows that $\partial_i A_0$, and hence F_{0i}, vanishes. Similarly, multiplication of (8b) A_k yields

$$0=\int \mathrm{d}^3x[\sqrt{{}^3g}g^{ij}g^{kl}F_{ik}F_{jl}]N \tag{10}$$

so that $F_{ij}=0$ as well and we are back to the trivial source-free Einstein case. Actually, the theorem also holds in stationary space-time [2].

For the Yang–Mills case, the above derivation fails (as it does for flat space!) because of the presence of (Yang–Mills) covariant divergences in (8) along with that of the usual A^2 nonlinear term in the definition of $F_{\mu\nu}$. It is still possible to derive the vanishing of $\vec{F}_{0i}$ by working in static gauge ($\partial_0\vec{A}_i=0$), where $\vec{F}_{0j}=\vec{D}_j\vec{A}_0$, since multiplication of (8a) by $\vec{A}_0$ and integration then yields the equivalent of (9). This suffices for the momentum density T^{0i} to vanish as required by (2). However, it is still to be shown that $\vec{F}_{ij}=0$; in flat space, this is obtained [3] from the vanishing of the self-stress, $\int \mathrm{d}^3x T^i_i$ as a result of $\int(\mathrm{d}^3x)x^i T^j_{i,j}=0$. This in turn follows from the existence of dilation invariance, which is related to tracelessness of $T_{\mu\nu}$($\xi^\mu=x^\mu$ is a conformal Killing vector, so that $(\xi^\mu T^\nu_\mu)_{,\nu}=0$, in flat space). This vector is not directly available in curved space, however. Although it should be possible to deduce $\vec{F}_{ij}=0$ directly from the equivalent of (8b), this has not yet been done even in flat space. It should be noted that the above argument excludes solutions in all but 4 space (5 space-time) dimensions, where instantons can be present [4]. Thus (8b) alone is not enough, and explicit use of the dimensionality must enter in the non-abelian case. In the absence of this type of derivation, one would have to look at the equivalent of the self-stress condition so as to include the 'stress-tensor' of the gravitational field. However, this would involve detailed consideration of the proof of positive energy in general relativity to establish that the gravitational contribution to the stress tensor' is non-negative and therefore that the Yang–Mills part also vanishes. The vanishing of $\vec{F}_{0i} \equiv D_i\vec{A}_0$ actually also implies that of $\vec{F}_{ij}$ in any dimensions. From $[D_i, D_j]\vec{A}_0 \equiv \vec{F}_{ij} x \vec{A}_0$ it follows that $\vec{F}_{ij}$ is parallel to $\vec{A}_0$ for any semisimple internal group. Now $\vec{A}_0$ is an 'isovector' (because its inhomogeneous transformation part vanishes within static gauges) and may be taken to define a direction in the internal space. Therefore $\vec{F}_{ij}$ is purely abelian and may be represented as the oridnary curl of some potential B_i in the $\vec{A}_0$ direction. The corresponding component of the field equation $D_i\vec{F}^{ij} = 0$ then reduces to an ordinary Maxwell one which implies that $\vec{F}_{ij}$, vanishes. Of course, once $\vec{F}_{ij}$ vanishes, it follows that $\vec{A}_0$ also does because now $D_i\vec{A}_0 \rightarrow \partial_i\vec{A}_0$. This argument carries over intact to the gravitationally coupled case. However, it does not handle systems (such as instantons) for which $\vec{A}_0 = 0$ *ab initio*. It is therefore still a challenge to prove directly that the field equations $D_i\vec{F}^{ij} = 0$ imply $\vec{F}_{ij} = 0$ in order to be able to establish the conjecture for Yang–Mills coupled to gravity even in four space–time dimensions.

Spin 1/2 – gravity coupling

In order to investigate the static behavior of the spin $\frac{1}{2}$ system, we must first go over to the second-order form of the Dirac equation, since the Laplace operator is the key to exclusion of static solutions. In flat space, this is trivial since

$$(\gamma.\nabla)(\gamma.\nabla)\psi = \nabla^2\psi = 0 \Rightarrow \psi = 0. \tag{12}$$

(The result clearly holds in presence of mass since $(\nabla^2 - m^2)$ is also a positive operator.) However, in curved space, the Dirac equation reads

$$\gamma^\mu D_\mu \psi \equiv (e^\mu_\alpha \gamma^a) D_\mu \psi = 0, \qquad D_\mu \equiv \partial_\mu + \tfrac{1}{2}\omega_{\mu ab}\sigma^{ab}. \tag{13}$$

Here e^μ_a are the (inverse) vierbeins and $\omega_{\mu ab}$ the Ricci coefficients. We consider first the torsion-free system, so that $\omega_{\mu ab}$ is not an independent variable, but the usual function of the vierbeins and their derivatives. The second-order operator obtained by iterating $\gamma.D$ has the well-known form

$$(\gamma.D)^2\psi \equiv (D^2 + \tfrac{1}{4}R)\psi \tag{14}$$

due to the Ricci identity. This follows from

$$\gamma^\mu\gamma^\nu D_\mu D_\nu = g^{\mu\nu} D_\mu D_\nu + \tfrac{1}{4}[\gamma^\mu, \gamma^\nu][D_\mu, D_\nu] \equiv D^2 - \tfrac{1}{2}\sigma^{\mu\nu} R_{\mu\nu ab}\sigma^{ab} \tag{15}$$

and the fact that only the anticommutator of the σ's contributes. There are now two additional terms to include in the static limit, namely $\frac{1}{4}$ R and the $\omega_0.\sigma$ part of the D_0, which does not vanish even though ∂_0 does; this is the unavoidable covariant derivative structure mentioned earlier. Fortunately, both can be dealt with. The scalar curvature vanishes by virtue of the Einstein equation, since it is proportional to the trace of the stress tensor of the spinor field, and the latter vanishes by virtue of the Dirac equation, as it does in flat space. We therefore have the equation

$$[\nabla^2 + g^{00}(\tfrac{1}{2}\omega_{0ab}\sigma^{ab})^2]\psi = 0 \tag{16}$$

in the static limit.

Now

$$\begin{aligned}(\omega_0,\sigma)^2 &\equiv \tfrac{1}{2}\omega_{0ab}\omega_{0cd}\{\sigma^{ab},\sigma^{cd}\} \\ &\equiv -\omega_{0ij}^2 + 2(\omega_{0i0})^2.\end{aligned} \tag{17}$$

The negative term vanishes for static geometries as is most easily seen in time gauge where e_{0i} vanishes. [The choice of time gauge is just a determination of three of the six antisymmetric components of $e_{\mu a}$, and so does not conflict with the static metric coordinate conditions]. Basically, all quantities with an odd number of time indices vanish in this gauge. An alternate, more elegant way to see that the ω contribution to (16) is positive is to use Schwinger's [5] remark that in the Dirac action, because of anticommutativity of the ψ's, the term $\bar{\psi}\gamma^\mu\sigma_{ab}\psi\omega_{\mu ab}$ only contributes through the anticommutator $\{\gamma^\mu,\sigma^{ab}\} = \varepsilon^{\mu abc}\gamma_5\gamma_c$. It therefore has the simple form $\bar{\psi}\gamma_5\gamma_c\psi{}^*\omega^c$, where ${}^*\omega^c$ is the totally antisymmetric part $\varepsilon^{\mu abc}\omega_{\mu ab}$. Thus, $D_0^2 \to (\gamma_5{}^*\omega^0)^2 = -({}^*\omega_0)^2 \leqslant 0$ and we have the elliptic operator

$$(\nabla^2 - ({}^*\omega_0)^2]\psi = 0 \tag{18}$$

which ensures that there are no static solutions. Indeed, in this case ψ (and not just its gradient) vanishes. Once ψ vanishes, we may go back to the original form of the source-free Einstein equations to establish flatness.

Consider now the case where torsion is present. This is equivalent to the previous model with an additional term $\sim |e|(\bar{\psi}\gamma_5\gamma^a\psi)^2 \sim \psi^4$. For this system, the Dirac equation reads $\gamma.D\psi = \psi^3$, and the trace of the stress tensor has the form $a\bar{\psi}\gamma.D\psi + b\psi^4$, so that on mass shell, $T^\mu_\mu \sim \psi^4$. On the other hand, multiplying the Dirac equation by $\gamma.D$ now yields

$$(\Box + \tfrac{1}{4}R)\psi + \gamma.D\psi^3 = 0. \tag{19}$$

For an anticommuting Majorana spinor ψ, any power of ψ greater than quartic vanishes. Consequently, both $\gamma.D\psi^3 \sim \psi^5$ and $\frac{1}{4}R\psi \sim \psi^5$ disappear from the second-order equation (as would be the case for any quartic self-interaction). We are then back to the case treated above, and there are again no static excitations in present of torsion either. Of course, these remarks (as well as all our derivations) hold only at the 'classical' level in which ψ is an anti-commuting c-number rather than an operator which obeys non-trivial anticommutation relations.

Finally, we remark that Birkhoff's theorem that there are no nonsingular spherically symmetric (but not necessarily time independent) excitations should be easily extendable in the presence of massless spin > 0 fields, since the spin 1 field has none and the stress tensor of the spin 1/2 field vanishes if it is spherically symmetric.

Supergravity

We close with some preliminary results on supergravity. In flat space, one merely iterates the Rarita–Schwinger operator

$$(R_0\psi)^\mu \equiv \varepsilon^{\mu\nu\alpha\beta}\gamma_5\gamma_\nu\partial_\alpha\psi_\beta \equiv \gamma_5\gamma_\nu{}^*f^{\mu\nu} \tag{20}$$

to get the desired second-order equation,

$$\begin{aligned}(R_0^2\psi)^\mu &= \varepsilon^{\mu\nu\alpha\beta}\gamma_5\gamma_\nu\partial_\alpha(R_0\psi)_\beta \\ &= \partial_\nu f^{\nu\mu} - \sigma^{\mu\nu}\partial_\alpha f^\alpha_\nu + \tfrac{1}{2}\partial_\mu\sigma.f\end{aligned} \tag{21}$$

where $f_{\mu\nu} \equiv \partial_\mu\psi_\nu - \partial_\nu\psi_\mu$ is the gauge invariant field strength, ${}^*f^{\mu\nu} \equiv \frac{1}{2}\varepsilon^{\mu\nu\alpha\beta}f_{\alpha\beta}$ and $\sigma_{\mu\nu} \equiv \frac{1}{4}[\gamma_\mu, \gamma_\nu]$. Using the first-order equations in the equivalent forms $\gamma_\nu{}^*f^{\mu\nu} = 0 = \gamma_\nu f^{\mu\nu}$, we finally obtain the Maxwell equation $\partial_\mu f^{\mu\nu} = 0$. A more elegant derivation starts with the operator $\gamma_\alpha\sigma^{\mu\nu}(R_0\psi)^\alpha$. This is a projection to the subspace of γ-orthogonal vectors, $\tilde{R}^\mu \equiv R^\mu - \frac{1}{4}\gamma^\mu\gamma.R$; it is clearly singular, since $\sigma_{\mu\nu}\gamma_\alpha\sigma^{\mu\nu} \equiv 0$. The trace part $\gamma.R$ is simply proportional to $\sigma.f$ or equivalently to $\gamma_5\sigma.{}^*f$ so the Rarita–Schwinger equation reduces to the pair

$$H^{\mu\nu} \equiv f^{\mu\nu} + \gamma_5{}^*f^{\mu\nu} = 0, \qquad \sigma^{\mu\nu}f_{\mu\nu} = 0 \tag{22}$$

with $\sigma{\cdot}H \equiv 0$ and $\gamma_\mu H^{\mu\nu} = \tilde{R}^\nu$. One may prove the absence of static solutions particularly simply from the first of (22) and the Bianchi identity $\partial_\mu{}^*f^{\mu\nu} \equiv 0$, without explicit use of the second-order equations: with the usual notation $f^{0i} = E^i, \frac{1}{2}\varepsilon^{ijk}f_{jk} \equiv B^i$, we have

$$\boldsymbol{E} + \gamma_5\boldsymbol{B} = 0 \tag{23a}$$

$$\boldsymbol{\nabla}\cdot\boldsymbol{B} \equiv 0 = \boldsymbol{\nabla}\times\boldsymbol{E} \tag{23b}$$

in the static case. Since $\boldsymbol{E}, \boldsymbol{B}$ are respectively curl- and divergence- free, it follows from (23a) that they separately vanish.

To obtain the second-order equations one may simply consider

$$\partial_\mu H^{\mu\nu} \equiv \partial_\mu f^{\mu\nu} = 0. \tag{24}$$

The second order form of $\sigma \cdot f = 0$ is included here because (by the cyclic identities on $f_{\alpha\beta}$

$$\partial_\mu \sigma^{\alpha\beta} f_{\alpha\beta} \equiv \sigma^{\alpha\beta} \partial_\alpha f_{\mu\beta} \equiv \tfrac{1}{2}(\gamma_\alpha \gamma_\beta - \eta_{\alpha\beta}) \partial^\alpha f^\beta_\mu$$

which is itself proportional to $\partial_\alpha f^{\mu\nu}$, using $\gamma_\mu f^{\mu\nu} = 0$. The absence of static solution again follows from (24), which states that $\nabla \cdot \boldsymbol{E} = \nabla x \boldsymbol{B} = 0$, together with (23b).

In supergravity [6], the situation becomes more complicated. The covariant forms of (22) are still the correct field equations (with $\partial_\mu \to D_\mu \equiv \partial_\mu + \frac{1}{2}\omega_{\mu ab}\sigma^{ab}$ in $f_{\mu\nu}$), and

$$H^{\mu\nu} \equiv (\sqrt{-g} g^{\mu\alpha} g^{\nu\beta} + \gamma_5 \tfrac{1}{2} \varepsilon^{\mu\nu\alpha\beta}) f_{\alpha\beta} = 0, \tag{25a}$$

$$e^\mu_a e^\nu_b \sigma^{ab} f_{\mu\nu} = 0. \tag{25b}$$

Note that $H^{\mu\nu}$ does not involve the vierbeins, $e_{\mu a}$ or e^ν_a and that D_μ commutes with $g_{\alpha\beta}$. The fundamental identity $\partial_\mu R^\mu \equiv 0$ of flat space reads

$$0 = \partial_\mu \tilde{R}^\mu + \tfrac{1}{4}\partial_\mu \gamma^\mu \gamma \cdot R \equiv \partial_\mu (\gamma_\nu H^{\mu\nu}) + \gamma \cdot \partial \sigma \cdot f \tag{25c}$$

and its implementation in supergravity will of course involve use of the Einstein and torsion equations, as can be seen by explicit consideration of the vierbein and D_μ dependence of (25c) in curved space. To reach the second-order form, we may take the divergence of (25a), using the above covariant derivative appropriate to spin 1/2 (rather than 3/2). This is because the ordinary divergence of a contravariant antisymmetric tensor density such as $(\sqrt{-g} f^{\mu\nu})$ or $*f^{\mu\nu}$ is a vector density. [The same holds when taking the divergence of the contravariant density $D_\mu {*f^{\mu\nu}}$ or $D_\mu(\sqrt{-g} f^{\mu\nu})$. We note, therefore, that

$$D_\mu D_\nu H^{\mu\nu} \equiv \tfrac{1}{2}[D_\mu, D_\nu] H^{\mu\nu} \equiv \tfrac{1}{4} R_{\mu\nu} \cdot \sigma H^{\mu\nu}$$

correctly vanishes by virtue of $H^{\mu\nu} = 0$.] Of course, we could have used any other operator (covariant or not!) including the spin 3/2 D_μ, but the net effect is the same since they would all differ by an affinity times the original Equation (25a). Although $D_\mu(\omega)$ does commute through all the explicit metrics in (25a), it does not commute with the D's in $*f^{\mu\nu}$. From the Ricci identity on a spinor,

$$[D_\mu, D_\nu]\psi = -\tfrac{1}{2} R_{\mu\nu} \cdot \sigma \psi \tag{26}$$

we find that [7]

$$D_\mu[\sqrt{-g}f^{\mu\nu} + \gamma_5 {}^*f^{\mu\nu}] = (D^2\psi_\nu - D_\mu D_\nu \psi^\mu) - \tfrac{1}{2}{}^*R^{*\mu\nu}\cdot\sigma\psi_\mu = 0 \tag{27}$$

where

$${}^*R^{*\mu\nu cd} \equiv \tfrac{1}{4}\varepsilon^{\mu\nu\alpha\beta}\varepsilon^{cdab}R_{\alpha\beta ab}$$

and we have used the identity $\gamma_5\sigma^{\mu\nu} \equiv \frac{1}{2}\varepsilon^{\mu\nu\alpha\beta}\sigma_{\alpha\beta}$. Thus, the second-order equation has the (covariant) Maxwell operator plus an additional term proportional to the *full* curvature (note that ${}^*R^* = -R +$ Ricci). It is no help to commute the mixed derivatives in the Maxwell operator, since the $D_\mu D_\nu \equiv D_\nu D_\mu + [D_\mu, D_\nu]$, commutator doubles, rather than cancels, the ${}^*R^*$ contribution. Thus, we are left with an equation, (27), which is considerably more complicated than its spin 1/2 counterpart. The ${}^*R^*\cdot\sigma$ curvature contribution contains a Weyl tensor part which is left undetermined by the Einstein equations, along with parts proportional to the Einstein tensor or equivalently to the spin 3/2 stress tensor. The latter may simplify by Fierz transformations, and indeed may even be proportional to the Rarita–Schwinger operator and so vanish. (The form is $(\bar{\psi}\gamma_5\gamma D\psi)(\sigma\psi)$.) Clearly, a demonstration that there are no static solutions in supergravity will have to deal with these complications. We hope to return to this elsewhere, but point out that the main apparent obstacle to establishing the result, namely the Weyl tensor, is in fact probably absent because the space is static and therefore contains no gravitational excitations, which that tensor describes.

Work supported in part by NSF PHY-78-09644.

Acknowledgements

I am grateful to Drs. M. Duff and J. H. Kay for useful discussions.

References

[1] Lichnerowicz, A., *Theories Relativistes de la Gravitation*, Masson, Paris, 1955.
[2] Thiry, Y., *J. Math. Pure Appl.* **30**, 275 (1951).
[3] Deser, S., *Phys. Lett.* **64B**, 463 (1976).
Coleman, S., 1975 Erice Lectures.
[4] Belavin, A. A., *et al., Phys. Lett.* **59B**, 85 (1975).
[5] Schwinger, J., *Particles, Sources and Fields*, Addison-Wesley, Reading, Mass., 1970.
[6] Freedman, D. Z., van Nieuwenhuizen, P., and Ferrara, S., *Phys. Rev.* **D13**, 3214 (1976).
Deser, S., Zumino, B., *Phys. Lett.* **62B**, 335 (1976).
[7] The second-order Rarita–Schwinger equation has also been derived by S. Christensen and M. Duff and by S. Hawking and C. Pope (to be published).

D. ARNAL, J. C. CORTET, M. FLATO AND D. STERNHEIMER

STAR-PRODUCTS: QUANTIZATION AND REPRESENTATIONS WITHOUT OPERATORS

ABSTRACT. In this contribution we briefly review the deformation (phase-space) approach to quantization in both quantum mechanics and quantum field theory. This leads to a new framework for group representations by star-products rather than operators; some examples of these are studied more in details, in particular induced star-representations of the Poincaré group.

INTRODUCTION

From the beginning of quantum theory attempts have been made to establish a bridge between classical and quantum mechanics. Since phase-space was introduced for the Hamiltonian formulation of classical mechanics, it was natural to look for a phase-space formulation of quantum mechanics. The probabilistic character of quantum mechanics invited people to try, inverting Weyl's quantization prescription [1], to interpret quantum mechanics as a statistical theory over phase-space.

A first attempt in this direction was done by Wigner [2] who introduced a distribution function (over phase-space) to stand for a quantum mechanical state; a more systematic attempt along these lines was made by Moyal [3] and later by many other authors. However, there were difficulties in these kind of approaches: for instance, the distribution functions were not positive, etc.

Another aspect of quantization, which was probably in the back of the mind of many authors, is its relation to deformations of classical theories. Some terminology and methodology are characteristic of this line of thought, such as the classical limit (when $\hbar \to 0$) or semi-classical (W.K.B., etc.) approximations (developments in series of powers of $\hbar$; for a comprehensive presentation of these approximations, see Voros [4]). However in these approaches the relation to mathematical deformation theory, though intuitive, was not very clear due to the essential difference in the formulations: functions over phase-space in classical mechanics, operators over functions on (e.g.) position space in quantum mechanics.

In the so-called geometric quantization [5], a more general and intrinsic

E. Tirapegui (Ed.), Field Theory, Quantization and Statistical Physics, 85–111.

attempt is made to explain the process of quantization (for the enveloping algebra of a Lie algebra of functions over a symplectic manifold). But in this case also there is a fundamental change in the nature of the observables.

On the other hand, the mathematical theory of deformations [6] (of Lie groups, Lie algebras or associative algebras) would have permitted us, even in the absence of experimental evidence, to predict the next level of a physical theory. For instance, starting from Galilean invariance of nonrelativistic theories, deformation theory could have given us the Poincaré group (and the deformation parameter $1/c$), i.e. the group theoretical aspect of the special theory of relativity, even if we ignored the Michelson and Morley experiment. But deformation theory was developed half a century after that experiment. Along the same lines, deformation theory of the Poincaré group gives us only the $(4+1)$ and $(3+2)$ De Sitter groups, connected to special solutions of general relativity which have been studied by many authors (special representations of which have been used for the construction of particle states which retain masslessness in the flat space limit [71]). Then the chain stops since we have a semi-simple group, stable under deformations.

Similarly, classical mechanics being based on the structure of Poisson brackets $\{u,v\}$ of two functions u and v over a symplectic manifold W, it is natural to look for deformations of this bracket and see if a 'new level' of formalism can be obtained in this way. Some deformations have been considered [8], e.g. in connection with open systems. Moreover since this bracket can be obtained as the commutator associated with the product law $u \cdot v + \nu\{u,v\}$ of two classical observables u and v, and since that product law (associative modulo ν^2) is an infinitesimal deformation of the associative algebra of functions over phase-space, one is led to the question of looking for this deformation at all orders (which we shall call a $*$-product) and for the corresponding bracket. It turns out that one such product and bracket had already been found by Groenewold [9] and Moyal [3] as the inverse Weyl transforms of the product and commutator of operators; it was rediscovered independently, from mathematical considerations of deformation theory, by Vey [10], who also showed the existence of nontrivial deformations of the Poisson bracket Lie algebra on any symplectic manifold with vanishing de Rham 3-cohomology; the same result for associative algebra deformations was recently obtained by Neroslavsky and Vlasov [11] and Lichnerowicz [12].

Using these deformations, it is possible to develop quantum mechanics in an autonomous manner on functions over phase-space. A similar approach is possible for field theory. Quantization can then be understood as a deformation of the algebra of observables, and not as a radical change in their nature. Thus, should the mathematical tools of deformation theory have existed early enough, it would have been possible to predict quantum mechanics at the time when Poisson brackets were introduced, before it was required by experimental evidence.

In contradistinction with Poisson brackets, the deformed brackets and products have a very small invariance group [13]: the affine symplectic in the case of the Moyal bracket. This has prompted a more general framework for quantization, based on star-products invariant under some Lie algebra [13]. This approach presents some similarity with the usual geometric quantization [5], with one basic difference: no passage to operators is required, since everything can be developed in an autonomous manner using star-products only (thus there is no need for polarizations, etc.). In the same spirit, star-products and star-exponentials based on orbits in the coadjoint representation of a Lie algebra enables a 'symplectic' theory of group representations (without operators) to be developed.

In the present paper, we shall enter slightly more into details of the subject matters noted above, and especially the 'induced star-representations theory' for semi-direct products.

1. Deformation Approach to Quantum Theories

1.1. *Quantum Mechanics*

Let W be a symplectic manifold (differentiable, connected, paracompact, of dimension $2n$, endowed with a closed 2-form F such that the symplectic volume element F^n is everywhere nonzero, or with the corresponding 2-tensor Λ having a vanishing Schouten auto-bracket). This is the phase-space of classical mechanics. One may also, for many purposes, take for W a Poisson manifold [13, 14] (of dimension m, with a 2-tensor Λ of rank $2n \leqslant m$). Then a *star-product* on W is a deformation of the associative algebra N_a of C^∞ -functions on W:

$$(1.1)\quad u *_\nu v = u \cdot v + \sum_{r=1}^{\infty} (\nu^r/r!)\, C_r(u,v), \qquad u,v \in N_a.$$

where the C_r's are 2-cochains for the Hochschild associative algebra cohomology, symmetric or skew-symmetric according to the parity (even or odd, respectively) of r, and where we take:

$$C_1(u,v) = i(\Lambda)(du \wedge dv) = \{u,v\} = P(u,v)$$

the Poisson bracket on W.

The parity hypothesis and the choice of C_1 show that the corresponding commutator:

$$(1.2)\quad [u,v]_\lambda = (2\nu)^{-1}(u *_\nu v - v *_\nu u), \qquad \lambda = \nu^2$$

is a deformation of the Poisson bracket Lie algebra on $N = C^\infty(W)$:

$$(1.3)\quad [u,v]_\lambda = \{u,v\} + \sum_{r=1}^{\infty} (\lambda^r/(2r+1)!)C_{2r+1}(u,v).$$

Both the bracket (1.3) and the product (1.1) are to be considered as maps from $N \times N$ to the spaces $E(N,\lambda)$ and $E(N,\nu)$ of formal series in λ and ν (respectively), with coefficients in N, and can be extended to these latter spaces. The hypothesis that (1.1) is a deformation means that we have associativity:

$$(1.4)\quad (u *_\nu v) *_\nu w = u *_\nu (v *_\nu w) \in E(N,\nu)$$

and this implies the Jacobi identity for the bracket (1.3) if it is defined by (1.2).

Jacques Vey (who disappeared untimely in December 1979) showed [10] that on any symplectic manifold with vanishing third Betti number $b_3(W) = \dim H^3(W, \mathbb{R}) = 0$ – a technical assumption which could certainly be weakened, according to him – there exists what we call a *Vey bracket*, namely a deformation (1.3) where the cochains C_r are differentiable, vanish on the constant functions, and have the same principal symbol (i.e. the same higher-order term in any coordinate system) as the rth power of the Poisson bracket bidifferential operator:

$$(1.5)\quad P^r(u,v)|_U = \Lambda^{i_1 j_1} \ldots \Lambda^{i_r j_r} \partial_{i_1 \ldots i_r} u \partial_{j_1 \ldots j_r} v$$

where $(x_i;\ i = 1, \ldots, 2n)$ is a local chart on $U \subset W$ on which the coefficients Λ^{ij} of Λ are constant and $\partial_i = \partial/\partial x_i$.

Later, Neroslavsky and Vlasov [11] (see also Lichnerowicz [12]) showed that under the same conditions on W there exists what we call a *Vey$*_\nu$-product* (1.1), i.e. a star-product where the C_r's have the above-mentioned properties. These Vey products and brackets can even be chosen [11] so that the class of C_3 in $H^2(N)$, the Chevalley–Eilenberg 2-cohomology of the Lie algebra N (with values in the adjoint representation), is the same as the class β of P^3 (in this case, the choice of C_3 determines uniquely C_2 and vice-versa).

In the particular case of $W = \mathbb{R}^{2n}$ with the usual symplectic structure defined by $F = \Sigma_{\alpha=1}^{n} \mathrm{d}p_\alpha \wedge \mathrm{d}q_\alpha$ in symplectic coordinates, taking $v = \frac{1}{2} i\hbar$ we obtain the Moyal bracket [3]. We recall that the Weyl quantization prescription [1] writes:

$$(1.6)\quad f(p_\alpha, q_\alpha) \to \Omega(f) = \hat{f} = \int \tilde{f}(\xi_\alpha, \eta_\alpha) \exp((\xi \cdot P + \eta \cdot Q)(i\hbar)^{-1}) \mathrm{d}\xi\, \mathrm{d}\eta$$

where $\tilde{f}$ is the inverse Fourier transform of the function f and P_α, Q_α are the generators corresponding to p_α, q_α in the Schrödinger representation of the canonical commutation relations Lie algebra H_n (this prescription can be extended to e.g. tempered distributions and has in general to be understood as an operator equality on some domain of differentiable or analytic vectors for this representation). Then in this prescription the product (resp. commutator) of two operators comes from the Moyal $*$-product or bracket which can be written in obvious notations based on (1.5):

$$(1.7)\quad u * v = \exp(\tfrac{1}{2} i\hbar P)(u, v),$$

$$(1.8)\quad M(u, v) = 2\hbar^{-1} \sin(\tfrac{1}{2}\hbar P)(u, v).$$

The same formulas hold on a flat Poisson manifold, when the P^r's are defined by a formula similar to (1.5) with partial replaced by covariant derivatives associated with a symplectic connection Γ having zero torsion and curvature [13].

The inverse of the Weyl transformation is usually called the Wigner correspondence. It maps operators to functions, and can in many cases be given by a trace formula (cf. e.g. [15]). More general examples of Weyl and Wigner correspondences have been introduced recently [16, 17], in relation with star-representations of Lie groups and Lie algebras of 'distinguished observables' to which we shall come back later; they are associated with more general symplectic manifolds (orbits in coadjoint representations) and with star-products of the pseudo-differential type. The main role in our point of view, of these Weyl and Wigner correspondences is to permit a comparison between our star-approach to quantum mechanics and group representations and their usual operatorial formulation. In the Moyal case of $\mathbb{R}^{2n}$ where quantization is usually done, they have been rather well studied, also from a completely rigorous point of view (cf. [4, 15], etc.). In more general cases their definition remains largely heuristic, with some examples treated more precisely ([16, 17]). This could be related to the fact that quantization in curved phase-spaces is usually not straightforward and might be one of the reasons why the autonomous star-approach to quantum mechanics should be preferred.

For the sake of clarity we shall indicate here how this autonomous approach is developed [13]. The basic tool is the star-exponential of a given Hamiltonian H (which corresponds to the evolution operator)

$$(1.9)\quad \operatorname{Exp}(tH) = \sum_{n=0}^{\infty} (n!)^{-1} (t/i\hbar)^n (H*)^n$$

where by $(H*)^n$ we mean the n-th $*$-power of the function H. This is a distribution (in t and on phase space) whose Fourier transform (in t) gives the $*$-spectrum of H, which is the same as the quantum mechanical spectrum of Ω (H) and is the support of this Fourier transform; the Fourier development gives also the functions corresponding to the eigenprojectors on the eigenspaces of Ω (H). Examples treated in this manner include [13] the harmonic oscillator and angular momentum, for which the Moyal $*$-product is appropriate; the hydrogen atom and Kepler problems require $*$-products on cotangent bundles to spheres which can be derived from Moyal $*$-products in higher dimensional flat spaces. The (one-dimensional) Kepler problem has been shown [17] to be related to (piecewise analytic) $*$-representations of $SL(2, \mathbb{R})$ and exhibits an example of a non-conventional Wigner correspondence between operators defined by integral kernels and functions on phase-space.

The deformation approach to quantum mechanics has further advantages from the conceptual point of view. First of all it is easy to check [13] that the Vey brackets are non-trivial deformations of the Poisson brackets (the class of their C_3 in $H^2(N)$ is non trivial). The equivalence of two deformations is defined as usual by the existence of formal series of operators (necessarily differential in the case of differentiable deformations):

$$(1.10)\quad T_\nu = I + \sum_{r=1}^{\infty} \nu^r T_r, \qquad T_\lambda = I + \sum_{r=1}^{\infty} \lambda^r T_r$$

such that, for the products $*_\nu$ and $*'_\nu$ (resp. brackets $[,]_\lambda$, $[,]'_\lambda$), we have, for u,v in N (or $E(N,\nu)$, resp. $E(N,\lambda)$):

$$(1.11)\quad \begin{aligned} T_\nu(u *_\nu v) &= (T_\nu u) *'_\nu (T_\nu v), \\ T_\lambda([u,v]_\lambda) &= [T_\lambda u, T_\lambda v]'_\lambda, \end{aligned}$$

and triviality means equivalence with the original product or bracket. Similarly the Vey products are nontrivial deformations of N_a.

We have thus seen the existence (at least when $b_3(W)=0$) and non-triviality of the autonomous formulation of quantum mechanics. We shall now give some results on the question of (mathematical) *uniqueness*.

The physical interpretation of this uniqueness, which includes the question of physical equivalence of various orderings (which are mathematically equivalent in the sense described above) will not be developed here (cf. for instance [18]).

It has been proved by Gutt [19] that, if F is exact and $\dim H^2(N(W))=1$ (e.g. in the case of $\mathbb{R}^{2n}$), all nontrivial deformations of the Poisson bracket Lie algebra N can be reduced to the Moyal bracket by equivalence and monomial change in the deformation parameter λ. Furthermore, if such a Lie algebra deformation comes from a star-product, the latter is unique (up to a limited change in the exponent relating the two deformation parameters λ and ν). In the same spirit, Lichnerowicz [12] showed that if $b_2(W)=\dim(H^2(W,\mathbb{R}))=0$, all Vey products and brackets are equivalent. It was shown before [13] that in the flat case the exponential and sine functions are the only formal functions of the Poisson bracket P giving rise to associative and Lie algebra (respectively) deformations.

The uniqueness of Vey products up to equivalence follows [12] from the fact that the equivalence to the order $2r-1$ ($r \geqslant 1$) extends immediately to the order $2r$, and that the second space $H^2(N)$ of differentiable Chevalley cohomology of N is generated by the images (in the map $\mu^{-1}: T^{*k}(W) \to T^k(W)$ defined by the symplectic structure F) of the classes $H^2(W)$ of closed 2-forms on W, and by the class β of the 2-cocycle S^3_Γ defined by a symplectic connection Γ and written in any canonical chart $\langle x^i \rangle$ of domain U as:

$$(1.12)\quad S^3_\Gamma(u,v) = \Lambda^{i_1 j_1}\Lambda^{i_2 j_2}\Lambda^{i_3 j_3}(\mathscr{L}(X_u)\Gamma)_{i_1 i_2 i_3}(\mathscr{L}(X_v)\Gamma)_{j_1 j_2 j_3}$$

where $\mathscr{L}(X_u)\Gamma$ is the Lie derivative of Γ by the Hamiltonian vector field $X_u = \mu^{-1}(\mathrm{d}u)$ defined by, i denoting the interior product:

$$\mathrm{d}u = -i(X_u)F, \qquad u,v \in N.$$

S^3_Γ has the same principal symbol as P^3, and any differentiable Chevalley

2-cocycle of the Vey type can be written:

$$(1.13)\quad C_3 = S^3_\Gamma + 3\partial T + A$$

where Γ is uniquely defined (by C_3), T being a differential operator of order $\leqslant 2$ and $\mu(A)$ a closed 2-form; then the corresponding C_2 can be written $C_2 = P^3 + \tilde{\partial} T$($\tilde{\partial}$: Hochschild coboundary operator). The equivalence of Vey products is then given by a formal automorphism T_ν (or a product of such T_ν); this will imply a passive physical equivalence; an active physical equivalence would require e.g. their realization as an 'inner automorphism' of the type:

$$(1.14)\quad T_\nu(u) = a * u * (a *')^{-1}$$

where $a, u \in E(N, \nu)$ and $(a*')^{-1}$ is the $*'$-inverse of a: such realization will give the same spectrum for the same classical observable u in both star-product quantizations.

1.2. *Field Theory*

A first, rather straight forward, generalization to the case of infinitely many degrees of freedom is to pass to the limit $n \to \infty$ in the quantization scheme based on the canonical commutation relations (C.C.R.) Heisenberg Lie algebra H_n. This is traditionally done by considering a discrete (box quantization) or continuous infinite set of C.C.R and looking for operatorial representations. But we have seen [13] that the autonomous 'Moyal approach' of deformed brackets and products can be equally successful, in particular when the phase-space $\mathbb{R}^{2n}$ with its usual symplectic structure is a natural one for the problem, so that an H_n-based quantization with deformed products and brackets (1.7) and (1.8) can be used.

One can therefore look for a space N of differentiable functions on some topological vector space with a basis, e.g. a Hilbert space with orthonormal basis $\{p_\alpha, q_\alpha; \alpha = 1, 2, \ldots\}$ on which a Poisson bracket operator defined by:

$$(1.15)\quad P(u, v) = \sum_{\alpha=1}^{\infty} (\partial_\alpha u \partial_{\bar{\alpha}} v - \partial_{\bar{\alpha}} u \partial_\alpha v)$$

where ∂_α (resp. $\partial_{\bar{\alpha}}$) denotes the derivative in the p_α (resp. q_α) direction and $u, v \in N$, would make sense. Then one could define its rth powers by a formula similar to (1.5) and, provided the space N is so chosen that they make sense, define the formal deformations of products and brackets by

(1.7) and (1.8). Then one could proceed along the same lines, e.g. using (1.9) to introduce a star-exponential that would be the analog in this approach of an evolution operator and correspond to Feynman's path integral, as we shall indicate later.

In the 'continuous' case, the most natural approach is to look for a symplectic structure on the space of solutions of the classical field equation under consideration, and use it in a similar way to that of the finite-dimensional (discrete) case. Such a symplectic structure, on the (local) solution manifold of a relativistic equation of the type $\Box\Phi = F(\Phi)$, where $\Box$ is the d'Alembertian and F a smooth function, has been defined by Segal [20], to which we refer for a precise formulation (that would take us too far here) and earlier references.

Considering now scalar valued functionals over the space of solutions of the classical field equations, i.e. over the space of initial conditions $\varphi(\vec{x}) = \Phi(\vec{x},0)$ and $\pi(\vec{x}) = ((\partial/\partial t)\Phi)(\vec{x},0)$ which plays here the role of phase-space and is endowed with the preceding symplectic structure, one can introduce the Poisson bracket of two such functionals Ψ_1 and Ψ_2 as the following expression (where δ denotes the Gateaux functional derivative, in the sense defined e.g. by Gelfand [21]):

$$(1.16)\quad P(\Psi_1,\Psi_2) = \int \mathrm{d}^3x\left(\frac{\delta\Psi_1}{\delta\varphi}\frac{\delta\Psi_2}{\delta\pi} - \frac{\delta\Psi_1}{\delta\pi}\frac{\delta\Psi_2}{\delta\varphi}\right).$$

This formula can then, in an obvious manner (at least formally) be extended to the rth powers of this Poisson bracket 'bidifferential' operator, whence by (1.7) and (1.8) the corresponding star-products and 'Moyal' (deformed) brackets. Eventually one can reach in this way a notion of star-exponential which could be made rigorous.

With the introduction of these star-products, brackets, and exponentials, the functionals over the phase-space of solutions of the classical field equation can now be considered as quantized fields, in the same way as for quantum mechanics: here also, quantization is to be understood as a deformation of the algebraic structure of the observables.

Now the star-exponential of the classical Hamiltonian, which describes the evolution of the quantized system in the phase-space deformation approach to quantization, is closely related to Feynman's path integral. In fact (see e.g. [22]), the path-integral is the Fourier transform, over momentum space, of the star-exponential function in quantum mechanics. A similar relation holds certainly in field theory – but is at present purely formal since already both sides are formally defined. However this relation should prove useful in several respects: one might get, via the field-theoretical star-exponential, a better understanding of the path-integral, and vice-versa. Moreover, since the star-exponential may be defined in a more general context than the particular Moyal quantization already in quantum mechanics (it can be defined in curved phase-spaces and in quantization procedures associated with various groups), one can expect the star-exponential approach in field theory to provide an extension of the path-integral approach to a more general context. In particular, constraints of various kinds can be built in the theory.

2. General remarks on star-products in quantization and in representation theory

For a star-product (1.1) on the function space N over a phase-space W in which the cochains vanish on constant functions (which we always assume), then obviously:

(2.1) $k*u = u*k = ku,\qquad \text{for all } k\in\mathbb{C} \text{ and } u\in N(W).$

Moreover the Moyal star-product (1.7) has a remarkable property, which we call invariance: whenever a is a polynomial of order $\leqslant 2$ in the usual coordinates of $W=\mathbb{R}^{2n}$, that is a belongs to the Lie algebra (for the Poisson bracket) $\mathscr{A}=sp(n,\mathbb{R}).H_n$, then (since the cochains other than P in (1.8) contain derivatives of order at least 3 in each variable) we have $M(a,u)=P(a,u)$ for all $u\in N$.

In other words:

(2.2) $a*u-u*a=i\hbar\{a,u\}, \forall a\in\mathscr{A}, u\in N.$

This implies the following two relations:

(2.2)′ $a*b-b*a=i\hbar\{a,b\},\qquad \forall a,b\in\mathscr{A}.$

(2.2)″ $\{a,u*v\}=\{a,u\}*v+u*\{a,v\},\qquad \forall a\in\mathscr{A},\qquad u,v\in N,$

which constitute (cf. [13, 16, 17]) together with (2.1) the definition of an $\mathscr{A}$-invariant star-product. (2.2)′ means that the Lie algebra $\mathscr{A}$ (under Poisson bracket) is preserved by the deformation, while (2.2)″ means that the infinitesimal canonical transformation (derivation of the Poisson Lie algebra N) X_a defined by $a\in\mathscr{A}$ by $X_a u=\{a,u\}, u\in N$, acts as a derivation (infinitesimal automorphism) also on the star-product algebra, i.e. that $\mathscr{A}$ is an invariance algebra for the star-product.

An immediate consequence of the invariance requirement (2.2) is that the classical time-evolution of an observable $u_t \in N$ under an Hamiltonian $a \in \mathscr{A}$, which is $\mathrm{d}u_t/\mathrm{d}t = \{a, u_t\}$, is the same as the quantum mechanical evolution:

$$\mathrm{d}u_t/\mathrm{d}t = M(a, u_t) = (i\hbar)^{-1}(a * u_t - u_t * a).$$

In particular, when the quantum time evolution is governed by an Hamiltonian $a \in \mathscr{A}$, the time evolution of the usual product of two observables u and v will be the product of the time evolutions of the two observables, as in the classical motion: $(uv)_t = u_t v_t$. (For a general Hamiltonian, one has to take the star-product of u and v). Thus, when $a \in \mathscr{A}$ governs the evolution, the classical and quantum evolutions of a coordinate function f in W will coincide, and quantum trajectories in W (not only in N) will have a meaning (independently of the choice of the coordinates).

All what precedes was said on the basis of our experience with the Moyal star-product, but formulated so that the generalization is already included in the formulation. Indeed, we say that a *star-product* (1.1) *is invariant under a Lie algebra* $\mathscr{A}$ *of 'distinguished observables'* if (2.1) and (2.2) hold. The phase-space will then be usually (in accordance with Souriau's identification) imbedded in the dual $\mathscr{A}^*$ of $\mathscr{A}$ as an orbit (or a collection of orbits) for the coadjoint representation of the corresponding Lie group. In particular the various orderings (standard, normal, ...) that are considered in the literature can be treated in our approach as H_n-invariant star-products; only the Weyl ordering admits in addition the symplectic Lie algebra as distinguished observables, with the important consequences for the harmonic oscillator, though in all cases the phase-space W is the same ($\mathbb{R}^{2n}$, considered as an orbit in H_n^*).

This approach has been shown to be useful in various physical problems. In particular the one-dimensional Kepler problem is naturally treated in terms of $so(2,1)$ invariant star-product [17]; $E(2)$-invariant star-products (related to a particle in a potential well) have been constructed [16]. The hydrogen atom has also been treated in the same spirit [13], though in this case the star-product used, which is natural for the geometry of the phase-space $W = T^*S^3$, exhibits a smaller invariance than expected from consideration of the phase-space as an orbit (it is $so(4)$-invariant and not $so(4,2)$-invariant).

We call this approach *star-product quantization*, associated with a Lie algebra $\mathscr{A}$ of distinguished observables. It has several advantages with respect to the usual quantization procedure, in particular due to a better utilization of the geometry of phase-space. The connection with the usual (operatorial) approach can be realized with the help of a generalized Weyl transformation Ω, which maps functions over phase-space into operators over some Hilbert space, or its inverse Ω^{-1}, a generalized Wigner transformation (mapping operators and their products into functions and their star-products). If we have an algebra $\mathscr{A}$ of distinguished observables, the Wigner transformation will have to be $\mathscr{A}$-invariant also [17], i.e. map the commutator of two operators into the Poisson bracket of the corresponding functions. Examples of nonconventional invariant Weyl and Wigner transformations have been given (cf. [16, 17]).

In the Kostant–Souriau [5] quantization approach, the Hilbert space and the operator formulation on it are realized with the help of a polarization on phase-space, which "takes away half of the dimension". In our case the corresponding notion, which can be defined in a general context, is that of star-polarization: it is a subspace of $N(W)$ stable and irreducible under the action of a Lie subalgebra $\mathscr{A}_0 \subset \mathscr{A}$ (often a maximal subalgebra, see below) by left star-multiplication: $u \to a*u, a \in \mathscr{A}_0$ (cf. [16], [23]). It may be of interest to note that in most cases (except for e.g. H_n) this left star-multiplication is expressed as an operator of the pseudo-differential type (an algebraic function of differential operators). In a similar way one can define and utilize the right star-multiplication: $u \to u*a$.

Now for a given element $a \in \mathscr{A}$ one can introduce a one-parameter group (for star-multiplication) by the star-exponential $\operatorname{Exp}(ta)$ defined by (1.9). For two elements $a, b \in \mathscr{A}$, if $c = c(a,b)$ is related to them by the Campbell–Hausdorff formula defined by the relation $e^a e^b = e^c$ in the connected and simply connected Lie group G with Lie algebra $\mathscr{A}$, we shall then have:

$$(2.3) \quad \operatorname{Exp}(a)*\operatorname{Exp}(b) = \operatorname{Exp}(c(a,b)).$$

Defining:

$$(2.4) \quad E(e^a) = \operatorname{Exp}(a)$$

we can thus extend the group law from the image of $\mathscr{A}$ by the exponential mapping in G to all of G and obtain what we call a star-representation of G ([13, 16]). In general, the star-exponential (1.9) is obtained – besides of the formal series aspect – as a distribution on phase space which coincides at least almost everywhere with a differentiable function in $N(W)$. Thus what will be called a *star-representation* of G will be a mapping $\mathscr{E}$ from a space D of test-functions on G into $N(W)$, or into a space of distributions on W, or into $N(W_0)$ where W_0 is a dense open set in W. To make this definition coherent we have to assume that the domain D and the kernel of $\mathscr{E}$ are stable under convolution, and to make it compatible with the notion of $\mathscr{A}$-invariant star-product we further assume that $\mathscr{E}$ is $\mathrm{ad}_{\mathscr{A}}$-invariant, that is:

(2.5) $\mathscr{E}\circ\mathrm{ad}_{\mathscr{A}}(a)=\mathrm{ad}^*_{\mathscr{A}}(a)\circ\mathscr{E}.$

The first condition ensures that:

(2.6) $\mathscr{E}(f*f')=\mathscr{E}(f)*\mathscr{E}(f')$

(where on the left-hand side we have the convolution of f and f' in D) defines a star-product on $\mathscr{E}(D)$. In the case we start with an $\mathscr{A}$-invariant star-product and define $E(g), g\in G$, by (2.4), we of course define the star-representation $\mathscr{E}$ by:

(2.7) $\mathscr{E}(f)=\int_G f(g)E(g^{-1})\mathrm{d}g$

The phase-space W is taken as an orbit of G in $\mathscr{A}^*$ under the coadjoint representation and elements of $\mathscr{A}$ are the restrictions of linear functions on $\mathscr{A}^*$, which permits to consider $E(e^a)$, hence $E(g)$, as functions on W.

We have here obtained the star-representation $\mathscr{E}$ with the help of the distribution E on G, the latter being an extension of the star-exponential Exp of elements of $\mathscr{A}$. This supposes some analyticity property of E near the identity of G, and gives what is called an *analytic star-representation.* However it may very well happen that a kernel E for a star-representation $\mathscr{E}$ exists and has a singularity at the identity. As long as the kernel $E(g,\xi)$ (where $g\in G$ and $\xi\in W$ or W_0) is analytic almost everywhere on G we can still define $\mathscr{E}$ by (2.7) for a large class of test-functions f, a star-product by (2.6) and derive the star-product for the Lie algebra in the neighbourhood of a regular point for E. In this case we shall speak of a *piecewise analytic star-representation*, an example of which, associated with a singular orbit in the coadjoint representation, has been found for $\mathrm{SL}(2,\mathbb{R})$ in connection with the Kepler problem [17].

The distribution on G defined by integrating $\mathscr{E}$ over W(with respect to a suitable quasi-invariant measure) can be called the *character* associated with the star-representation $\mathscr{E}$. It is a character on G, associated with the operatorial representation T of G corresponding to $\mathscr{E}$ in a suitable invariant Wigner Ω^{-1} transformation: $\mathscr{E}=\Omega^{-1}\circ T$, or $E(g)=\Omega^{-1}(T(g))$. The comparison between star and operatorial group representations can thus be carried out by consideration of the characters. Moreover since we know that characters of group representations are analytic functions on a dense subset $G' \subset G$, and do have singularities (often including at the origin, at least for most one-parameter subgroups) it follows that the notion of piecewise analytic star-representation is the most appropriate. In particular it has been mentioned [13] that, when H is the harmonic Hamiltonian on $W = \mathbb{R}^2$, $\operatorname{Exp}(tH)$ was the one-parameter subgroup associated with the compact generator H in a star-representation of the two-fold covering of SL (2, $\mathbb{R}$), which decomposes into two irreducible components. The character for each of these components is singular at the origin on the above-mentioned compact one-parameter group, but the singularities cancel out (on this subgroup) so that $\operatorname{Exp}(tH)$ is analytic in t for $|t| < \pi$. This is another remarkable feature of the harmonic oscillator Hamiltonian, which was helpful in computing $\operatorname{Exp}(tH)$ but could be misleading in the general theory of star-representations.

Finally we shall mention that, exactly as for operatorial representations, we can define and construct induced representations. This will be made more explicit in the following section, together with the geometric structure of the orbits in the case of semi-direct products with a vector normal subgroup. In this section we shall also deal with the notion of contraction of groups and star-representations in our geometrical framework.

3. Contraction and Induction of Representations in Phase-Space Approach. Examples

In the quantization procedure as defined in 2, a Lie algebra $\mathfrak{g}$ of distinguished observables is given in terms of functions on the phase-space W so that

$$[u,v] = \{u,v\} \ \forall u,v \in \mathfrak{g}.$$

where $[\,,]$ denotes the bracket associated with the $*$-product defining the quantization. This $*$-product defines a morphism from the associative universal enveloping algebra $\mathscr{U}(\mathfrak{g})$ of $\mathfrak{g}$ into $(E(N,\lambda),*)$. Moreover $E(N,\lambda)$ is a bi-module for the following structure:

$$\mathfrak{g} \times E(N,\lambda) \to E(N,\lambda) \qquad E(N,\lambda) \times \mathfrak{g} \to E(N,\lambda)$$

and

$$(u,v) \to u*v \qquad (v,u) \to v*u$$

and the relevance of Lie group and Lie algebra theory is obvious in this context.

The following two examples are typical of both points of view of quantum mechanics and group theory.

Example 1: The manifold W is an orbit of the coadjoint representation of a given Lie group G. The elements of the Lie algebra $\mathfrak{g}$ of G are represented by functions on W. A 'theory of orbits' can be developed by considering all $*$-products on W such that $\mathfrak{g}$ is an algebra of distinguished observables, sufficiently regular to be exponentiated to G (or to its universal covering). Such a theory has evidently some analogies with the Kostant–Souriau theory (see for instance [5]). However the various possible $*$-products on each orbit allow for richer constructions.

For instance Fronsdal shows [16] in the case $\mathfrak{g} = su(2)$ that for each integer or half integer l each orbit carries a $*$-product such that every factor of the representation:

$$u \in E(N, \lambda) \to \operatorname{Exp}(a) * u, \qquad (a \in \mathfrak{g})$$

is equivalent to $D(l)$.

Example 2: Let G be a Lie group of transformations of a manifold V. The Lie algebra $\mathfrak{g}$ is then represented by functions on the cotangent bundle $W = T^*(V)$ of V. The representations of G obtained by $*$-products on W are 'quasi-regular' representations of G.

We consider here two different applications of $*$-products in this field:

(i) The notion of contraction of Lie algebra is well-known. A geometrical interpretation can generally be given for instance when the parameter of the contraction is the curvature of some homogeneous space. However, the definition of this notion is not obvious in the universal enveloping algebra. Moreover although the contractions may be useful for construction of representations, mathematical theories in this matter are somewhat confused and unsatisfactory, and in any case the geometrical interpretation is completely lost at this level. We consider here contractions as restrictions of $*$-products on submanifolds defined by appropriate constraints.

(ii) The theory of induced representations for inhomogeneous Lie groups is well known and intensively used since Mackey and Bruhat. Concerning Lie algebras two different notions exist (induced and produced representations by Blattner and some other authors, see for instance [24]). These notions are fruitful especially in the theory of simple $\mathfrak{g}$-modules. Nevertheless the carrier spaces of representations are not spaces of functions on orbits and the theory does not permit a notion of local induced representations containing global induced representations of Lie groups and including those obtained by cut-off of the homogeneous space (only partially integrable). We consider here the Poincaré group which is generic for many problems on induction for inhomogeneous groups in addition to its physical interest. We show for representations with non-zero mass how it is possible to define a global notion of induced $*$-products and to interpret it geometrically. In this way we quantize an elementary relativistic system (with or without spin).

3.1. *Contractions* [25]

3.1.1. *Structural contractions*

Consider the symplectic vector spaces $W = \mathbb{R}^{2n} = \{(q,p);\ q,p \in \mathbb{R}^n\}$ provided with the Moyal $*$-product (1.7). The Lie algebra $so(n)$ is naturally realized as the space of functions on W spanned by:

$$(3.1)\quad M_{j,k} = q_j p_k - q_k p_j, \qquad 1 \leqslant j < k \leqslant n.$$

Of course:

$$\{M_{jk}, M_{j'k'}\} = [M_{jk}, M_{j'k'}] \text{ (Moyal bracket).}$$

The restrictions of the M_{jk}'s to the symplectic subspace $\mathbb{R}^{2n-2}$ defined by $q_n = -c$ and $p_n = 0$ are:

$$(3.2)\quad \begin{aligned} M_{jk} &= q_j p_k - q_k p_j, \qquad && 1 \leqslant j < k \leqslant n-1, \\ M_{jn} &= c p_j, && 1 \leqslant j \leqslant n-1. \end{aligned}$$

These restrictions generate for Poisson or Moyal bracket on $\mathbb{R}^{2n-2}$ the Lie algebra $\mathscr{E}(n-1)$ of the inhomogeneous rotation group in $\mathbb{R}^{n-1}$, and we have thus performed the usual contraction $so(n) \to \mathscr{E}(n-1)$. More precisely, if k_1 and k_2 are the two second-class constraints on $\mathbb{R}^{2n}$ given by $k_1 = q_n + c, k_2 = p_n$, the associated Dirac bracket [8] is:

$$[u,v]_D = \{u,v\} + C(u,v), \qquad u,v \in N(\mathbb{R}^{2n})$$

where

$$C(u,v) = (\{u,k_1\}\{k_2,v\} - \{u,k_2\}\{k_1,v\})\{k_1,k_2\}^{-1}.$$

The restriction of the Dirac bracket on the submanifold defined by k_1, k_2 is nothing but the Poisson bracket on $\mathbb{R}^{2n-2}$. The contraction of the Lie algebra $so(n)$ to $\mathscr{E}(n-1)$ is given by a formal deformation of the Poisson bracket on $\mathbb{R}^{2n}$ to the Dirac bracket relative to the constraints k_1, k_2.

Now there exists an injective morphism from $\mathscr{U}(so(n))$ (resp. $\mathscr{U}(\mathscr{E}(n-1))$) into the space $\mathbb{C}[p,q]$ of polynomial functions on $\mathbb{R}^{2n}$ (resp. $\mathbb{R}^{2n-2}$) endowed with the Moyal $*$-product (1.7). We can *define a 'contraction map' from* $\mathscr{U}(so(n))$ to $\mathscr{U}(\mathscr{E}(n-1))$ by extension of the deformation to the corresponding Moyal $*$-products.

3.1.2. *Contractions of representations*

In the above realization of $\mathscr{E}(n-1)$, the obtained representation is evidently reducible. The Casimir element

$$C = \sum_{j=1}^{n-1} (M_{jn}*)^2 = \sum_{j=1}^{n-1} c^2 p_j^2, \qquad (c \geqslant 0)$$

has a continuous spectrum. Its spectral resolution is given by the Laplace transform:

$$\operatorname{Exp}(-tC) = \int_0^\infty \exp(-t\lambda)\delta(\lambda - c)\mathrm{d}\lambda, \qquad (\operatorname{Re}(t) > 0).$$

This fact suggests restricting the representation to the cotangent bundle $T^*(S_{n-2})$ of the sphere $\Sigma_{j=1}^{n-1} p_j^2 = R^2$ in order to obtain a representation with scalar element C.

For this purpose we have to construct a $*$-product on $T^*(S_{n-2})$.

For further considerations we give a general procedure [26] of construction of a $*$-product on the cotangent bundle W defined by the equations in $\mathbb{R}^{2n}$:

$$(3.3) \quad \sum_{i=1}^{n} \varepsilon_i p_i^2 = R^2 (R > 0), \qquad \sum_{i=1}^{n} \varepsilon_i p_i q_i = 0$$

where

$$\varepsilon_i = +1 \text{ if } 1 \leqslant i \leqslant p, \qquad \varepsilon_i = -1 \text{ if } p < i \leqslant n.$$

The group $\{(\rho,\sigma)/\rho > 0, \sigma \in \mathbb{R}\}$ acting on the open set $V = \{(p,q) \in \mathbb{R}^{2n};\ \Sigma_{i=1}^n \varepsilon_i p_i^2 > 0\}$ by:

$$V \ni (p_i, q_i) \to (\rho p_i, \rho^{-1}(q_i + \sigma \varepsilon_i p_i))$$

defines an equivalence relation and W as quotient. There exists a bijective map between $N(W)$ and the subspace of $N(V)$ of functions invariant with respect to this action:

(3.4) $N(W)\ni u \to \hat{u}(p_i, q_i)$

$$= u\left(\frac{Rp_i}{\sqrt{\Sigma\varepsilon_j P_j^2}}, \frac{\sqrt{\Sigma\varepsilon_j P_j^2}}{R}\left(q_i - \frac{\Sigma\varepsilon_j P_j q_j}{\Sigma P_j^2}\varepsilon_i p_i\right)\right)\in N(V).$$

The formula:

(3.5) $u * v = \hat{u} \hat{*} \hat{v}, \qquad u, v \in N(W),$

where $\hat{*}$ denotes the Moyal $*$-product on $\mathbb{R}^{2n}$, defines a $*$-product on W, invariant with respect to the inhomogeneous Lie algebra $iso(p,q)$, $q = n - p$.

Let us denote $W_n = T^*(S_n)$. Then W_{n-2} is obtained from W_{n-1} by the Dirac constraints $k_1 = q_n + c, k_2 = p_n$ and the Poisson bracket on W_{n-2} is the restriction of the Dirac bracket on W_{n-1}. Since the Lie algebra $so(n)$ (resp. $\mathscr{E}(n-1)$) is generated by the functions M_{ij} on W_{n-1} (resp. W_{n-2}) for Poisson or $*$-bracket the above procedure of contraction in $\mathbb{R}^{2n}$ is applicable here. By extension to the corresponding universal enveloping algebras (generated by the M_{ij}'s for the $*$-product on W_{n-2} or W_{n-1}) we have a contraction map from $\mathscr{U}(so(n))$ onto a quotient of $\mathscr{U}(\mathscr{E}_{n-1})$. The restriction of the Dirac deformation from $T^*(S_{n-1})$ to the submanifold $T^*(S_{n-2})$ defined by the constraints $k_1 = q_n + c$, $k_2 = p_n$ realizes the contraction of the representation of $so(n)$ on $T^*(S_{n-1})$ to a representation of $\mathscr{E}(n-1)$ on $T^*(S_{n-2})$ with scalar Casimir element C.

These results are easily generalizable to the contraction from $so(p,q)$ to $iso(p-1,q)$ or $iso(p,q-1)$.

3.1.3. *Case* $n = 3$

In the obtained representation of SO(3) on $T^*(S_2)$ each representation $D(l)$ (with $l \in \mathbb{N}$) appears. In fact the Casimir element $\Sigma(M_{ij}*)^2$ has the spectrum $\{n(n+1)\hbar^2\}$ (see [13]). In the contracted representation of $\mathscr{E}(2)$ on $T^*(S_1)$ we propose to diagonalize $M_{12} = z = q_1 p_2 - q_2 p_1$.

Let us define $\theta \in [0, 2\pi[$ by $p_1 = R\cos\theta$, $p_2 = R\sin\theta$. Then the Fourier development of the distribution $\operatorname{Exp}(tM_{12})$ gives

$$M_{12} = \sum_{k\in\mathbb{Z}} k\hbar\pi_k$$

where π_k is the unique solution of the equation:

$$z * f = f * z = kf.$$

Moreover $\pi_0(z) = (\hbar/2z)\sinh(2z/\hbar)$.
Define $\mathbb{S} = \{f \in E(N,\lambda); f * z = 0\}$. Then:

$$f \in \mathbb{S} \Leftrightarrow f = \varphi(\theta) * \pi_0$$

and

$$M_{12} * (\varphi * \pi_0) = \left(i\hbar \frac{\partial \varphi}{\partial \theta}\right) * \pi_0,$$

$$M_{13} * (\varphi * \pi_0) = (cR \cos\theta \cdot \varphi) * \pi_0,$$

$$M_{23} * (\varphi * \pi_0) = (cR \sin\theta \cdot \varphi) * \pi_0,$$

$$C = c^2 R^2 = (M_{13} *)^2 + (M_{23} *)^2.$$

For $c \neq 0$ we obtain the usual irreducible representation of $E(2)$ with non null Casimir; for $c = 0$ the representation is a sum of characters, i.e. the usual standard result.

3.2. *Induction for Inhomogeneous Classical Groups*

3.2.1. *Coadjoint representation. Notations*

Let G be a real Lie group, $\mathfrak{g}$ its Lie algebra. G acts on the dual $\mathfrak{g}^*$ of $\mathfrak{g}$ by:

$$(g.\xi)(X) = \xi(\mathrm{Ad}^{-1} g(X)), \qquad \forall \xi \in \mathfrak{g}^*, \qquad g \in G, \qquad X \in \mathfrak{g}.$$

Let $\xi \to X_\xi$ be the vector field defined by differentiation of this action on $\mathfrak{g}^*$.

The tangent space at ξ to the orbit $\mathcal{O}_\xi$ of ξ in $\mathfrak{g}^*$ is generated by the set $\{X_\xi, X \in \mathfrak{g}\}$. Moreover $\mathcal{O}_\xi$ is endowed with a canonical symplectic structure by the 2-form defined by:

$$\omega_\xi(X_\xi, Y_\xi) = -\xi([X, Y]), \qquad X, Y \in \mathfrak{g}.$$

The elements X, Y of $\mathfrak{g}$ are realized as functions on $\mathcal{O}_\xi$ and $[X, Y] = \{X, Y\}$ where $\{,\}$ denotes the Poisson bracket. If G_ξ denotes the stabilizer of $\xi, \mathcal{O}_\xi \simeq G/G_\xi$ and

$$\mathfrak{g}_\xi = \{X \in \mathfrak{g}; \xi([X, Y]) = 0, \forall Y \in \mathfrak{g}\}.$$

3.2.2. *Mackey's orbits for semi-direct products*

If $G_0 = G.T$ is the semi-direct product of a Lie group G by $T = \mathbb{R}^n$, we denote by $\mathfrak{g}_0, \mathfrak{g}, \mathfrak{t}$ the corresponding Lie algebras and by $\mathfrak{t}^\perp, \mathfrak{g}^\perp$ the orthogonal in $\mathfrak{g}_0^*$ of $\mathfrak{t}$ and $\mathfrak{g}$. $\mathfrak{g}^\perp$ is isomorphic to $\mathfrak{t}^*$ and then $\mathfrak{g}_0 = \mathfrak{t}^\perp + \mathfrak{t}^*$. The elements of $\mathfrak{g}_0^*$ are denoted $\zeta = (\eta, \xi)$, $\eta \in \mathfrak{t}^\perp$, $\xi \in \mathfrak{t}^*$. G acts on $\mathfrak{t}^*$ by restriction of the coadjoint action.

The corresponding orbit Ω_ξ of ξ in $\mathfrak{t}^*$ is called Mackey's orbit of ξ. The tangent space θ_ξ to Ω_ξ at ξ is the subspace of the elements X_ξ in $\mathfrak{t}^*$ defined by $X_\xi = \lim_{t\to 0} t^{-1}(e^{tX}\xi - \xi)$. The cotangent space θ_ξ^* is the quotient $\mathfrak{t}^{**}/\theta_\xi^\perp \simeq \mathfrak{t}/\theta_\xi^\perp$. If we denote by P_ξ the quotient projection, the duality is:

$$\langle P_\xi(T) | X_\xi \rangle = X_\xi(T);\ \forall X \in \mathfrak{g}, \forall T \in \mathfrak{t}.$$

The map Φ from $T^*(\Omega_{\xi_0})$ into $\mathfrak{g}_0^*$ defined by:

$$\Phi\begin{pmatrix} P_\xi(T) \\ \xi \end{pmatrix} = e^T(0,\xi), \qquad \forall \xi \in \Omega_{\xi_0}$$

is a symplectomorphism from $T^*(\Omega_{\xi_0})$ to the orbit $\mathcal{O}_{(0,\xi_0)}$ of $(0,\xi_0)$ in $\mathfrak{g}_0^*$. Moreover G acts canonically on $T^*(\Omega_{\xi_0})$ and $\mathcal{O}_{(0,\xi_0)}$ and Φ intertwins these actions.

If H_{ξ_0} is the stabilizer (little group) of ξ_0 and Y_{ξ_0} any orbit of the coadjoint representation of H_{ξ_0}, we define a bundle Y of little group orbits on Ω_{ξ_0} as a bundle:

$$Y \xrightarrow{\rho} \Omega_{\xi_0} \text{ with fiber } Y_{\xi_0}.$$

The set of bundles of little group orbits is in bijection with the set of orbits of the coadjoint representation of G_0 ([27]).

More precisely if $\mathcal{O}$ is the orbit for the coadjoint representation of the point $\zeta_0 = (\eta_0, \xi_0)$, $\rho(\zeta) = \zeta|\mathfrak{t}$ defines a bundle projection from $\mathcal{O}$ onto $\mathfrak{t}^*$ and $\{\rho(\zeta), \zeta \in \mathcal{O}\}$ is the Mackey orbit Ω_{ξ_0}.

In a neighbourhood of ζ_0, $\mathcal{O}$ is diffeomorphic to an open set in $Y_{\xi_0} \times T^*(\Omega_{\xi_0})$ and then we can construct a local symplectomorphism between $\mathcal{O}$ and $Y_{\xi_0} \times T^*(\Omega_{\xi_0})$ with the product symplectic structure.

This local product structure suggests to introduce 'induced' $*$-products as products of $*$-products on the Mackey orbits by $*$-products on the orbits of the coadjoint representation of the little group.

3.2.3. *Polarizations*

Given a symplectic manifold W endowed with a $*$-product invariant with respect to $\mathfrak{g}$, we define a representation of $\mathfrak{g}$ in $E(W,\lambda)$ by:

(3.6) $\mathfrak{g} \ni X \to T_X$ where $T_X(f) = X * f, \qquad f \in N(W)$.

Rather than to isolate irreducible subspaces in $E(N,\lambda)$, the duality between induced and coinduced representations suggests to introduce the following notion of polarization: If $\mathcal{O}_\zeta$ is the orbit of ζ in the coadjoint representation of a given Lie group G, the associative $*$-algebra $(\mathbb{P}, *)$ generated by the coordinate functions with respect to some $\mathfrak{g}$-invariant $*$-product is isomorphic to a quotient of the enveloping algebra $\mathcal{U}(\mathfrak{g})$. Let $\mathbb{P}'$ be the algebraic dual space of $\mathbb{P}$ and $u \to u^+$ the antilinear canonical antiautomorphism in $\mathcal{U}(\mathfrak{g})$.

Let $\mathfrak{k}_\zeta$ be a maximal subalgebra in $\mathfrak{g}^{\mathbb{C}}$ such that:

$$\zeta([\mathfrak{k}_\zeta, \mathfrak{k}_\zeta]) = 0$$

We call polarization in ζ with respect to $\mathfrak{k}_\zeta$ the space $\mathbb{S}$ of elements φ in $\mathbb{P}'$ such that:

$$(3.7) \quad \varphi(R * k) = \zeta(k^+)\varphi(R), \qquad \forall k \in \mathfrak{k}_\zeta, \qquad \forall R \in (\mathbb{P}, *)$$

The representation defined by:

$$(3.8) \quad X \in \mathfrak{g} : \varphi \to X * \varphi = T^t_{X^+}(\varphi), \qquad \forall \varphi \in \mathbb{S}$$

is called the representation associated with $\mathbb{S}$.

In fact these representations are the so-called algebraic induced representations.

In most cases of physical interest the elements of $\mathbb{P}$ are polynomial and then $\mathbb{S}$ is a space of formal series.

3.2.4. *Representations of the Poincaré group*

Each point X of the Minkowski space E_4 is denoted by $X = \begin{pmatrix} x \\ t \end{pmatrix}$ where $x \in \mathbb{R}^3$. The Poincaré group $SO_0(3,1)$. T_4 is represented by the matrices

$$a = \begin{pmatrix} L & C \\ 0 & 1 \end{pmatrix} \text{ where } L \in SO_0(3,1), \qquad C \in T_4.$$

Its Lie algebra $\mathcal{P}$ is then the set of the matrices

$$Z = \begin{pmatrix} j(\omega) & \beta & \gamma \\ \bar{\beta} & 0 & \varepsilon \\ 0 & 0 & 0 \end{pmatrix} \qquad \text{where } \omega, \beta, \gamma \in \mathbb{R}^3, \varepsilon \in \mathbb{R} \text{ and}$$

$j(\omega)(x) = \omega \wedge x, \forall x \in \mathbb{R}^3$.

The dual $\mathcal{P}^*$ can be described as the set of pairs $\xi = (M, P)$ where

$M = \begin{pmatrix} j(l) & g \\ \bar{g} & 0 \end{pmatrix}$, $\quad P = \begin{pmatrix} p \\ E \end{pmatrix}$ with the duality

$$\xi(Z) = \langle l,\omega\rangle - \langle g,\beta\rangle + \langle p,\gamma\rangle - E\varepsilon$$

Then $a\cdot\xi = (LM\bar{L} + C\bar{P}\bar{L} - LP\bar{C}, LP)$.

Each orbit is characterized by the invariants of mass and spin:

$$\bar{P}\cdot P = m^2,\ \bar{W}W = -m\sigma \text{ where } W(\xi) = \begin{pmatrix} p \wedge g + l.E \\ \langle l,p\rangle \end{pmatrix}$$

(*i*) *Case* $m > 0, \sigma = 0$: We are here concerned with the orbit $\mathcal{O}_{\xi_0}$ of the point $\xi_0 = \left(0, \begin{pmatrix} 0 \\ m \end{pmatrix}\right)$.

$$\xi \in \mathcal{O}_{\xi_0} \Leftrightarrow \xi = (Q\bar{P} - P\bar{Q}, P) \text{ with } \bar{P}P = m^2, \bar{P}\cdot Q = 0.$$

$\mathcal{O}_{\xi_0}$ is the cotangent bundle $T^*(\Omega_{\xi_0})$ to the hyperboloid (Mackey orbit) $\bar{P}.P = m^2$.

Let $\mathbb{P}$ be the space of polynomial functions on $T^*(\Omega_{\xi_0})$ generated by the functions: $p_i, p_4, (p \wedge q)_i, (pq_4 - p_4 q)_i$, $i = 1,2,3$. The elements Z of $\mathcal{P}$ are represented by:

$$Z = (p \wedge q).\omega - (p_4 q - pq_4).\beta + p.\gamma - \varepsilon p_4 \in \mathbb{P}.$$

Moreover $\mathbb{P}$ is stable with respect to the $*$-product (3.5) and is isomorphic to a quotient of $\mathcal{U}(\mathcal{P})$.

We define the representation T of $\mathcal{U}(\mathcal{P})$ by:

$$\mathrm{T}_u(\mathrm{R}) = u * R,\ \forall u \in \mathcal{U}(\mathcal{P}), R \in \mathbb{P}.$$

More precisely, $(\mathbb{P}, *)$ is isomorphic to the quotient of the enveloping algebra $\mathcal{U}(\mathcal{P})$ by the ideal of elements vanishing on $T^*(\Omega_{\xi_0})$. T_u is a representation of $\mathcal{U}(\mathcal{P})$ with mass m and spin 0. Here $\mathfrak{h}_{\xi_0} = su(2) + \mathbb{R}^4$ and there is only one possible polarization given by the space $\mathbb{S}_{m,0}$ of formal series in the variables M_{i4} $(i = 1,2,3)$.

Let $U_{m,0}$ be the unitary irreducible representation with mass m and spin 0 in $L^2(\Omega_{\xi_0})$. We can write

$$\Omega_{\xi_0} = \{e^{t_1 M_{14} + t_2 M_{24} + t_3 M_{34}} \xi_0\}.$$

Then for all $\xi \in \Omega_{\xi_0}$ there exists a unique l_ξ such that $\xi = e^{l_\xi}.\xi_0$. For each analytic vector f of $U_{m,0}$ let us define φ_f in $\mathbb{S}_{m,0}$ by:

$$\varphi_f((l_\xi *)^n) = (\mathrm{d}U_{m,0}(-l_\xi)^n f)(\xi_0), \qquad \forall n \geqslant 0, \qquad \forall \xi \in \Omega_{\xi_0}.$$

Then for each u in $\mathscr{U}(\mathscr{P})$ we have:

$$\varphi_{\mathrm{d}U_{m,0}(u)f} = u * \varphi_f$$

i.e. there exists an intertwining operator φ between the $*$-representation associated with the polarization and the usual representation $\mathrm{d}\,U_{m,0}$ in $L^2(\Omega_{\xi_0})$ on its analytic vectors.

(*ii*) *Case* $m > 0, \sigma \neq 0$: We are concerned with the orbit of the point:

$$\zeta_0 = \left(M_0 = \sigma \begin{pmatrix} j(J_0) & 0 \\ 0 & 0 \end{pmatrix}, P_0 = m\, I_0 \right) \text{with } J_0 = \begin{pmatrix} 0 \\ 0 \\ 1 \\ 0 \end{pmatrix}, I_0 = \begin{pmatrix} 0 \\ 0 \\ 0 \\ 1 \end{pmatrix}.$$

This orbit can be parametrized by the coordinates:

$$\begin{pmatrix} p \\ p_4 \end{pmatrix} \begin{pmatrix} q \\ q_4 \end{pmatrix} \begin{pmatrix} s \\ 0 \end{pmatrix} \text{ where } p_4^2 - p^2 = m^2, \qquad p_4 q_4 - pq = 0, s^2 = 1.$$

In fact it is possible to prove that the orbit $\mathcal{O}_{\zeta_0}$ characterized by the mass m and the spin σ is symplectomorphic to the symplectic product of the orbit $T^*(\Omega_{\xi_0})$ of mass m and spin 0 by the orbit S_2 of the coadjoint representation of the little group $SU(2)$ with spin σ. A complete classification of sufficiently regular $*$-products on S_2 is given in [16]. Let $*_2$ be the $*$-product of this classification on S_2 such that the Casimir element is

$$\sum_{i=1}^{3} (\sigma s_i *_2)^2 = \hbar^2 l(l+1) \qquad l\hbar = \sigma.$$

This $*$-product gives the corresponding representation of SU(2) with spin σ and is defined on the space $\mathbb{P}_\sigma$ of polynomial functions in the s_i's with degree $\leqslant 2l$.

If we denote by $*_1$ the $*$-product constructed in the case $\sigma = 0$ on $T^*(\Omega_{\zeta_0})$, the product of $*_1$ and $*_2$ is defined by:

$$(u_1 \otimes u_2) * (v_1 \otimes v_2) = (u_1 *_1 v_1) \otimes (u_2 *_2 v_2) \begin{cases} u_1, v_1 \in \mathbb{P} \\ u_2, v_2 \in \mathbb{P}_\sigma \end{cases}.$$

This $*$-product is invariant with respect to $so(3,1)$ and there exists an isomorphism Φ from $(\mathbb{P} \times \mathbb{P}_\sigma, *)$ onto a quotient of the enveloping algebra $\mathscr{U}(\mathscr{P})$ such that:

$$Z \in \mathscr{P} \to \Phi(Z) = \sigma s \wedge \omega - (p \wedge q) \cdot \omega$$

$$- \frac{\sigma}{m + p_4} (p \wedge s).\beta - (p_4 q - p q_4)\beta + p\gamma - p_4 \varepsilon.$$

The stabilizer of ζ_0 being $so(2).\mathbb{R}$, the only (equivalent) possible polarizations are the spaces $\mathbb{S}_{m,\sigma}$ of linear forms on $\mathbb{P} \times \mathbb{P}_\sigma$ such that

$$\varphi(R*Z) = \zeta_0(Z^+)\varphi(R) \quad \begin{cases} R \in \mathbb{P} \otimes \mathbb{P}_\sigma \\ Z \in \{M_{12}, M_{23} + iM_{14}, T_j (j = 1,2,3,4)\} \\ \text{or } \{M_{12}, M_{23} - iM_{14}, T_j (j = 1,2,3,4)\}. \end{cases}$$

Moreover $\mathbb{S}_{m,\sigma} = \mathbb{S}_{m,0} \otimes \mathbb{S}_\sigma$ where $\mathbb{S}_\sigma$ is the space of polynomial functions in the variables $s_1 \pm is_2$ with degree $\leqslant 2l$.

On the other hand, let us consider the unitary irreducible representation $U_{m,\sigma}$ with mass m and spin σ of Poincaré in the Hilbert space $L^2(\Omega_{\xi_0}, \mathrm{d}^3p/p_4) \otimes \mathbb{S}_\sigma$.

To each analytic vector of the form $f \otimes \varphi_2$ we associate the form $\varphi_{f \otimes \varphi_2} = \varphi_f \otimes \varphi_2$. Then for each analytic vector $\Sigma_i f^i \otimes \varphi_2^i$ of $U_{m,\sigma}$ and each $u \in \mathscr{U}(\mathscr{P})$:

$$\varphi_{\mathrm{d}U_{m,\sigma}(u)}\left(\sum_i f^i \otimes \varphi_2^i\right) = u * \varphi_{\sum f^i \otimes \varphi_2^i}$$

This proves that the polarized $*$-representation is intertwined by φ with the differential of $U_{m,\sigma}$ on its analytic vectors.

(*iii*) *Case* $m^2 < 0$: First, we consider the orbit $\mathcal{O}_{\xi_0}$ of the point

$$\xi_0 = \left(0, \begin{pmatrix} m \\ 0 \\ 0 \\ 0 \end{pmatrix}\right).$$

As in the case $\sigma = 0, m > 0$, $\mathcal{O}_{\xi_0}$ is the cotangent bundle $T^*(\Omega_{\xi_0})$ to the hyperboloid $\bar{P}P = -m^2$. The space $\mathbb{P}$ of polynomial functions on $T^*(\Omega_{\xi_0})$ generated by the functions $p_i, p_4, (p \wedge q)_i, (pq_4 - p_4 q)_i$, is stable with respect to the star product defined by (3.5). The representation T of $\mathscr{U}(\mathscr{P})$ defined by

$$T_u(R) = u * R, \qquad u \in \mathscr{U}(\mathscr{P}), \qquad R \in \mathbb{P},$$

is a representation of $\mathscr{U}(\mathscr{P})$ with mass m and 'spin' 0 (that means induced by the trivial one of the little group). Thus, we consider the orbit $\mathcal{O}_{\zeta_0}$ of the point. $\zeta_0 \left(M_0, \begin{pmatrix} m \\ 0 \\ 0 \\ 0 \end{pmatrix}\right)$

$$M_0 = \begin{pmatrix} j \begin{pmatrix} 0 \\ 0 \\ l_1 \end{pmatrix} \begin{pmatrix} 0 \\ 0 \\ g_3 \end{pmatrix} \\ 0\,0 - g_3\,0 \end{pmatrix}$$

with either $l_1 = 0$, or $g_3 = 0$ or $l_1 = \pm g_3$.

Let Ω'_{ξ_0} be the open submanifold defined by $p^2 + p_1^2 \neq 0$ and $p_4 \neq 0$ and $\mathcal{O}'_{\zeta_0}$ the corresponding submanifold of $\mathcal{O}_{\zeta_0}$. One can show that $\mathcal{O}'_{\zeta_0}$ is the symplectic product of $T^*(\Omega'_{\xi_0})$ by the orbit of M_0 under the action of the little group $SL(2,\mathbb{R})$.

Thus the same procedure as in the $m^2 > 0$ case is available here. Taking an invariant $*$-product on the little group orbit, we define by direct product a $*$-product on $\mathcal{O}'_{\zeta_0}$ corresponding to a representation of $\mathcal{P}$, induced by a representation of $SL(2,\mathbb{R})$, associated with the Mackey orbit Ω_{ξ_0}.

(*iv*) *Case* $m^2 = 0$: In this case, there does not seem to exist a similar construction of an invariant $*$product on the cotangent bundle of the light cone. A possible approach to that problem is to define directly a geometrical $*$-product. The existence and classification of such Vey $*$-products was shown by Lichnerowicz, and Neroslavsky and Vlasov [11]. Moreover in the physically interesting situation (discrete helicity), the orbits of the coadjoint representation are diffeomorphic to $T^*(\Omega_{\xi_0})$ and there is no difficulty with the little group orbits since they are trivial in this case (and the question of the factorization does not appear).

There is also no difficulty in constructing star-products (e.g. by lifting from the Moyal product on $\mathbb{R}^6$): the only problem lies with the Poincaré (and not only Euclidean) invariance requirement. SO(3)-invariant star products, corresponding to the zero-mass representations of Poincaré, have recently been constructed [28].

Finally, let us mention that the experience of the previous case ($m^2 < 0$), shows that a good notion of 'fibered product' of star-product would be relevant for induced star-representations in more general situations.

4. Conclusion and Open Problems

The main motivation of the series of works briefly reviewed in the present contribution was twofold:

(1) To be able to get a better understanding of the meaning of the classical limit ($\hbar \to 0$) in different quantum theories, and in particular to create a device natural for applications in different approximation schemes

of quantum theories (semi-classical approximations, WKB etc.). In doing so one gains enormous conceptual (and we believe also eventually technical) advantages: We deal with *scalar-valued* quantities instead of operator-valued ones (even though in general some non-locality in phase-space is introduced to the star-product, translating the fact that we are dealing with quantum theories); we have directly a phase-space formulation of physical theories (something which is very natural from the viewpoint of the foundation of these theories, and which we know from canonical field theories how important it is); we incorporate in the very formulation of the quantized theory the geometry of phase-space (which is difficult to achieve in the usual formulation and is of special importance in the presence of constraints) and thereby obtain more general quantization procedures adapted to the phase-space geometry.

(2) The existence and importance of symmetries in physical problems made it natural to try to obtain the analog of unitary group representations in our phase-space approach. This lead to a generalization of the notion of quantization (with invariant star-products) and, with star-representations considered per se, brought in turn a possible approach to the problem of group representations. This point of view is somehow related to the Kostant–Souriau geometric quantization procedure, but goes quite beyond it (in particular because it is fully autonomous already on the phase-space level). Being based on a phase-space structure, our approach must also be related to the famous Gelfand–Kirillov conjecture, which identifies the enveloping field of a Lie algebra with that of a generalized Heisenberg algebra and hints at an underlying structure of flat Poisson manifold, where the Moyal product and bracket are also defined [13].

Evidently there is still a lot to be done in this new approach to both quantum theories and group representations. One should build as many as possible interacting field models and try to compare their quantization schemes in the standard and in the new formulation (sine-Gordon, nonlinear sigma model, nonlinear Schrödinger, etc.) At the same time, one should try to develop representation theory in the star formalism in a systematical way: study of characters, 'unitarity' criteria for representations (which might look simpler in this approach), exhaustive study of the dual of a Lie group (especially in the semi-simple case, which looks promising here), etc. These goals, we hope, will be achieved, at least partially, in the coming years.

Physique-mathématique, Université de Dijon

References

[1] Weyl, H., *Gruppentheorie und Quantemmechanik*, Hirzel-Verlag, Leipzig, 1928; *Z. Phys.* **46**, 1 (1927).

[2] Wigner, E. P., *Phys. Rev.* **40**, 749 (1932).

[3] Moyal, J. E., *Proc. Cambridge Phil. Soc.* **45**, 99 (1949).

[4] Voros, A., Thèse Orsay No. 1843, (1977).

[5] Kostant, B., *Lecture Note in Math.* **170**, 87 (1970).
Souriau, J-M, *Structure des Systèmes Dynamiques*, Dunod, 1970.

[6] Gerstenhaber, M., *Ann. Math.* **79**, 59 (1964).

[7] Fronsdal, C., Flato, M., *Lett. Math. Phys.* **2**, 421 (1978).

[8] Flato, M., Lichnerowicz, A., and Sternheimer, D., *J. Math. Phys.* **17**, 1754 (1976).

[9] Groenewold, H. J., *Physica* **12**, 405 (1946).

[10] Vey, J., *Comment. Math. Helvet.* **50**, 421 (1975).

[11] Neroslavsky, M., Vlasov, A., (to be published).

[12] Lichnerowicz, A., *Lett. Math. Phys.* **3**, 495 (1979).

[13] Bayen, F., Flato, M., Fronsdal, C., Lichnerowicz, A., Sternheimer, D., *Ann. Phys.* **111**, 61, 111 (1978).
see also: Flato, M., Lichnerowicz, A., Sternheimer D., *C. R. Acad. Sc. Paris* **283**, 19 (1976).

[14] Lichnerowicz, A., *J. Diff. Geom.* **12**, 253 (1977).

[15] Agarwal, G. S., and Wolf, E., *Phys. Rev.*, **D2**, 2161 (1970).
Mehta, C. L., *J. Phys. A (Proc. Phys. Soc.)* ser. 2, **1**, 385 (1968).
Maillard: J-M (to be published).

[16] Fronsdal, C., *Rep. on Math. Phys.* **15**, 111 (1979).

[17] Fronsdal, C., *J, Math. Phys.*, **20**, 2226 (1979).

[18] Flato, M., Sternheimer, D., in J. Wolf M. Cahen and M. De Wilde (eds.), *Harmonic Analysis and Representations of Semisimple Groups*, D. Reidel, 1980.

[19] Gutt, S., *Lett. Math. Phys.* **3**, 297 (1979).

[20] Segal, I., *Symposia Math.* **XIV**, 99 (1974).

[21] Gelfand, I. M., Fomin, F. V., *Calculus of Variations*, (1963), Prentice-Hall.

[22] Pankaj Saran, *Phys. Rev.*, **D**. 15 (in press).

[23] Huynh, T. V., *Lett. Math. Phys.* **4**, (1980).

[24] Dixmier, J., *Algèbres Enveloppantes*, Gauthier-Villars (1974).

[25] Arnal, D., Cortet, J-C, *J. Math. Phys.* **20**, 556 (1979).

[26] Lichnerowicz, A., *Lett. Math. Phys.* **1**, 505 (1977).

[27] Rawnsley, J. H., *Math. Proc. Camb. Phil. Soc.* **78**, 345 (1975).

R. GASTMANS

HIGH ENERGY TESTS OF QUANTUM ELECTRODYNAMICS

ABSTRACT. Developments in the calculation of radiative corrections for $e^+e^- \to e^+e^-$, $\mu^+\mu^-$, and $\gamma\gamma$ are discussed, and a review is made of the comparison between QED and recent experimental results on these reactions.

1. INTRODUCTION

Quantum electrodynamics has played and continues to play a very important role in elementary particle physics for several reasons.

First, it is the most successful theory ever created by the human mind, in the sense that its validity has been verified over an incredible domain of roughly 25 orders of magnitude, from 3×10^{-15} cm [1] to 3×10^{10} cm [2]. These results were obtained by either studying low energy phenomena with very high precision (parts per million), or by performing high energy experiments with a typical 5% accuracy.

Secondly, it is the first gauge theory to be studied in detail, and one now knows how important gauge theories are for the understanding of all the interactions of the elementary particles: the electromagnetic, the weak, the strong, and the gravitational interactions. It is also the simplest gauge theory, from the mathematical point of view, and, as such, it provided the theorists with very valuable opportunities for trying out new ideas or techniques without running into the complexities of the other gauge theories.

Last but not least, with the advent of high energy electron-positron colliding beam facilities and all the exciting new physics they have provided, quantum electrodynamics has proven to be very useful in the understanding of these phenomena, which are mostly associated with hadronic channels. Indeed, the theory must be used to determine the luminosities of the e^+e^- beams and to calculate the various backgrounds for processes involving hadron states coupled to the leptons. But, to be certain that these backgrounds and luminosities are calculated correctly, one must be sure about the validity of the theory, and this is why high energy tests of quantum electrodynamics continue to be of vital importance. This

Onderzoeksleider N.F.W.O., Belgium

E. Tirapegui (Ed.), Field Theory, Quantization and Statistical Physics, 113–121.

article is devoted to a review of the tests which have been carried out over the last couple of years.

Traditionally, one considers the processes,

$$e^+e^- \to e^+e^-,$$

$$e^+e^- \to \mu^+\mu^-,$$

and

$$e^+e^- \to \gamma\gamma,$$

as tests of quantum electrodynamics, because of their simplicity in terms of lowest-order Feynman diagrams. To compare theory and experiment at the percent level, it is necessary to calculate these processes to order α^3, where α is the fine structure constant, i.e. one has to evaluate the virtual corrections to the lowest-order amplitude and the bremsstrahlung contribution. As is well-known, the bremsstrahlung has to be taken into account since final states with an extra soft photon simulate elastic scattering and any experiment, no matter how precise it is, always measures the sum of these contributions. Obviously, the precise amount of inelastic scattering which has to be taken into account depends crucially on the selection criteria which the experimentalist has decided to take. The bremsstrahlung contribution must therefore be recalculated for every experiment, in contradistinction to the virtual corrections, which are independent of the experimental set-up.

In Section 2, we review the latest developments in the calculation of the radiative corrections, and, in Section 3, we present a comparison of the most recent experiments with the theory. Our conclusion will be that QED in the high energy regime looks as good as ever!

2. Calculation of radiative corrections

Of the three QED processes mentioned in Section 1, mu-pair production is the simplest one. Explicit formulae for the virtual corrections and the soft bremsstrahlung can be found in References 3 and 4. The correctness of the formulae has been checked by Passarino and Veltman [5], who have developed a numerical method for calculating any one-loop Feynman diagram. These authors have moreover extended the analysis of the radiative corrections to include all the weak interaction effects for $e^+e^- \to \mu^+\mu^-$ as predicted by the Weinberg model. To be complete, these corrections have to be supplemented by the contribution of hadronic vacuum polarization effects [6], which are generally important at high energies, and the contribution of the virtual τ-lepton loop in the photon propagator, which is easily obtained from the well-known expression for the electron loop.

To compare theory and experiment, one still has to evaluate the hard photon corrections. This is done in several steps. First, one calculates the bremsstrahlung cross section for $e^+e^- \to \mu^+\mu^-\gamma$. With the standard Feynman techniques, this can lead to very cumbersome and lengthy formulae, but if one can neglect lepton masses compared to the beam energy, reasonably compact results are obtained [7]. This limit is somewhat tricky as, in the zero mass limit, some lepton propagators blow up when the photon is emitted parallel to one of the leptons. These effects only occur in the square of amplitudes associated with individual Feynman diagrams, not in interference terms, and can therefore be treated without too much difficulty.

Secondly, one has to choose a set of integration variables in which to express the experimental selection criteria. These are usually threshold cuts, below which the muons cannot be detected, or acollinearity cuts, which require the outgoing particles to be back-to-back to a given precision. These criteria are then translated into allowed regions of phase space which are accessible for the undetected photon. Eidelman and Kuraev [8] emphasized the importance of a 'symmetrical' approach to the problem of hard photon radiative corrections in e^+e^- collisions. They introduced an 'average' scattering direction, which is determined by the bisectrix of the angle between the momentum of one outgoing muon, say $\vec{q}_+$, and the opposite of the momentum of the other muon, $-\vec{q}_-$. In the phase space analysis of Reference 7, it was found that the average scattering direction could be conveniently defined by the vector $\vec{q}_+ - \vec{q}_-$. In both cases, the photon variables are then defined with respect to this average scattering axis, which coincides with the usual scattering direction in the case of elastic scattering.

Finally, one has to integrate the bremsstrahlung cross-section over the relevant region of phase space. In Reference 8, analytical expressions are obtained corresponding to selection criteria (small polar and azimuthal acollinearity angles), which are easily treated using their variables. In Ref. 4, use is made of numerical integration techniques to accommodate the various experimental cuts. Also, special attention is given to the peaking regions, where the emitted photon travels parallel to one of the charged particles.

For $e^+e^- \to e^+e^-$ (Bhabha scattering), the analysis proceeds in the same way. The analytical expressions for the corrections due to virtual and real soft photons can be found in References 3 and 9. In Reference 3, the effects of virtual electron and muon loops are included, but not of the τ-loop. For hadronic vacuum polarization effects, we refer again to Reference 6. For small angle Bhabha scattering, which is often used to determine the beam luminosity, both these effects are however small.

In the zero mass limit, the bremsstrahlung cross section for $e^+e^- \to e^+e^-\gamma$ takes again a simple form [7]. This is achieved by writing it as a fraction, whose denominator is the product of all the different denominators appearing in the various amplitudes associated with the Feynman diagrams. It turns out that no squared denominator appears in the final answer when all diagrams are combined, although they are present when an individual diagram is squared. Moreover, the resulting numerator of the final expression factorizes into two factors, one of which also appears in the cross section for $e^+e^- \to \mu^+\mu^-\gamma$, provided one performs the analogous manipulations in this case. Why this is so is not understood.

The results of the integration over phase space are given in analytical form in References 8 for the same selection criteria as for mu-pair production, and numerically in Reference 9 for various experimental set-ups.

For $e^+e^- \to \gamma\gamma$, finally, the radiative corrections at the one-loop level involve only an extra virtual photon, and the analytical expression for the sum of the virtual and real soft photon contribution can be found in References 3, 10–12. The bremsstrahlung cross section for $e^+e^- \to \gamma\gamma\gamma$ in the zero mass limit [7] has again the property that no squares appear in the denominator once all 6 diagrams are combined. As explained in Reference 12, one has to be careful to introduce the proper combinatorial factors when integrating over the phase space: in certain regions, photons can become indistinguishable, in which case one should not count the events twice. To obtain numerical estimates of the radiative corrections, one proceeds otherwise in exactly the same way as for mu-pair production and Bhabha scattering [12].

3. Comparison of Theory and Experiment

In the last years, most experiments testing QED for the reaction $e^+e^- \to \mu^+\mu^-$ have been done at SLAC with the SPEAR electron-positron storage ring. One experiment [13] took data at center-of-mass energies of 3.0, 3.8 and 4.8 GeV, and found for

$$R(2E) = \sigma_{\exp} / \sigma_{\text{QED}}$$

$R(3.0 \text{ GeV}) = 0.95 \pm 0.04$, $R(3.8 \text{ GeV}) = 1.05 \pm 0.03$, and $R(4.8 \text{ GeV}) = 1.01 \pm 0.03$. This experiment also produced an angular distribution in the range $|\cos\theta_+| \leqslant 0.7$ at $E_{\text{c.m.}} = 4.8$ GeV, and found very good agreement with theory.

Another experiment [14] measured $e^+e^- \to \mu^+\mu^-$ at angles close to 90° at a center-of-mass energy of 5.2 GeV, and found $R(5.2 \text{ GeV}) = 0.89 \pm 0.11$.

The data at the highest energies were taken by Himel *et al.* [15], who measured the process in the center-of-mass energy range 5.8–7.4 GeV. Their selection criteria were: a maximum acollinearity between the muons of 10° and a muon momentum threshold of one half the beam energy. They observed an angular distribution, reproduced in Figure 1, in the range $|\cos\theta_+| \leqslant 0.6$ in very good agreement with second order QED. They also measured a forward backward asymmetry

$$A = 0.013 \pm 0.010,$$

whereas QED predicts

$$A = 0.0155 \pm 0.0008.$$

It is well-known that the existence of a neutral intermediate vector boson, which couples to the leptons, can also produce an asymmetry in this process [16]. The agreement between QED and the experiment therefore puts

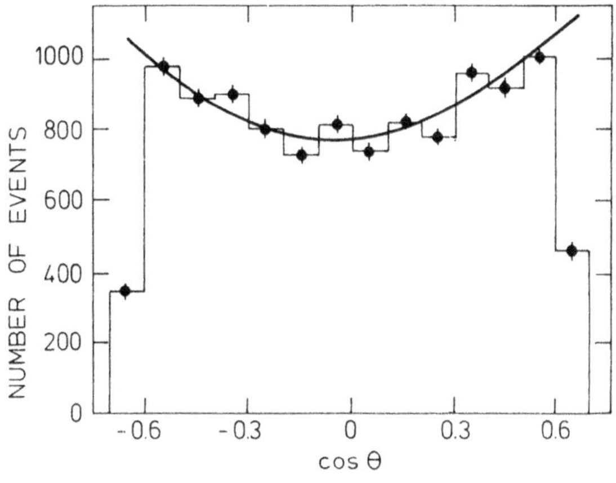

Fig. 1. Angular distribution of the μ^+ in $e^+e^- \to \mu^+\mu^-$ at an average center-of-mass energy of 6.8 GeV, as measured at SPEAR and presented in Reference 15 (Courtesy The American Physical Society). The solid line is the QED prediction, normalized to the data.

a limit on the ratio of the mass of the vector boson and the strength of its coupling to the leptons. The Weinberg model passes this test, but at least one other model could be ruled out by this experiment [15].

Just a passing remark about $e^+e^- \to \tau^+\tau^-$. Due to the short lifetime of the τ, the direct process $e^+e^- \to \tau^+\tau^-$ itself is unfortunately not a good reaction for testing QED. However, this reaction is topologically the same as $e^+e^- \to \mu^+\mu^-$, and the cross-section to order α^3 can therefore be obtained from the μ -pair production one by simply replacing $m_\mu \leftrightarrow m_\tau$. In one experiment [17] these radiative corrections were applied to $e^+e^- \to \tau^+\tau^-$ in order to determine accurately the τ-mass from the observed energy dependence of the cross section for $e^+e^- \to e^\pm + (\mu^\pm) +$ one charged track near threshold.

Bhabha scattering experiments have been performed at various places over the last five years or so. An experiment at CEA [18] finds that R(4 GeV) = 0.93 ± 0.10 and R(5 GeV) = 1.03 ± 0.09. It also measures the angular distribution in good agreement with QED. Two Stanford experiments find R(5.2 GeV) = 1.02 ± 0.05 [14], R(7 GeV) = 1.05 ± 0.04 and R(7.4 GeV) = 0.96 ± 0.03 [19]. Another experiment with SPEAR [13] at $E_{c.m.} = 4.8$ GeV produces an angular distribution in the range $50^\circ < \theta_+ < 130^\circ$ which is in remarkable agreement with QED. The highest energy data on $e^+e^- \to e^+e^-$ come from recent PETRA experiments, which observed the reaction at center-of-mass energies of 13 and 17 GeV. One experiment [1], using the MARK-J detector, measured the beam luminosity by looking at forward-angle Bhabha scattering (11.5°–26°), and selected the events according to the criteria of a maximum acollinearity of 20° and a total energy in the showers of at least 8 GeV. For the comparison with QED, the effects of the virtual τ-loops were included, as well as the hadronic vacuum polarization effects. They find an excellent agreement between theory and experiment as can be seen from the angular distribution, $s\, d\sigma/d\Omega$ versus θ, which is roughly independent of $s = 4E^2$ (see Figure 2). The other PETRA experiments, using the TASSO detector [20] or the PLUTO detector [21], also find an angular distribution at large scattering angles consistent with the small angle data.

Also the reaction $e^+e^- \to \gamma\gamma$ has been tested several times in the last years. At CEA, a first experiment [22] yielded $R(4\text{ GeV}) = 1.05^{+0.23}_{-0.21}$, and a second one [23] R(5 GeV) = 1.05 ± 0.23. With the DORIS colliding

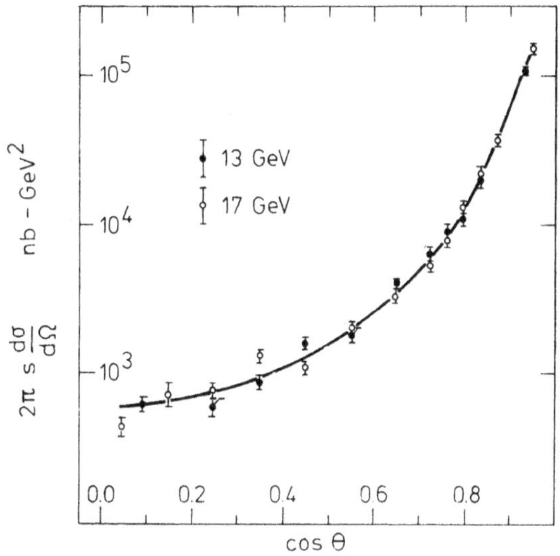

Fig. 2. Angular distribution for $e^+e^- \rightarrow e^+e^-$ at center-of-mass energies of 13 and 17 GeV, as measured at PETRA and presented in Reference 1 (Courtesy The American Physical Society). The solid line is the QED prediction, normalized to the small angle data.

beam facility, the DASP-collaboration [24] studied this reaction in the vicinity of the $J/\Psi(3100)$ resonance. They measured at six different energies in the range 3.08 GeV $< E_{c.m.} <$ 3.10 GeV, and every time the QED prediction fell within the errors. This experiment also produced an angular distribution in agreement with theory. Finally, with SPEAR several experiments were performed. A Stanford University group [14] measured at 5.2. GeV, and found R(5.2 GeV) = 1.05 ± 0.08. Another experiment at SLAC [25] observed $e^+e^- \rightarrow \gamma\gamma$ at angles close to 90°, and measured R eight times in the energy range 6.2–7.4 GeV. Here, too, the values of R oscillate around the reference value 1.

Recently, the process was studied in the neighborhood of the Υ and Υ' resonances with the e^+e^- storage ring DORIS using the DESY-Heidelberg detector [26]. They observed 244 photon-pairs, where QED predicts 271, and an angular distribution in agreement with QED. However, by studying $e^+e^- \rightarrow \gamma\gamma\gamma$ for 9.41 GeV $< E_{c.m.} <$ 9.51 GeV, they found six events in a region of phase space where QED only predicts 1.0 ± 0.4 events. It is not yet clear how this excess of events should be interpreted.

The Crystal Ball detector at SLAC is very well suited for the observation of γ-rays in e^+e^- collisions, and one can expect in the near future precise QED tests for $e^+e^- \to \gamma\gamma$. So far, only preliminary results [27] were reported at the $\psi'(3684)$, which show an angular distribution in the range $|\cos\theta| < 0.5$ in agreement with QED.

Incidentally, $e^+e^- \to \gamma\gamma$ may prove to be the most important process for testing QED at even higher energies, such as for LEP where $E_{\text{c.m.}} = 200$ GeV is envisaged. Bhabha scattering and mu-pair production are expected then to be very much influenced by the weak interaction effects, even at the level of the lowest-order cross-section, whereas this is not the case for pair-annihilation. Moreover, the effects of the weak interactions for the virtual corrections were calculated by Capdequi-Peyranère *et al.* [28] and found to be less than 10^{-3}, i.e. effectively negligible.

The conclusion of all these comparisons between theory and experiment is obvious: QED continues to describe the interaction between electrons, muons, and photons (and τ's) correctly at the highest energies, and the challenge remains of finding out where the real limits of applicability of the theory are.

Institute for Theoretical Physics, University of Leuven, Belgium

References

[1] Barber, D., *et al.*, *Phys. Rev. Lett.* **42**, 1110 (1979).
[2] Hollweg, J. V., *Phys. Rev. Lett.* **32**, 961 (1974).
[3] Berends, F. A., and Gastmans, R., in A. Donnachie and G. Shaw (eds.), *Electromagnetic Interactions of Hadrons*, Plenum Publ. Co., New York, 1978, Vol. 2, pp. 471–522.
[4] Berends, F. A., Gaemers, K. J. F., and Gastmans, R., *Nucl. Phys.* **B57**, 381 (1973); **B63**, 381 (1973).
[5] Passarino, G., and Veltman, M., University of Utrecht preprint, February 1979.
[6] Berends, F. A., and Komen, G. J., *Phys. Lett.* **63B**, 432 (1976).
[7] Berends, F. A., Gastmans, R., and Wu, T. T., in preparation.
[8] Eidelman, S. I., and Kuraev, E. A., *Phys. Lett.* **80B**, 94 (1978).
[9] Berends, F. A., Gaemers, K. J. F., and Gastmans, R., *Nucl. Phys.* **B68**, 541 (1974).
[10] Haris, I., and Brown, L. M., *Phys. Rev.* **105**, 1656 (1957).
[11] Tsai, Y. S., *Phys. Rev.* **137B**, 730 (1965).
[12] Berends, F. A., and Gastmans, R., *Nucl. Phys.* **B61**, 414 (1973).
[13] Augustin, J.-E., *et al.*, *Phys. Rev. Lett.* **34**, 233 (1975).
[14] Beron, B. L., *et al.*, *Phys. Rev.* **D17**, 2187 (1978).
[15] Himel, T., *et al.*, *Phys. Rev. Lett.* **41**, 449 (1978).
[16] Cabibbo, N., and Gatto, R. *Phys. Rev.* **124**, 1577 (1961);
Godine, J., and Hankey, A., *Phys. Rev.* **D6**, 3301 (1972);
Love, A., *Lett. Nuovo Cim.* **5**, 113 (1972);

Cung, V. K., Mann, A. K., and Paschos, E. A., *Phys. Lett.* **41B**, 355 (1972);
Khriplovich, I. B., *Sov. J. Nucl. Phys.* **17**, 298 (1973);
Dicus, D. A., *Phys. Rev.* **D8**, 338 (1973); 890 (1973);
Brown, R. W., *et al.*, *Phys. Lett.* **43B**, 403 (1973);
Budny, R., *Phys. Lett.* **45B**, 340 (1973).

[17] Bartel, W., *et al.*, *Phys. Lett.* **77B**, 331 (1978).

[18] Newman, H., *et al.*, *Phys. Rev. Lett.* **32**, 483 (1974).

[19] O'Neil, L. H., *et al.*, *Phys. Rev. Lett.* **37**, 395 (1976).

[20] Brandelik, R., *et al.*, *Phys. Lett.* **83B**, 261 (1979).

[21] Berger, Ch., *et al.*, *Phys. Lett.* **81B**, 410 (1979).

[22] Hanson, G., *et al.*, *Lett. Nuovo Cim.* **7**, 587 (1973).

[23] Law, M. E., *et al.*, *Lett. Nuovo Cim.* **11**, 5 (1974).

[24] Braunschweig, W., *et al.*, *Phys. Lett.* **53B**, 491 (1975).

[25] Hilger, E., *et al.*, *Phys. Rev.* **D15**, 1809 (1977).

[26] Bienlein, J. K., *et al.*, preprint DESY 79/24.

[27] Bloom, E. D., in J. Trân Thanh Vân (ed.), *Current Hadron Interactions*, Proceedings of the XIVth Rencontre de Moriond, Les Arcs, March 1979. Editions Frontières, Dreux, 1979, Vol. II, p. 188.

[28] Capdequi-Peyranère, Grunberg, G., Renard, F. M., and Talon, M., *Nucl. Phys.* **B149**, 243 (1979).

L. GOMBEROFF and E. K. MASCHKE

NON-IDEAL EFFECTS ON THE STABILITY OF A CYLINDRICAL CURRENT-CARRYING PLASMA

ABSTRACT. The stability of a cylindrical plasma, limited by fixed boundaries and having a constant current distribution and constant density, is considered. For parallel wave numbers $k_{\parallel} \simeq 0$, non-ideal effects are shown to play an important role. Solutions of the linearized equations including viscosity, thermal conductivity and resistivity are obtained. It is shown that viscosity is always stabilizing, as expected on physical grounds, but the range of unstable modes remains unchanged. Resistivity, on the other hand, reduces both the growth rates and the unstable domain of incompressible modes, provided that viscosity is not negligible. If it is, then resistivity plays an equivalent role to viscosity. When the pressure gradient attains a critical value, there exists a marginal mode with zero parallel wave number which characterizes the onset of large-scale steady convection in the plasma.

1. Introduction

The stability of a current-carrying cylindrical plasma limited by fixed boundaries (perfectly conducting walls) and placed in a strong longitudinal magnetic field, has been extensively studied within the frame of ideal magnetohydrodynamics (MHD) theory. The first analytical results concerning the growth rates of instabilities in a system with shearless magnetic field are due to Tayler [1] and Shafranov [2] who solved the ideal MHD equations under the assumption that the plasma is incompressible. In this case, it was found that the growth rate as a function of parallel wave vector, $k_{\parallel}$, has two maxima located symmetrically on both sides of $k_{\parallel} = 0$, and it vanishes in the limit $k_{\parallel} = 0$. Subsequent numerical [3,4] and analytical [5,6] studies have shown that the compressibility of the plasma has the effect of changing the shape of the unstable spectrum in such a way that the growth rate has a single maximum about $k_{\parallel} = 0$, with the unstable range of $k_{\parallel}$ values remaining unchanged. As a matter of fact, it turns out that there is no linear solution of the MHD equations closed to $k_{\parallel} = 0$, if the plasma is incompressible. This result will be proved in the following section.

A new situation arises when non-ideal effects like viscosity, thermal conductivity, or resistivity are considered. Even when the corresponding coefficients are small, such non-ideal effects play an important role for

E. Tirapegui (Ed.), Field Theory, Quantization and Statistical Physics, 123–145.

perturbations with $k_{||} = 0$. This is related to the fact that the linearized ideal MHD equations for incompressible perturbed motion are singular at $k_{||} = 0$.

In the present article the effect of viscosity, thermal conductivity, and resistivity of the unstable motion of a cylindrical plasma with constant density confined by a shearless magnetic field, is investigated. To this end, in Section 2 a summary of the ideal MHD situation is presented, showing, in particular, the singular behaviour of the perturbed velocity at $k_{||} = 0$. In Section 3 the effect of perpendicular viscosity is considered by solving the linearized perturbation equations for the case where the adiabaticity index is infinite. It is then shown that the singularity at $k_{||} = 0$ is removed and furthermore, in the presence of thermal conductivity, a marginal mode with $k_{||} = 0$ exists even if the plasma is compressible (γ finite). This mode appears when the pressure gradient attains a critical value, which, in fact characterizes the onset of large-scale steady convection in the plasma [7]. In Section 4, resistivity is incorporated into the scheme, and it is shown that on top of reducing further the growth rate, it also decreases the range of unstable modes. It is also shown that resistivity and perpendicular viscosity play equivalent roles, in the sense that everything that was found for the combined effect of thermal conductivity and perpendicular viscosity, when resistivity is negligible, applies equally well for thermal conductivity and resistivity when perpendicular viscosity can be neglected. Finally, in Section 5, a discussion of the results is given.

2. Ideal MHD Motion

2.1. *Basic Equations and Equilibrium*

The plasma is described by the usual equations of ideal magnetohydrodynamics:

$$\rho\left(\frac{\partial v}{\partial t} + v.\nabla v\right) = \frac{1}{4\pi}(\nabla \times \boldsymbol{B}) \times \boldsymbol{B} - \nabla p, \tag{1a}$$

$$\frac{\partial \rho}{\partial t} + \nabla.(\rho v) = 0, \tag{1b}$$

$$\frac{1}{\gamma}\left(\frac{\partial p}{\partial t} + v.\nabla p\right) = -p\nabla.v, \tag{1c}$$

$$\boldsymbol{E} + v \times \boldsymbol{B} = 0, \tag{1d}$$

$$\frac{\partial \boldsymbol{B}}{\partial t} = -\nabla \times \boldsymbol{E}, \tag{1e}$$

$$\nabla . \boldsymbol{B} = 0, \tag{1f}$$

$$\nabla \times \boldsymbol{B} = 4\pi \boldsymbol{j}. \tag{1g}$$

The system consists of a cylindrical current-carrying plasma in a strong longitudinal magnetic field. The equilibrium magnetic field is given by

$$B_z^{(0)} = B_0, \qquad B_\theta^{(0)} = B_I(r/a), \tag{2}$$

where B_0 and B_I are constants, and a is the radius of the cylinder.

From Equation (1a) it follows that the zeroth-order equilibrium pressure is

$$p^{(0)} = p_0 - \frac{B_I^2}{4\pi}\left(\frac{r}{a}\right)^2 \tag{3}$$

where p_0 is a constant.

The rotational transform is constant and, therefore, the magnetic field is shearless:

$$q = \frac{2\pi r B_z^{(0)}}{L B_\theta^{(0)}} = \frac{2\pi a B_0}{L B_I}$$

where L is the length of the cylinder.

2.2. *Incompressible Modes* ($\gamma = \infty$)

In this subsection a summary of the dispersion relation for incompressible modes is given. Most of the material presented here is well known and amply discussed in the literature. One of the reasons for including it here is because a similar method of solution will be used for the non-ideal MHD system treated in the following sections.

The density is assumed to be nearly constant, $\rho \simeq \rho_0$ and the motion incompressible, i.e. $\nabla . \boldsymbol{v} = 0$. Under these condition the set of Equations (1) can be satisfied only for $\gamma = \infty$.

The stability analysis is carried out by linearizing Equations (1) in the usual way, assuming that all perturbed quantities behave like

$$f^{(1)}(r,\theta,z) = f^{(1)}(r)\exp(im\theta + ikz + \omega t).$$

Defining the displacement vector by $(\partial\xi/\partial t) = \omega\xi = v$, the linearized equation of motion (1a) becomes, after using Maxwell's equations

(1e)–(1g), and Ohm's law, Equation (1d),

$$\{\hat{\omega}^2 + (m - nq)\}\xi = -\nabla\hat{p}^{(1)} - 2i(m - nq)(\xi_\theta\hat{e}_r - \xi_r\hat{e}_\theta). \qquad (4)$$

In Equation (4),

$$\hat{\omega} = \sqrt{\frac{4\pi\rho_0 a^2}{B_I^2}}\,\omega\,; n = -\frac{kL}{2\pi} \qquad (5)$$

and $\hat{p}^{(1)}$ is the total perturbed pressure

$$\hat{p}^{(1)} = \frac{4\pi}{B_I^2}\left(p^{(1)} + \frac{(\boldsymbol{B}.\boldsymbol{B})^{(1)}}{8\pi}\right). \qquad (6)$$

From Equation (4) it follows that

$$\begin{aligned} \xi_r &= \frac{a^2}{\Lambda_0}\left(\frac{\partial\hat{p}^{(1)}}{\partial r} + \frac{m}{r}\sigma_0\hat{p}^{(1)}\right), \\ \xi_\theta &= \frac{ia^2}{\Lambda_0}\left(\frac{m}{r}p^{(1)} + \sigma_0\frac{\partial\hat{p}^{(1)}}{\partial r}\right), \\ \xi_z &= \frac{ia^2 k}{\Lambda_0}(1 - \sigma_0^2)\hat{p}^{(1)}, \end{aligned} \qquad (7)$$

where

$$\sigma_0 = \frac{2(m - nq)}{\hat{\omega}^2 + (m - nq)^2} \qquad (8)$$

and

$$\Lambda_0 = [\hat{\omega}^2 + (m - nq)^2](\sigma_0^2 - 1).$$

Using the incompressibility condition, $\nabla.\xi = 0$, Equation (7) reduces to

$$\nabla^2\hat{p}^{(1)} + k^2\sigma_0^2\hat{p}^{(1)} = 0 \qquad (9)$$

whose solution, regular at $r = 0$, is

$$p^{(1)} = \alpha J_m(k\sqrt{\sigma_0^2 - 1}\,r) \qquad (10)$$

Imposing the boundary condition $\xi_r(r = a) = 0$, and using Equations (7) and (10) yields the dispersion relation. For $|ka| \ll 1$, it reduces to

$$ka\sqrt{\sigma_0^2 - 1} = z_m \qquad (11)$$

where z_m is the first zero of $J_m(z)$. From Equations (8) and (11) it follows

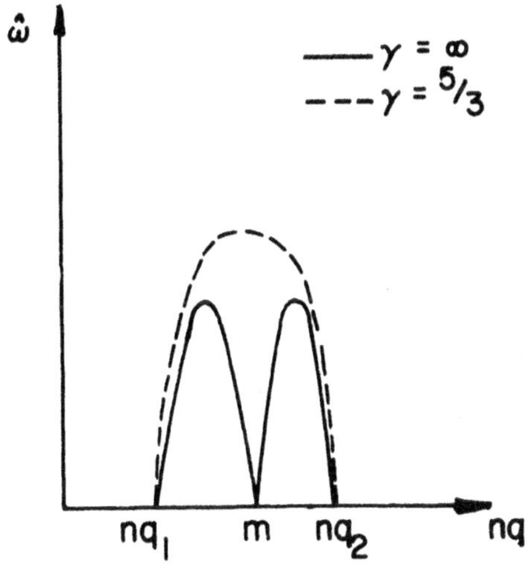

Fig. 1. Growth rate $\hat{\omega}$ as a function of nq for the incompressible ($\gamma = \infty$) and compressible ($\gamma = 5/3$) case in ideal MHD.

that

$$\hat{\omega}^2 \simeq \frac{2|m - nq|\,|k|a}{z_m} - (m - nq)^2. \tag{12}$$

This result was first given in Reference 2. The exact expression for $\hat{\omega}^2$ has the form of Equation (12) with z_m replaced by a function $z(m,k)$. The spectrum given by Equation (12) is shown in Figure 1.

Notice that $\hat{\omega} = 0$ for $nq = m$, that is, $\hat{\omega} = 0$ for rational values of q. However, for $nq \to m$

$$\hat{\omega}^2 \simeq \frac{2|m - nq|\,|k|a}{z_m} \tag{13}$$

and

$$\Lambda_0 \simeq \frac{2|m - nq|\,z_m}{|k|a} \tag{14}$$

so that

$$\left(\frac{\hat{\omega}}{\Lambda_0}\right) \simeq \left(\frac{ka}{z_m}\right)^{3/2} [2|m - nq|]^{-1/2} \tag{15}$$

which goes to infinity as nq goes to m.

Thus, since $v = \omega\xi$ from Equations (7) it follows that as nq approaches m, a finite pressure perturbation leads to an infinite velocity. In other words, there is no linear solution to the problem if the plasma is assumed incompressible. As it will be shown in the following sections, this singular behaviour does not occur when non-ideal effects are taken into account. In ideal MHD, however, the singular behaviour indicates that the incompressibility assumption must be abandoned and therefore, compressible modes are important in the neighborhood of rational values of q. This has indeed been found in References (5) and (6). On the other hand, the effect of compressibility remains negligible near the marginally unstable modes, i.e. near the values nq_1 and nq_2 (see Figure 1) as it will be discussed in the following subsection.

2.3. *Compressible Modes*

Consider first values of q in the neighborhood of the stability limit, where ω vanishes (in the incompressible case these limits occur for $nq_{1,2} = m \mp 2 |k|a/z_m$).

For sufficiently small $\omega \neq 0$ the linearized equation of motion (1a) becomes

$$\frac{iB_I}{4\pi a}(m - nq)\boldsymbol{B}^{(1)} - \frac{2B_I}{4\pi a}(B_\theta^{(1)}\hat{e}_r - B_r^{(1)}\hat{e}_\theta) - \frac{B_I^2}{4\pi a^2}\nabla\hat{p}^{(1)} = 0(\omega). \quad (16)$$

On the other hand, Equations (1d) and (1e) yield

$$\boldsymbol{B}^{(1)} = \frac{iB_I}{a}(m - nq)\xi - \boldsymbol{B}^{(0)}\nabla.\xi. \quad (18)$$

From Equation (16) the components of $\boldsymbol{B}^{(1)}$ can be expressed in terms of $\hat{p}^{(1)}$ in the same way as the components of ξ were written in Equation (7). Using $\nabla.\boldsymbol{B}^{(1)} = 0$ it is found that $\hat{p}^{(1)}$ satisfies Equation (9) and thus is again given by Equation (10). Therefore, in the limit $\omega = 0$, the perturbed magnetic field $\boldsymbol{B}^{(1)}$ is identical to the one obtained in Section 2.1, that is, under the assumption $\nabla.\xi = 0$ Consequently, Equation (16) implies that in the limit $\omega = 0$, ξ is given by Equations (7) and $\nabla.\xi = 0$.

Notice that ξ is finite at $\omega = 0$ (with $m \neq nq$) whereas $v = 0$ in this limit.

By making use of the fact that $\nabla.\xi$ is small in the neighborhood of the stability limits, the dispersion relation for compressible modes can be determined near these limits. To a first approximation, the term proportional to $\nabla.\xi$ in Equation (18) can be neglected and the resulting expression may be used to eliminate $\boldsymbol{B}^{(1)}$ from the linearized Equation (1a). The resulting equation yields

$$\nabla.\xi = \frac{1}{\Lambda_0}(\nabla^2 \hat{p}^{(1)} + k^2 \sigma^2 \hat{p}^{(1)}). \tag{19}$$

This equation can now be combined with Equation (1c) to eliminate $\nabla.\xi$. The result is

$$\nabla^2 \hat{p}^{(1)} + \left[k^2 \sigma_0^2 - \frac{\delta \hat{\omega}^2 \sigma_0^2 nq}{a^2(m - nq)} \right] \hat{p}^{(1)} = 0 \tag{20}$$

where

$$\delta = \frac{B_I^2}{\gamma} \frac{1}{4\pi p^{(0)}}. \tag{21}$$

To obtain Equation (20), the pressure $p^{(1)}$ has to be expressed in terms of the total pressure $\hat{p}^{(1)}$. To this end, Equation (18) is inserted into Equation (6) after setting $\nabla.\xi = 0$, to obtain

$$p^{(1)} = -\frac{B_I^2}{4\pi\Lambda_0}\left\{ [\hat{\omega}^2(1 - \sigma_0^2) - m\sigma_0^2(m - nq)]\hat{p}^{(1)} - (m - nq)\sigma_0 r \frac{\partial \hat{p}^{(1)}}{\partial r} \right\}. \tag{22}$$

Inserting Equation (22) into the linearized Equation (1c) yields

$$\nabla.\xi = \frac{1}{\gamma p^{(0)}} \frac{B_I^2 r^2}{4\pi\Lambda_0} \left\{ [\hat{\omega}^2(1 - \sigma_0^2) - (m - nq)m\sigma_0^2 + 2m\sigma_0] \frac{\hat{p}^{(1)}}{r} - [(m - nq)\sigma_0 - 2] \frac{1}{r} \frac{\partial \hat{p}^{(1)}}{\partial r} \right\}. \tag{23}$$

Equation (23), together with Equation (19), leads to Equation (20).

Assuming now $p_0 \gg B_I^2/4\pi$ in Equation (3), so that $p^{(0)}$ varies slowly and δ may be considered as constant, the solution of Equation (20) is then

$$\hat{p}^{(1)} = \alpha J_m(br) \tag{24}$$

where

$$b^2 = k^2(\sigma_0^2 - 1) - \frac{\delta \hat{\omega}^2 \sigma_0^2 nq}{a^2(m - nq)}. \tag{25}$$

The boundary condition $\xi_r(r=a)=0$ yields the following dispersion relation

$$\hat{\omega} = -\left[(m-nq)^2 + \frac{2\delta nq\,ka}{z_m}\right] + \left\{\frac{4(m-nq)^2(ka)^2}{z_m} + 4\delta nq(m-nq)^2\frac{ka}{z_m} + \frac{4\delta^2(nq)^2}{z_m^2}\right\}^{1/2} \tag{26}$$

From this expression it follows that compressibility, as expected, has the effect of increasing the growth rate without changing the unstable domain of q-values. This is illustrated in Figure 1.

The effect of compressibility in the central region of the spectrum ($k_{\|} \approx 0$) where $\nabla.\boldsymbol{\xi}$ is not small compared to other perturbed quantities, has been calculated analytically by Friedberg [5] and by Takeda *et al.* [3, 4]. The result is illustrated in Figure 1. It is seen that the doubled-humped spectrum of the incompressible case becomes a single hump peaked at $nq = m$, when compressibility is taken into account.

3. Effect of perpendicular viscosity and thermal conductivity

3.1. *Incompressible Modes*

In this section the effect of viscosity and thermal conductivity on the stability of the equilibrium characterized by Equations (2) and (3) is investigated. Instead of the full viscosity tensor [8], only the term corresponding to viscosity perpendicular to the magnetic field is considered. The effect of other terms will be discussed at the end of this section.

The equations describing the system are

$$\rho\left(\frac{\partial \boldsymbol{v}}{\partial t} + \boldsymbol{v}.\nabla\boldsymbol{v}\right) = \frac{1}{4\pi}(\nabla\times\boldsymbol{B})\times\boldsymbol{B} - \nabla p - \mu_{\perp}\nabla\times(\nabla\times\boldsymbol{B}), \tag{27a}$$

$$\frac{\partial\rho}{\partial t} + \nabla.(\rho\boldsymbol{v}) = 0, \tag{27b}$$

$$\frac{1}{\gamma}\left(\frac{\partial p}{\partial t} + \boldsymbol{v}.\nabla p - \frac{2}{3}K\nabla^2 p - S_0\right) = -p\nabla.\boldsymbol{v}, \tag{27c}$$

$$\boldsymbol{E} + \boldsymbol{v}\times\boldsymbol{B} = 0, \tag{27d}$$

$$\frac{\partial\boldsymbol{B}}{\partial t} = -\nabla\times\boldsymbol{E}, \tag{27e}$$

$$\nabla . \boldsymbol{B} = 0, \tag{27f}$$

$$\nabla \times \boldsymbol{B} = 4\pi \boldsymbol{j}. \tag{27g}$$

Here $\mu_\perp$ is the perpendicular viscosity coefficient, K is the heat conductivity, and S_0 is a constant heat source which maintains the equilibrium pressure profile, i.e.,

$$S_0 = \tfrac{2}{3} K B_I^2 / 4\pi a^2. \tag{28}$$

The plasma is assumed to be incompressible, that is, γ is set equal to infinity and Equation (27c) is replaced by

$$\nabla . \boldsymbol{v} = 0. \tag{29}$$

Assuming constant density and linearizing Equation (27a) for perturbations of the form used in the previous section, the following equation for the displacement $\boldsymbol{\xi}$ is obtained:

$$\left\{ \rho_0 \omega^2 + \frac{B_I^2}{4\pi a^2}(m - nq)^2 \right\} \boldsymbol{\xi} = -\nabla\left(p^{(1)} + \frac{\boldsymbol{B}^{(0)} . \boldsymbol{B}^{(1)}}{4\pi} \right) - \frac{2 B_I^2 i}{4\pi a^2}(m - nq)(\xi_\theta \hat{e}_r - \xi_r \hat{e}_\theta) - \mu_\perp \omega \nabla \times (\nabla \times \boldsymbol{\xi}) \tag{30}$$

In the ideal case treated in Section 2, the solution given by Equations (7) and (10), has the property that $\nabla \times \boldsymbol{\xi}$ is proportional to $\boldsymbol{\xi}$. This can be obtained in a straight forward way by applying the curl to Equation (14). This property will be assumed to hold also in the presence of viscosity. Thus, writing

$$\nabla \times \boldsymbol{\xi} = \beta \boldsymbol{\xi}, \tag{31}$$

where β is a constant to be determined later, Equation (30) reduces to

$$\{\hat{\omega}^2 + \hat{\mu}_\perp \hat{\omega} \beta^2 + (m - nq)^2\} \boldsymbol{\xi} = -\nabla \hat{p}^{(1)} - 2i(m - nq)(\xi_\theta \hat{e}_r - \xi_r \hat{e}_\theta) \tag{32}$$

where $\hat{\omega}$ and $\hat{p}^{(1)}$ are defined in Equations (5) and (6), and

$$\hat{\mu}_\perp = \sqrt{\frac{4\pi}{a^2 \rho_\theta B_I^2}} \mu_\perp . \tag{33}$$

Equation (32) can be solved for the components of $\boldsymbol{\xi}$, as was done in the previous section. Using the incompressibility condition, $\nabla . \boldsymbol{\xi} = 0$, yields the following equation for $\hat{p}^{(1)}$

$$\nabla^2 \hat{p}^{(1)} + k^2 \sigma^2 \hat{p}^{(1)} = 0 \tag{34}$$

where

$$\sigma = \frac{2(m - nq)}{\hat{\omega}^2 + \hat{\mu}_\perp \hat{\omega} \beta^2 a^2 + (m - nq)^2}. \tag{35}$$

The constant β is determined by taking the curl of Equation (32). The result is

$$\beta = k\sigma. \tag{36}$$

The solution of Equation (34) regular at $r = a$ is

$$\hat{p}^{(1)} = \alpha J_m(k\sqrt{\sigma^2 - 1}\, r) \tag{37}$$

Defining

$$\Lambda = [\hat{\omega}^2 + \hat{\mu}_\perp \hat{\omega} k^2 a^2 \sigma^2 + (m - nq)^2](\sigma^2 - 1) \tag{38}$$

the components of ξ are given by expressions of the form of Equations (7) with σ_0, Λ_0 replaced by σ, Λ. In particular,

$$\xi_r = \frac{a}{\Lambda}\left(\frac{\partial \hat{p}^{(1)}}{\partial r} + \frac{m}{r}\sigma \hat{p}^{(1)}\right). \tag{39}$$

The boundary condition $\xi_r(r = a) = 0$, leads to

$$0 = m\sigma J_m(ka\sqrt{\sigma^2 - 1}) + ka\sqrt{\sigma^2 - 1}\, J'_m(ka\sqrt{\sigma^2 - 1}). \tag{40}$$

For $|ka| \ll 1|$, this reduces to $J_m(ka\sqrt{\sigma^2 - 1}) = 0$.
Hence,

$$|ka|\sqrt{\sigma^2 - 1} = z_m \quad or \quad \sigma \simeq \mp \frac{z_m}{|ka|} \tag{41}$$

where z_m is again the first zero of $J_m(z)$.

From Equation (35) it follows that for unstable modes, $\omega > 0$, the sign of σ must be equal to the sign of $m - nq$. Thus, for $|ka| \ll 1|$

$$\sigma \simeq - \frac{z_m}{|ka|} \frac{(m - nq)}{|m - nq|} \tag{42}$$

and

$$\hat{\omega} = -\tfrac{1}{2} z_m^2 \mu_\perp^2 + \left\{ (\tfrac{1}{2} z_m^2 \mu_\perp^2)^2 + \frac{2|ka|}{z_m} |m - nq| - (m - nq)^2 \right\}^{1/2}. \tag{43}$$

In the last expression the sign of the square root has been chosen so that $\hat{\omega}$ is real and positive. Otherwise, from Equation (35) it follows that $\hat{\omega}$ is complex and the boundary condition is not satisfied.

For $\mu_\perp = 0$ the dispersion relation given by Equation (43) reduces to expression (12). As expected on physical grounds, viscosity has the effect of reducing the growth rate of all unstable modes, but the form of the spectrum as a function of $k_{||} = (m - nq)$ is the same as for $\mu_\perp = 0$. In particular, ω vanishes for $nq = m$. However, in contrast to the situation encountered in Section 2 for the ideal case, now, as $m - nq \to 0, \hat{\omega}$ and Λ behave as

$$\hat{\omega} \simeq \frac{2|ka|\,|m-nq|}{z_m^3 \hat{\mu}_\perp},$$

$$\Lambda \simeq \frac{2|m-nq|z_m}{|ka|}, \tag{44}$$

so that

$$\left(\frac{\hat{\omega}}{\Lambda}\right) \simeq -\frac{(ka)^2}{z_m^2 \hat{\mu}_\perp}. \tag{45}$$

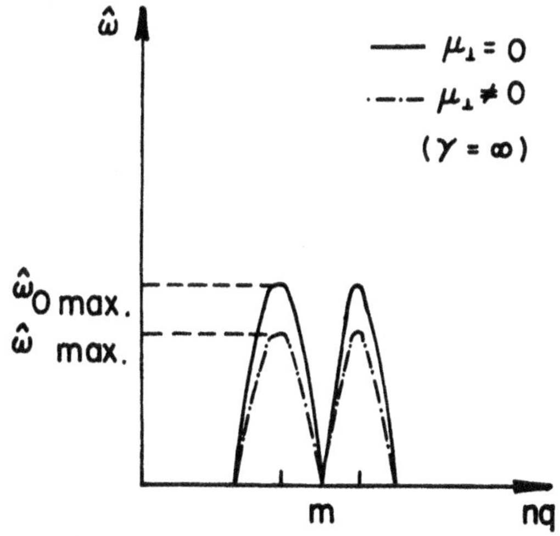

Fig. 2. Effect of viscosity on the growth rate $\hat{\omega}$ in the incompressible case ($\gamma = \infty$), with $\hat{\omega}_{0,\max}$ and $\hat{\omega}_{\max}$ defined in Equation (93) and (94).

That is, the ratio between $\hat{\omega}$ and Λ remains finite as $k_{\|} \to 0$ in contrast to the ratio $(\hat{\omega}/\Lambda_0)$ of Equation (15). Therefore, from $v_r = \omega\xi_r$ and expression (39) it follows that in the limit $nq = m$ a finite pressure perturbation leads to a finite velocity v_r; the same is true for v_θ and v_z. Hence, viscosity has the effect of removing the singularity found in Section 2, for rational q-values. Due to this fact, a solution of the complete system of Equations (27) for rational values of q can be obtained, as it will be seen in the next subsection. The effect of viscosity on the unstable spectrum is shown in Figure 2.

3.2. *A Marginal Mode for Finite* γ

In the case of a marginally stable state, i.e. $\omega = 0$, Equation (27e) implies that $\boldsymbol{E} = -\nabla\phi$, so that the linearized Equation (27d) takes the form

$$v \times \boldsymbol{B}^{(0)} = \nabla\phi^{(1)}. \tag{46}$$

From the components of this vector equation it follows that

$$v_r(m - nq) = 0, \tag{47a}$$

$$\nabla . v = i(m - nq)\left[\frac{v_\theta}{r} + \frac{B_I}{aB_0}v_z\right]. \tag{47b}$$

For marginal states occuring at $nq \neq m$ (for instance at the limits nq_1 and nq_2 of the unstable region in Figure 1) relation (47a) implies $v_r = 0$. In fact, as it has been shown in Section 2.3, all three components of v vanist at the limits nq_1, nq_2.

On the other hand, for marginal states occurring at $nq = m$, relation (47a) can be satisfied for $v_r \neq 0$, but (47b) requires $\nabla . v = 0$. This is the situation to be investigated now. Specifically, it will be shown that for rational q, there exists a marginal mode satisfying the complete set of Equations (27), including Equation (27c) with $K \neq 0$ and arbitrary finite γ.

Thus, assuming

$$\omega = 0 \quad \text{at} \quad k_{\|} = 0 \tag{48}$$

according to Equation (47b) $\nabla . v = 0$, and Equation (27c) reduces to

$$v_r \frac{\mathrm{d}p^{(0)}}{\mathrm{d}r} = \frac{2}{3} K \nabla^2 p^{(1)}. \tag{49}$$

In the preceding subsection a solution of Equations (27a, b) and (27e, f, g) has been obtained for arbitrary values of $m - nq$ and the expression for ξ_r has been given in Equation (39). From that equation and Equation (45) it follows that

$$\lim_{nq \to m} (\omega \xi_r) = \left(\frac{B_I^2}{4\pi \rho_0 a^2} \right)^{1/2} \frac{2(ka)^2 a^2}{z_m^3 \hat{\mu}_\perp} \left[\frac{\partial \hat{p}^{(1)}}{\partial r} + \frac{m}{r} \sigma \hat{p}^{(1)} \right]. \tag{50}$$

Since v_r is a function of $\hat{p}^{(1)}$, it is convenient to express the right-hand side of Equation (49) in terms of $\hat{p}^{(1)}$. To this end, from Equations (27e, f) and (45) it follows that (for $\nabla . \boldsymbol{v} = 0$)

$$\boldsymbol{B}^{(1)} = \lim_{nq \to m} \left\{ i \frac{m - nq}{\omega} \frac{B_I}{a} \boldsymbol{v} \right\} = \frac{i z_m^3 B_I \hat{\mu}_\perp}{2ka^2} \boldsymbol{v}^{(1)}. \tag{51}$$

Introducing this expression into Equation (6) and using the fact that $\hat{p}^{(1)}$ satisfies Equation (34), yields

$$\nabla^2 \hat{p}^{(1)} = - \frac{B_I}{4\pi} \frac{\hat{\omega}}{\Lambda} \frac{\hat{\mu}_\perp \sigma^3 k^2}{\rho_0 a} \times \left[\left(2ka\sigma(\sigma^2 - 1) + \frac{m}{2}(ka)^2 \sigma^4 \right) \hat{p}^{(1)} + \tfrac{1}{2}(ka)^2 \sigma^3 r \frac{\partial \hat{p}^{(1)}}{\partial r} \right]. \tag{52}$$

Using the value of the equilibrium pressure, Equation (3), and introducing Equations (50) and (52) into Equation (49), it is found that such an equation is satisfied provided (see Appendix A)

$$\mu_\perp K = \frac{3(ka)^2}{z_m^6} \left. \frac{\mathrm{d}p^{(0)}}{\mathrm{d}r} \right|_{r=a}. \tag{53}$$

Thus, when this relation between the pressure gradient and the physical constants is attained, a marginal solution of the complete system (27) exists for arbitrary γ.

This state is actually the analog in a plasma of the well-known stationary convection in ordinary hydrodynamics. Thus, it is possible to define the plasma anolog of the Rayleigh number used in hydrodynamics [9],

$$\mathscr{R} = \left(\frac{1}{r_c} \left| \frac{\mathrm{d}p^{(0)}}{\mathrm{d}r} \right|_{r=a/2} a^4 \right) \bigg/ \mu_\perp K \tag{54}$$

where

$$\frac{1}{r_c} = \frac{1}{a} \left(\frac{B_I}{B_0} \right)^2 \tag{55}$$

is the curvature of the helical field lines at $r = a$. The curvature plays here the same role as gravity does in the hydrodynamic stability problem.

Condition (53) for the existence of a marginal solution may then be written in the form

$$\mathscr{R} = \mathscr{R}_{\text{crit}}, \qquad \mathscr{R}_{\text{crit}} = \frac{1}{6}\frac{z_m^6}{m^2}. \tag{56}$$

As it has been shown in the case of a purely azimuthal magnetic field [7], Equation (56) defines in fact the critical 'Rayleigh number' for the onset of steady convection. In the same way as in Reference 7, it can be shown that for $\mathscr{R} > \mathscr{R}_{\text{crit}}$ the nonlinear Equations (27) have stationary solutions with finite convection velocity $\mathfrak{v}$. The nonlinear problem will not be treated here, but, instead, this section will be concluded with some remarks on the flow pattern of the solutions of the linearized equations, and the possible influence of parallel viscosity.

The flow pattern is most easily recognized by writing the velocity in terms of the flux function. Since $\nabla . \mathfrak{v} = 0$, $\mathfrak{v}$ is given by

$$\mathfrak{v} = \nabla \times \boldsymbol{G}. \tag{57}$$

Taking $G_r = 0$ and using the helical symmetry of the perturbation, the flux surfaces are found to be described by

$$\Phi(r, \theta, z) = m G_z - kr G_\theta = \text{const.} \tag{58}$$

As it can be easily seen, Φ can be expressed in terms of v_r as follows

$$\Phi = \text{Re}\{-irv_r\} = A\left\{J_m(z_m r/a) + \frac{r z_m}{m\sigma a} J'_m(z_m r/a)\right\}\sin(m\theta + kz) \tag{59}$$

The flux surfaces $\Phi = \text{const.}$ define convection cells which have the form of helically twisted tubes, as shown in Figure 3, for the case $m = 1$. The centre of each tube is determined by the extremes of Φ, that is,

$$\frac{\partial \Phi}{\partial r} = 0, \qquad \frac{\partial \Phi}{\partial \theta} = 0. \tag{60}$$

Hence,

$$m\theta + kz = (2n + 1)\frac{\pi}{2} \qquad n = 0, 1, 2, \ldots$$

$$J'_m(z_m r/a) + \frac{z_m}{m\sigma}\frac{\mathrm{d}}{\mathrm{d}r}[r J'_m(z_m r/a)] = 0 \tag{61}$$

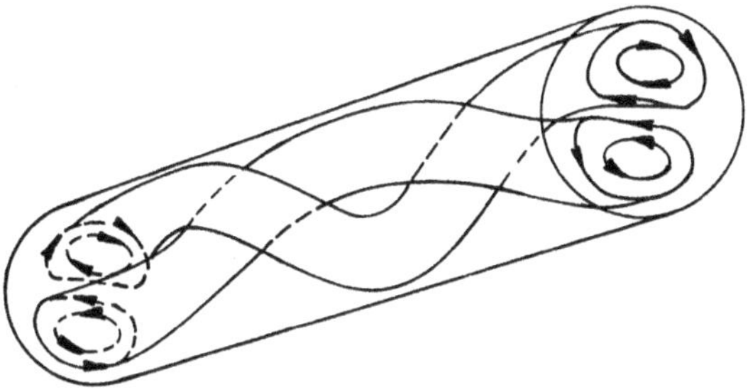

Fig. 3. Qualitative picture of helical convection cells ($m = 1$).

On the centre line of each tube, defined by Equations (61), $\boldsymbol{v}$ is given by

$$v_r = 0, \qquad v_\theta = (B_\theta/B_z)v_z, \qquad v_z = (-1)^{n+1}A(ka)J_m(z_m r/a) \quad (62)$$

Thus, on the centre line there is a nonvanishing flow velocity parallel to $\boldsymbol{B}^{(0)}$. This means that the component of $\boldsymbol{v}$ parallel to $\boldsymbol{B}^{(0)}$ is large and has large derivatives. It is, therefore, necessary to consider the influence of parallel viscosity on the perturbed motion.

Using the parallel viscosity part of the viscosity tensor as given in Reference 8, the corresponding force term is given by

$$\boldsymbol{F}_0 = \nabla(\mu_{||}H) - 3\boldsymbol{h}[\nabla\cdot(\mu_{||}H\boldsymbol{h})] - 3\mu_{||}H(\boldsymbol{h}.\nabla)\boldsymbol{h} \qquad (62)$$

where

$$\boldsymbol{h} = \boldsymbol{B}^{(0)}/B^{(0)}$$

and

$$H = -\,\boldsymbol{h}.[(\boldsymbol{h}.\nabla)\boldsymbol{v}] + \tfrac{1}{3}\nabla.\boldsymbol{v}.$$

Assuming $\mu_{||} = \text{const.}$, $\nabla.\boldsymbol{v} = 0$, and $m - nq \simeq 0$, expression (62) takes the form

$$\boldsymbol{F}_0 = -\,\mu_{||}\nabla\left(\frac{B_I}{B^{(0)}}\frac{rv_r}{a^2}\right) - 3\mu_{||}\left(\frac{B_r}{B^{(0)}}\right)^4\frac{rv_r}{a^4}\hat{e}_r. \qquad (63)$$

The first term in this expression is a gradient and, therefore, it can be combined with the pressure gradient and, consequently, it has no consequence in the solution of the problem. The other term in $\boldsymbol{F}_0$ vanishes on the boundary, $r = a$, and on the centre line of each helical convection cell. For other parts of the plasma volume, this term has to be compared to the perpendicular viscosity force $-\mu_\perp \nabla \times (\nabla \times v) = -\mu_\perp \beta^2 v$. By doing this, it is found that the parallel viscosity can be neglected if

$$\frac{\mu_{\parallel}}{\mu_\perp}\left(\frac{B_I}{B^{(0)}}\right)^4 \frac{3}{z_m} < 1. \tag{64}$$

Since $(\mu_{\parallel}/\mu_\perp)$ is large (the assumption $\omega_i \tau_i \gg 1$ is used in the derivation of the viscosity tensor [8]) the above condition is rather restrictive for small values of m. For sufficiently large m, and $(B_I/B^{(0)}) \ll 1$, the condition is easily satisfied since for large $m, z_m \sim m$.

4. Effect of Resistivity

The equations describing the system are the same as Equations (27), except for (27d) which now becomes

$$\boldsymbol{E} + v \times \boldsymbol{B} = \eta \boldsymbol{j} \tag{65}$$

where η is the resistivity.

The zeroth-order equilibrium requires an electric field $\boldsymbol{E}^{(0)} = \eta \boldsymbol{j}^{(0)}$ (if $v^{(0)} = 0$). This electric field can, in principle, produce charge separation and thereby destroy quasi-neutrality, which has been used in the derivation of the equation of motion. However, as it can be easily shown, if the quantity $4\pi\rho c^2/B^2$ is large enough, charge separation will not be significant and quasi-neutrality still holds. That such a quantity is indeed large for controlled thermonuclear plasma and laboratory plasmas in general, can easily be checked.

As in the previous sections the plasma is assumed to be incompressible, that is, γ is set equal to infinity and Equation (27e) replaced by $\nabla . v = 0$.

In the ideal case treated in Section 2, Equations (1d) and (1e) together with Equations (2) yield

$$\boldsymbol{B} = i\frac{B_I}{a}(m - nq)\boldsymbol{\xi}. \tag{66}$$

Since, as pointed out before, $\nabla \times \xi \sim \xi$, the last equation implies that $\nabla \times \boldsymbol{B} \sim \boldsymbol{B}$. It will be assumed that when non-ideal effects are included these properties are still true, i.e.

$$\nabla \times \xi = \beta \xi$$

and

$$\nabla \times \boldsymbol{B} = \beta' \boldsymbol{B} \tag{67}$$

where β and β' are constants to be determined later.

Thus, Equations (27e) and (65) yield

$$\omega \boldsymbol{B}^{(1)} = \frac{iB_I}{a}(m - nq)\boldsymbol{v} - \frac{\eta}{4\pi}\nabla \times (\nabla \times \boldsymbol{B}^{(1)}) \tag{68}$$

which due to (67) can be written in the following form

$$\left(\omega + \frac{\eta}{4\pi}\beta'^2\right)\boldsymbol{B}^{(1)} = \frac{iB_I}{a}(m - nq)\boldsymbol{v}^{(1)}. \tag{69}$$

Taking the curl of Equation (69) leads to

$$\beta = \beta'. \tag{70}$$

Replacing Equation (69) in the linearized Equation (27a) yields

$$\left\{\rho_0\omega + \mu_\perp\beta^2 + \frac{(m/r)B_\theta^{(0)} + kB_z^{(0)}}{4\pi(\omega + \eta\beta^2/4\pi)}\right\}\boldsymbol{v}$$
$$= -\frac{B_I}{4\pi}\nabla\hat{p}^{(1)} - \frac{i}{4\pi}\frac{2B_\theta^{(0)}}{r}\frac{(m/r)B_\theta^{(0)} + kB_z^{(0)}}{\omega + \eta\beta^2/4\pi}(v_\theta\hat{e}_r - v_r\hat{e}_\theta). \tag{71}$$

Introducing the definitions

$$\boldsymbol{\xi} = \frac{\boldsymbol{v}}{\omega + \eta\beta^2/4\pi}, \qquad \hat{\eta} = \sqrt{\frac{\rho_0}{4\pi a^2 B_I^2}}\,\eta,$$
$$\hat{\omega} = \sqrt{\frac{4\pi\rho_0 a^2}{B_I^2}}\,\omega, \qquad \hat{\mu}_\perp = \sqrt{\frac{4\pi}{a^2\rho_0 B_I^2}}\,\mu_\perp \tag{72}$$

Equation (71) reduces to

$$\{(\hat{\omega} + \hat{\mu}_\perp\beta^2)(\hat{\omega} + \hat{\eta}\beta^2) + (m - nq)^2\}\boldsymbol{\xi}$$
$$= -\nabla\hat{p}^{(1)} - 2i(m - nq)(\xi_\theta\hat{e}_r - \xi_r\hat{e}_\theta). \tag{73}$$

Solving for the components of $\boldsymbol{\xi}$ in terms of $\hat{p}^{(1)}$ leads to

$$\xi_r = \frac{a}{\Lambda_1}\left(\frac{\partial\hat{p}^{(1)}}{\partial r} + \frac{m}{r}\sigma_1\hat{p}^{(1)}\right),$$

$$\xi_\theta = \frac{ia}{\Lambda_1}\left(\frac{m}{r}\hat{p}^{(1)} + \sigma_1\frac{\partial\hat{p}^{(1)}}{\partial r}\right),$$

$$\xi_z = \frac{ika}{\Lambda_1}(1 - \sigma^2)\hat{p}^{(1)}, \tag{74}$$

where

$$\sigma_1 = \frac{2(m - nq)}{(\hat{\omega} + \hat{\mu}_\perp \beta^2)(\hat{\omega} + \hat{\eta}\beta^2) + (m - nq)^2} \tag{75}$$

and

$$\Lambda_1 = \{(\hat{\omega} + \hat{\mu}_\perp \beta^2)(\hat{\omega} + \hat{\eta}\beta^2) + (m - nq)^2\}(\sigma_1^2 - 1). \tag{76}$$

The constant β can be determined by taking the curl of Equation (73). The result is

$$\beta = k\sigma_1. \tag{77}$$

Taking the divergence in Equations (74) and setting it equal to zero, gives the following equation for $\hat{p}^{(1)}$:

$$\nabla^2 \hat{p}^{(1)} + k^2 \sigma_1^2 \hat{p}^{(1)} = 0. \tag{78}$$

This again is Bessel's equation, whose solution regular at the origin, $r = 0$, is

$$\hat{p}^{(1)} = \alpha J_m(k\sqrt{\sigma_1^2 - 1}r) \tag{79}$$

The boundary condition $\xi_r(r = a) = 0$, i.e.

$$m\sigma_1 J_m(z_m r/a) + ka\sqrt{\sigma_1^2 - 1}J'_m(z_m r/a) = 0 \tag{80}$$

and the condition $|ka| \ll 1$ yields the dispersion relation

$$ka\sqrt{\sigma_1^2 - 1} = z_m \tag{81}$$

where z_m is the first zero of $J_m(z)$.

Using Equation (75), Equation (81) leads, finally, to

$$\hat{\omega} = -\tfrac{1}{2}z_m^2(\hat{\mu}_\perp + \hat{\eta}) + \left\{\tfrac{1}{2}z_m^4(\hat{\mu}_\perp - \hat{\eta}_\perp)^2 + \frac{2|k|a}{z_m}|m - nq| - (m - nq)^2\right\}^{1/2}. \tag{82}$$

It is interesting to notice that if either $\hat{\mu}_\perp = 0$ or $\hat{\eta} = 0$ the range of unstable $(m - nq)$-values remains unchanged. If, on the other hand, both $\hat{\mu}_\perp \neq 0$ and $\hat{\eta} \neq 0$ then the growth rate as well as the unstable region are reduced. In particular, the neighborhood of $k_{\parallel} \simeq 0$ is stable. For fixed m and n, the limits of the unstable region are given by

$$|m - nq| = \frac{|ka|}{z_m} \pm \left\{\left(\frac{ka}{z_m}\right)^2 - \frac{z_m^4 \hat{\mu}_\perp \hat{\eta}}{2}\right\}^{1/2}. \tag{83}$$

The results are illustrated in Figure 4.

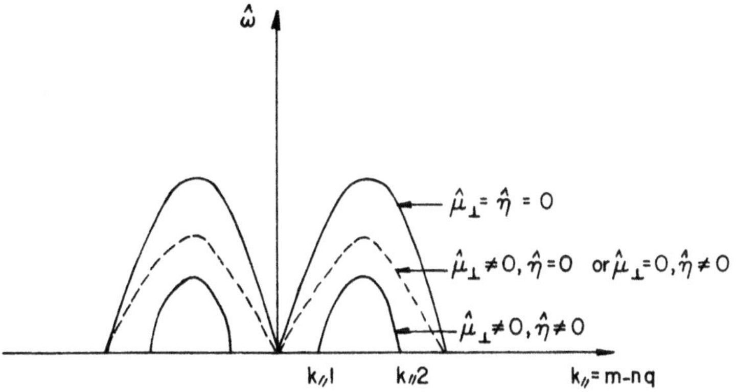

Fig. 4. Effect of resistivity on the growth rate $\hat{\omega}$ in the incompressible case ($\gamma = \infty$), with $k_{\parallel 1}$ and $k_{\parallel 2}$ given by Equation (83).

Another point worth mentioning is that viscosity and resistivity play equivalent roles as long as one of them is zero. For example, if viscosity is zero, resistivity will still give rise to stationary convection for a critical value of the product of thermal conductivity and resistivity, just as discussed in the previous section. It is not, however, a priori clear that such a state will be an exact solution of the set of Equations (27) with $\mu_\perp = 0$ and Equation (27d) replaced by Equation (65). The reason being that for the case when resistivity is not zero, Equation (40) no longer holds and, therefore, the incompressible character of the state is not guaranteed. Nevertheless, it is still true that such states fulfil the condition $\nabla . \boldsymbol{v} = 0$, as it will now be shown.

The linearized equation of motion (27a) with $\mu_\perp = 0$, can be written in the following way

$$\omega \rho_0 \boldsymbol{v} = -\nabla \hat{p}^{(1)} + \frac{1}{4\pi} \times \left\{ i\left(\frac{m}{r} B_\theta^{(0)} + k B_z^{(0)}\right) \boldsymbol{B}^{(1)} - \frac{2B_\theta^{(0)}}{r}(B_\theta^{(1)}\hat{e}_r - B_r^{(1)}\hat{e}_\theta)\right\}. \tag{84}$$

Since the state under consideration is characterized by $\omega = 0$ and $q = m/n$ Equation (84) for these particular values reduces to

$$B_\theta^{(1)}\hat{e}_r - B_r^{(1)}\hat{e}_\theta = -\frac{B_I}{2a}\nabla \hat{p}^{(1)}. \tag{85}$$

In terms of vector components, Equation (85) gives

$$B_\theta^{(1)} = -\frac{B_I}{2a}\frac{\partial \hat{p}^{(1)}}{\partial r},$$

$$B_r^{(1)} = \frac{B_I}{2a}\frac{im}{r}\hat{p}^{(1)}. \tag{86}$$

From these equations it follows that

$$\frac{1}{r}\frac{\partial}{\partial r}(rB_r^{(1)}) + \frac{1}{r}\frac{\partial B_\theta^{(1)}}{\partial \theta} = 0. \tag{87}$$

Since on the other hand, $\nabla . \boldsymbol{B}^{(1)} = 0$ the last equation yields

$$\frac{\partial B_z^{(1)}}{\partial z} = 0. \tag{88}$$

In other words,

$$B_z^{(1)} = 0. \tag{89}$$

From Equations (27e), (65), and (67) it follows that

$$(\omega + \eta\beta^2/4\pi)\,\boldsymbol{B}^{(1)} = i\left(\frac{m}{r}B_\theta^{(0)} + kB_z^{(0)}\right)\boldsymbol{v} - \boldsymbol{B}^{(0)}\nabla . \boldsymbol{v}. \tag{90}$$

Setting $\omega = k_{\|} = 0$ yields

$$(\eta\beta^2/4\pi)\,\boldsymbol{B}^{(1)} = -\boldsymbol{B}^{(0)}\nabla . \boldsymbol{v}. \tag{91}$$

Since this mode, according to the equation of motion (85), is characterized by $B_z^{(1)} = 0$, Equation (91) implies that $\nabla . \boldsymbol{v} = 0$. This completes the proof that indeed η and $\mu_\perp$ play equivalent roles provided that one of them is zero.

5. Conclusion

As discussed in Section 2.2, the incompressible modes obtained from the linearized MHD equations behave quite differently at the limits of the unstable domain of nq and at the point $m = nq$, which is inside this domain. For a given pressure perturbation, the perturbed velocity vanishes at the limits of the unstable domain, (points $nq_{1,2}$ in Figure 1), whereas it diverges at $nq = m$. Near the point $nq = m$, the compressibility of the plasma becomes important, but also non-ideal effects play a significant role. The effect of compressibility is to increase the growth rate without changing the instability domain.

When dealing with non-ideal effects, (Sections 3 and 4) only the perpendicular term of the viscosity tensor has been considered. This allows a complete solution of the MHD equations in the case of incompressible modes ($\gamma = \infty$). Viscosity, as expected on physical grounds, is always stabilizing, but it does not change the range of unstable modes (Figure 2). As a consequence of the stabilizing effect of viscosity, the dependence of the maximum growth rate on the azimunithal mode number m is modified. Indeed, with the position of the maximum growth rate given by

$$nq = m/\left(1 \pm \frac{1}{z_m}\frac{2\pi a}{qL}\right) \tag{92}$$

the corresponding $\hat{\omega}$-value is given by

$$\hat{\omega}_{\max} = -\frac{z_m}{2}\mu_\perp + \sqrt{\left(\frac{z_m}{2}\mu_\perp\right)^2 + \hat{\omega}^2_{0,\max}} \tag{93}$$

where

$$\hat{\omega}_{0,\max} = \frac{m}{z_m}\,\frac{2\pi a/qL}{1 \pm 2\pi a/z_m qL} \tag{94}$$

is the maximum growth rate for $\mu_\perp = 0$. Since m/z_m is a monotonically increasing function of m, approaching unity for $m \to \infty$, the growth rate without viscosity, $\omega_{0'\max}$, increases steadily with m. This result is typical of MHD modes of the interchange type. In contrast, for $\mu_\perp \neq 0$ the growth rate is largest for some finite m and tends to zero for $m \to \infty$. This behavior is illustrated in Figure 5 (the curve $\hat{\omega}_{\max}$ corresponds to $\hat{\mu}_\perp = 10^{-3}(2\pi/L)$).

As pointed out in Section 3.2, the simple perpendicular viscosity term will not represent correctly the full viscosity tensor, except for particular domains of plasma parameters or for modes with sufficiently high mode number m. It can, however, be argued that the qualitative features of the present solutions will remain when other terms, specifically, the parallel viscosity, are taken into account.

In Section 4, the effect of resistivity has been shown to reduce both the growth rates as well as the unstable domain of incompressible modes. It was also shown that resistivity and perpendicular viscosity play equivalent roles provided that one of them is negligible. In particular, they give rise to the onset of stationary convection in the plasma when the product of thermal conductivity with resistivity (or perpendicular viscosity)

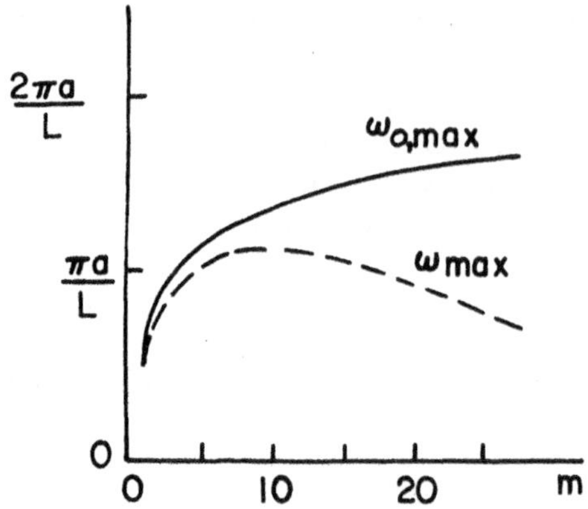

Fig. 5. Dependence of the maximum growth rate on the azimuthal wave number m, with and without viscosity.

attains a critical value, which depends on the pressure gradient. Marginal modes of this type have also been shown to exist in other geometries for both shearless [7] and sheared magnetic fields [10]. When the magnetic field has shear, non-ideal effects are important in a layer near the moderational surface where $k_{\parallel} = 0$ so that the methods of solution are quite different from those for the shearless equilibrium field treated here.

Appendix A

From Equations (22) (with σ_0, Λ_0 replaced by σ, Λ) and (45), it follows that as $nq \to m$ and $\hat{\omega} \to 0$

$$p^{(1)} = \frac{B_I^2}{4\pi a} \frac{\hat{\omega}}{\Lambda} \frac{\hat{\mu}_{\perp} \sigma^3 k^2}{2\rho_0} \left[m\sigma^2 \hat{p}^{(1)} + \sigma r \frac{\partial \hat{p}^{(1)}}{\partial r} \right]. \tag{A2}$$

Since $\hat{p}^{(1)}$ *satisfies Bessel's equation*

$$\nabla^2 \hat{p}^{(1)} = -k^2 \sigma^2 \hat{p}^{(1)} \tag{A2}$$

and

$$\nabla^2 \left(r \frac{\partial \hat{p}^{(1)}}{\partial r} \right) = 2 [\nabla^2 \hat{p}^{(1)} + k^2 \hat{p}^{(1)}] + r \frac{\partial}{\partial r} (\nabla^2 \hat{p}^{(1)})$$

$$= 2k^2 (1 - \sigma^2) \hat{p}^{(1)} - k^2 \sigma^2 r \frac{\partial \hat{p}^{(1)}}{\partial r} \tag{A3}$$

the Laplacian of Equation (A1) is

$$\nabla^2 p^{(1)} = -\frac{B_I^2}{4\pi a}\frac{\hat{\omega}}{\Lambda}\frac{\mu_\perp \pi^3 k^2}{\rho_0} \times \left[2k^2(\sigma^2-1)\hat{p}^{(1)} + \frac{m}{2}k^2\sigma^4\hat{p}^{(1)} + \frac{k^2\sigma^3}{2}\gamma\frac{\partial \hat{p}^{(1)}}{\partial r}\right]. \quad \text{(A4)}$$

Assuming $ka \ll 1$, Equation (A4) can be approximated by

$$\nabla^2 p^{(1)} = -\frac{B_I^2}{4\pi a}\frac{\hat{\omega}}{\Lambda}\frac{k^4\sigma^6\hat{\mu}_\perp}{2\rho_0}\left[m\sigma\hat{p}^{(1)} + r\frac{\partial \hat{p}^{(1)}}{\partial r}\right]. \quad \text{(A5)}$$

Replacing Equations (A5) and (39) into Equation (49) leads to Equation (53).

Dept. of Physics and Astronomy,
Tel Aviv University, Israel.
Association EURATOM-CEA sur la Fusion,
Centre d'Etudes Nucléaires, France.

References

[1] Tayler, R. J., *Proc. Phys. Soc. (London)* **B70**, 1049 (1957).
[2] Shafronov, V. D., *Plasma Physics and the Problem of Controlled Thermonuclear Reactions*, Pergamon Press, New York, 1972, Vol. 4.
[3] Ohta, M., Shimomura, Y., and Takeda, T., *Nucl. Fusion* **12**, 271 (1972).
[4] Takeda, T., Shimomura, Y., Ohta, M., and Yoshikawa, M., *Phys. Fluids* **15**, 2193 (1972).
[5] Freidberg, J. P., *Phys. Fluids* **13**, 1812 (1970).
[6] Goedbloed, J. P., and Hagebeuk, H. J. L., *Phys. Fluids* **15**, 1090 (1972).
[7] Maschke, E. K., and Paris, R. B., *Proc. Vth Conf. on Plasma Physics and Controlled Nuclear Fusion Research*, Tokyo 1974, I.A.E.A. Vienna, 1975.
[8] Braginskii, S. I., in M.A. Leontovich (ed.), *Reviews in Plasma Physics* Vol. 1, p. 205, Consultant Bureau, New York, (1965).
[9] Chandrasekhar, S., *Hydrodynamic and Hydromagnetic Stability*, Clarenton Press, Oxford, 1961.
[10] Dagazian, R. Y., and Paris, R. B., *Phys. Fluids* **20**, 9171 (1977).

LEOPOLD HALPERN

BROKEN SYMMETRY OF LIE GROUPS OF TRANSFORMATION GENERATING GENERAL RELATIVISTIC THEORIES OF GRAVITATION

ABSTRACT. Invariant varieties of suitable semisimple groups of transformations can serve as models of the space-time of the universe. The metric is expressible in terms of the basis vectors of the group. The symmetry of the group is broken by introducing a gauge formalism in the space of the basis vectors with the adjoint group as gauge group. The gauge potentials are expressible in terms of the basis vectors for the case of the De Sitter group. The resulting gauge theory is equivalent to De Sitter covariant general relativity. Group covariant generalizations of gravitational theory are discussed.

1. Introduction and Summary

The present article, following previous work of the author [3–5] derives general relativistic gravitational theories from a kind of gauge formalism which is applied to the theory of semisimple groups of transformations. All quantities which occur in the theory are generalizations obtained from group theory, mainly generators and structure constants. The metric of space-time can also be expressed this way, becoming thus a function of the dynamical variables of the theory.

Dirac's generalization of matter field equations to other invariance groups than the Poincaré group [1] inspired the author to construct a De Sitter covariant formulation of general relativity [10]. Preliminary considerations along this line were already made by Lubkin [19] without the author's knowledge. The author's work starts from a De Sitter covariant version of Pauli's [13] formulation of the principle of equivalence [10]. There are no doubt as many different formulations of this modified principle of equivalence possible, as of the original one, but they are largely equivalent.

The gauge group in the present approach is the narrowest one can choose, namely the adjoint group of the invariance group. In the example of the De Sitter group, the general formalism can be simplified by solving for the Lagrangian multipliers resulting in a De Sitter covariant general relativity. The formalism allows for generalizations to higher dimensions than space-time alone. It suggests a generalization of the law of motion

E. Tirapegui (Ed.), Field Theory, Quantization and Statistical Physics, 147–159.

and the unification with other fields, especially electromagnetism. The notation follows Reference 2, yet block letters are used for basis indices.

2. Minimal invariant varieties as models of the space-time of the universe

Consider a continuous group of transformations G_r with r essential parameters, acting on a n-dimensional space V_n. The rank of the matrix of base vectors: (ξ^i_A) $(i = 1 \dots n, A = 1 \dots r)$ of G_r be $q < r < n$ so that the group is intransitive and there exist q-dimensional invariant varieties.

One can, in general, construct a metric of V_n such that G_r is a group of motion and each of the family of invariant varieties is a q-dimensional Riemannian subspace V_q imbedded in V_n. The case $q = 4$ and signature $+2$ makes such a V_4 of suitable extensions a candidate for a model of space time of a universe. One can then introduce a coordinate system in V_n such that everywhere:

$$(1) \quad \xi^m_S = 0, \qquad \frac{\partial \xi^i_R}{\partial x^m} = 0, \qquad g_{mm} = \pm 1, \qquad g_{im} = 0, \qquad (i \neq m),$$

$$(S = 1 \dots r, \quad m = q+1, \dots, n, \quad i = 1 \dots n, \quad R = 1 \dots r)$$

because the generators $X_R = \xi^1_R \partial/\partial x^1$ of G_r act only within each invariant variety and because Killing's equations:

$$(2) \quad \frac{\partial g_{ik}}{\partial x^l} \xi^l_S + g_{lk} \frac{\partial \xi^l_S}{\partial x^i} + g_{il} \frac{\partial \xi^l_S}{\partial x^k} = 0$$

are satisfied. We shall consider here only semisimple groups for which a nonsingular matrix:

$$(3) \quad \gamma_{RS} = C^U_{RV} C^V_{SU}$$

can be formed out of the structure constants C^S_{RT}.

The author [3–5] has expressed the contravariant metric tensor as:

$$(4) \quad g^{ik} = C \xi^i_R \gamma^{RS} \xi^k_S,$$

$$(3a) \quad \gamma^{RS} \gamma_{ST} = \delta^R_T$$

in the coordinates satisfying Equations (1). C depends only on $x^m (m > q)$ and the ξ_R only on x^i $(i \leqslant q)$. For a given minimal invariant variety V_4, C is thus constant. Equation (4) satisfies, in general, Killings equations.

The metric of V_4 is thus expressed in terms of generators and structure constants of G_r. Field equations can now be constructed which consist only of group covariant expressions. Lie derivatives should replace all ordinary derivatives.

The Lagrangian density of a scalar field becomes:

(5) $\mathscr{L} = \sqrt{g}\gamma^{RS}\xi^i_R(\partial\phi^*/\partial x^i)\xi^k_S(\partial\phi/\partial x^k).$

A Lie derivative of spinors on the V_4 of a G_r has also been constructed. We shall not discuss it here [7].

Having achieved that task for the field equations, we would like to express the equation of motion of a classical body as well in a G_r covariant way. The time-like geodesics of V_4 however, do, not, in general, coincide with group trajectories. Agreement with the well established results of general relativity can only be achieved if trajectories exist which approximate these geodesics well enough. We consider in the following only the case where every geodesic of V_4 is a group trajectory. This situation is further explored in the next section.

3. GEODESICS AND GROUP TRAJECTORIES

The basis vectors ξ_R of G_r are determined only up to a nonsingular r-dimensional linear transformation C. The theory of Lie groups of transformations is formulated covariantly with respect to such linear transformations. The transformation C can be chosen such that at a given ordinary point P_0 of V_4, $q = 4$ of the basis vectors, let us say $\xi_B(B = 1 \ldots 4)$, are orthonormal and the remaining $\xi_M(M = q+1 \ldots r)$ vanish there. We denote such a system of basis vectors as special at P_0. A special system is of course not uniquely determined – apart from rotations, vectors ξ_M $(M > q)$ can arbitrarily be added to the ξ_B.

THEOREM. *The group trajectory of a base vector ξ is a geodesic through a point P_0, if a special system of base vectors exist there such that:*

$$(6) \quad [\xi_B, \xi]^i g_{ik}\xi^k = \left(\xi^m_B \frac{\partial}{\partial x^m}\xi^i - \xi^m \frac{\partial}{\partial x^m}\xi^i_B\right) g_{ik}\xi^k = 0,$$

$$(B = 1 \ldots q = 4)$$

Proof (1) The above condition of Equation (6) is satisfied at every point of the trajectory, if it is satisfied at P_0, because none of its points is preferred. Indeed at any point P_1 of the trajectory the linear transformation of the base vectors by the adjoint group of G_r with the group element which brings P_1 into P_0 along the trajectory, produces such a system at P_1. We shall show this for an infinitesimal displacement along the trajectory. We can always choose ξ itself as one of the vectors (let us say $\xi^i_1 = \delta^i_1$) of the special system pointing in the 1-direction. At a neighbouring point P_1 of the path,

$$(7) \quad x^i = x^i_0 + \delta^i_1 \delta\tau,$$

The components of all vectors differ by

$$(7a) \quad \delta\xi^i_R = \frac{\partial \xi^i_R}{\partial x^1}\delta\tau, \qquad (R = 1 \ldots r),$$

the transformation of the adjoint group alters them by

$$(7b) \quad \delta\xi^i_R = C^S_{R1}\xi^i_S \delta\tau = \left(\frac{\partial \xi^i_1}{\partial x^k}\xi^k_R - \frac{\partial \xi^i_r}{\partial x^1}\right)\delta\tau$$

clearly a coordinate transformation with:

$$(7c) \quad \frac{\partial x'^i}{\partial x^k} = -\frac{\partial \xi^i_1}{\partial x^k}$$

restores then all the values of the transformed components to their values at P_0. The transformation of the adjoint group leaves the structure constants unchanged because of Jacobi's identities so that even the metric (4) is the same as at P_0 after the transformations and (6) is satisfied at P_1.

(2) To prove that the trajectory is a geodesic if and only if Equation (6) is fulfilled at every one of its points we remember that every ξ in V_4 is the symbol of the group of motion and contract Killing's Equation (2) with ξ^i and ξ^k to obtain:

$$(2a) \quad \xi^i\xi^k \frac{\partial g_{ik}}{\partial x^m}\xi^m_R + 2g_{mk}\frac{\partial}{\partial x^i}(\xi^m_R)\,\xi^i\xi^k = 0$$

with the help of (6) we obtain from this

$$(7) \quad \frac{\partial}{\partial x^m}(\xi^i g_{ik}\xi^k)\xi^m_R = 0 \text{ implying} \frac{\partial}{\partial x^j}(\xi^i g_{ik}\xi^k) = 0,$$

$$(j = 1 \ldots 4).$$

One can thus choose the parameter such that:

$$(7a) \quad \dot{x}^k = \xi^k, \ddot{x}^k = \frac{\partial \xi^k}{\partial x^m}\xi^m$$

and

$$\Gamma^k_{im}\dot{x}^i\dot{x}^m = \tfrac{1}{2}g^{kj}\left(2\frac{\partial g_{ij}}{\partial x^m} - \frac{\partial g_{im}}{\partial x^j}\right)\xi^i\xi^m.$$

Because of (7):

$$\text{(7b)}\quad -\frac{\partial g_{im}}{\partial x^k}\xi^i\xi^m = 2g_{im}\xi^i\frac{\partial \xi^m}{\partial x^j}$$

and because of (2):

$$\text{(2b)}\quad \frac{\partial g_{ij}}{\partial x^k}\xi^i\xi^k = -\left(\xi^i g_{kj}\frac{\partial \xi^k}{\partial x^i} + \frac{\partial \xi^k}{\partial x^j}g_{ik}\xi^i\right)$$

together (7a), (7b), and (2b) show that the equation of the geodesic is satisfied.

$$\text{(8)}\quad g_{jk}(\ddot{x}^k + \Gamma^k_{im}\dot{x}^i\dot{x}^m) = 0.$$

We obtain immediately the Corollary:

COROLLARY *Every geodesic through P_0 is a group trajectory if a special system of base vectors exist so that* (6) *is fulfilled for any two vectors $\xi = \xi_A, \xi_B$ $(A, B = 1 \ldots 4)$. Each vector of the subspace spanned by $\xi_1 \ldots \xi_4$ has a geodesic as trajectory.*

We consider here only groups which fulfil this condition. Is the possibility of motion of macroscopic bodies along nongeodesic time-like group trajectories in conflict with experience? The author has repeatedly pointed out that this need not necessarily be so [4, 5].

Consider a special system of base vectors at P_0 in coordinates which result in a Minkowski metric there. The geodesic motion of a macroscopic body can be the trajectory of a vector $C^R\xi_R$ which $C^R = 0$ for $R > 4$, $\gamma_{RS}C^RC^S = 1$. One may attribute to that motion a volume in phase space proportional to:

$$\text{(9)}\quad \sum_{A=1}^{3}\frac{(C^A)^2}{\gamma_{RS}C^RC^S}.$$

Consider now the same relations in case of a nongeodesic trajectory of the same initial velocity, where $C^R \neq 0$ also for $R > 4$. The phase space volume according to Equation (9) will shrink relative to the geodesic case, in the limit the more, the smaller the ratio of the radii of curvature of the trajectory and the universe (geodesic motion) is. We can only observe a radius of curvature far smaller than that of the universe, which would correspond to such a small relative volume in phase space that we have practically no chance to encounter it among a limited number of samples – just as we do not encounter a macroscopic quantity of gas in vacuum, that will contract. The example given here is a possible generalized law of motion which is in the spirit of the group theoretical approach; it can be studied best in case of the De Sitter group treated in later sections. The equations of the generalized free motion are nonlinear and of higher order. They are briefly stated for the De Sitter group.

4. Breaking of the group symmetry by the gauge formalism

We have been able to express the metric of space-time in terms of quantities belonging to the group of transformations itself; it is therefore suggestive to describe the breaking of the symmetry of space due to local inhomogenous matter distributions in terms of a symmetry breaking of the group of transformation which acts on space. (Not of the abstract group which is to remain intact to allow action on localized quantities. Even the group of transformation should still function locally on path segments).

The author has suggested to use the covariance of the group theory with respect to linear transformations of the space of base vectors (see the first part of Section 3) to establish a formalism of the gauge type, which can describe the symmetry breaking of space and (in a still rather artificial way) relate it to the presence of matter [3–5].

We start by performing at every point of V_4 independent linear transformations of the space of base vectors. The transformations affect thus all indices with capital block letters. We are able to uphold our formalism in spite of this manoeuvre if we only replace derivatives of quantities with block indices by invariant derivatives. Suppose ξ_u transforms.

$$(10) \quad \xi_u(x) \to \xi'_u(x) = S_u^v(u^\alpha(x))\xi_v(x)$$

by some subgroup Γ of $GL(r)$ with canonical parameters u^α.* The invariant derivative is defined in a well known way [8] with a potential

*To distinguish Γ from G_r we use Greek indices instead of roman capital indices.

$A_k^\alpha(x)$ and the generators G_α of Γ:

$$\text{(10a)} \quad \xi_{u\cdot k} = \frac{\partial}{\partial x^k}\xi_u + A_k^u(G_\alpha)_u^v \xi_v.$$

A transformation S of Γ transforms the potentials inhomogenously with respect to the adjoint group of Γ. Infinitesimally:

$$\text{(10b)} \quad \delta A_k^\alpha(x) = C_{\beta\gamma}^\alpha A_k^\beta \delta u^\gamma(x) - \frac{\partial}{\partial x^k}\delta u^\gamma(x)$$

so that $\xi_{u.k}$ transforms in the same way as ξ_u:

$$\text{(10c)} \quad \xi_{U\cdot k} \to S_U^V \xi_{V\cdot k}.$$

The condition that the potential can be transformed to zero at all points x simultaneously by a transformation $S_U^V(x)$ is:

$$\text{(10d)} \quad F_{ik}^\alpha(x) = \frac{\partial}{\partial x^i}A_k^\alpha - \frac{\partial}{\partial x^k}A_i^\alpha + C_{\beta\gamma}^\alpha A_i^\beta A_k^\gamma = 0.$$

A transformation $S_U^V(u^\alpha(x))$ in our unperturbed V_4 of G_r produces potentials which satisfy Equation (10d). The commutation relations are now of the form:

$$\text{(11)} \quad \xi_R^m \xi_{U\cdot m}^i - \xi_U^m \xi_{R\cdot m}^i = C'^T_{RU} \xi_T^i,$$

where $\xi, \xi_{.k}$ are defined in Equations (10a) and (10c) and

$$\text{(11a)} \quad C'^T_{RU} = S_R^Y S_u^z (S^{-1})_V^T C_{YZ}^V$$

depends on x but

$$\text{(11b)} \quad C'^T_{RU\cdot k} \equiv 0.$$

Killings equations assume the form:

$$\text{(11c)} \quad \frac{\partial g_{ik}}{\partial x^m}\xi_S^m + g_{mk}\xi_{S.i}^m + g_{im}\xi_{S\cdot k}^m = 0.$$

To break the symmetry of V_4 we abandon Equations (10) and (10d) keeping, however, Equations (10a,b,c), (11), and (11a,b,c) for a potential which has now to be determined from field equations. Also the ξ_R and the $S_u^v(x)$ have to be obtained from field equations and from Equations (11) and (11a). The generalized Killing Equations (11c) imply furthermore that the metric is of the same form as Equation (4) yet formed with the γ'^{RS} out of the primed structure 'constants'.

To obtain consistent field equations we have to form invariants out of the A_k^α, the ξ_R^i, and the S_u^v and add Equations (11) with a Lagrangian multiplier, so that we can vary independently with respect to all the unknowns. We may in simpler cases solve Equations (11) eliminate thereby some of the unknowns and avoid the multipliers. The metric can always be constructed from the solutions.

The following two sections will provide examples for the procedures.

5. The Adjoint Group as Gauge Group

Every group G_r has a 'natural' group of linear transformations acting on the space of its base vectors: The adjoint group of G_r has the same structure constants as G_r and thus an isomorphic law of composition of the group parameters [2]. We have seen moreover in Section 3 that the structure constants of G_r are not altered by the transformations of its adjoint group. The $S_U^V(x)$ which were only auxiliary variables, serving to exhibit the full invariance properties of the theory, are constants here.

The adjoint group is not transitive; this leads us to an additional restriction: We want to avoid solutions for the ξ_R^i which the gauge group even for one single point fails to transform to a set of base vectors of the unperturbed V_4. This means we must be able to transform the base vectors at every point into a special system (see Section 3) in which for $A = 1 \ldots q = 4$ they form a vierbein whereas those with $M > 4$ vanish. This because as we saw in Section 3, we can also always introduce a special system at one given point by a constant linear transformation and then transform this property to any other point of V_4 by a transformation of the adjoint group. The constant linear transformation fixes the components of the structure constants.

Thanks to our requirement we can introduce even in the general case a gauge for which at every point the first four base vectors form a vierbein:

$$(12)\quad \xi_A^i = h_A^i (A, B = 1 \ldots 4), \qquad g_{ik}\xi_A^i \xi_B^k = \eta_{AB},$$
$$\xi_M^i = 0 \quad (M > 4).$$

Written explicitly Equations (11) in such a gauge are:

$$(13a)\quad h_A^m\left(\frac{\partial}{\partial x^m}h_B^i + A_m^E C_{BE}^D h_D^i\right) - h_B^m\left(\frac{\partial}{\partial x^m}h_A^i + A_m^E C_{AE}^D h_d^i\right) = C_{AB}^D h_D^i,$$

$$(13b)\quad h_A^m\left(\frac{\partial}{\partial x^m}\xi_M^i + A_m^E C_{ME}^D h_D^i\right) = C_{AM}^D h_D^i,$$

$$(A, B, D, E, = 1 \ldots 4, M = 5 \ldots r)$$

The Greek indices in Equation (10) are replaced for the adjoint group by block indices because structure constants are the same as for G_r.

The equations in this form are much simplified. They will be solved for the De Sitter group in the next section.

6. The De Sitter Groups as an Example

The De Sitter and anti De Sitter groups are five-dimensional orthogonal groups with signatures $+3$ and $+1$, respectively. The basis vectors are simplest expressed in five-dimensional Cartesian form, labeled by double indices which we denote here by two Greek letters:

$$(14) \quad \xi^i_{[\alpha,\beta]} = X^m(\eta_{m\beta}\delta^i_\alpha - \eta_{m\alpha}\delta^i_\beta), \qquad (i, m, \alpha, \beta = 1 \ldots 5).$$

If we use the labels only as symbols or in summation we shall write however block letters as before. The structure constants are:

$$(14a) \quad C^{[\varepsilon,\phi]}_{[\alpha,\beta][\gamma\delta]} = \tfrac{1}{2}[\eta_{\alpha\gamma}(\delta^\varepsilon_\beta\delta^\phi_\beta - \delta^\varepsilon_\delta\delta^\phi_\beta) - (\gamma \leftrightarrow \delta)] - \tfrac{1}{2}[(\alpha \leftrightarrow \beta)].$$

The geodesic condition Equation (6) is fulfilled here

$$(14b) \quad \gamma_{[\alpha,\beta][\delta,\varepsilon]} = 4(\eta_{\alpha\varepsilon}\eta_{\beta\delta} - \eta_{\alpha\delta}\eta_{\beta\varepsilon}).$$

double index summations are performed over each of the two letters.

The minimal invariant varieties are generalized spheres:

$$(14c) \quad \eta_{ik}X^iX^k = \pm R^2, \qquad (i, k = 1 \ldots n = 5).$$

(Upper sign De Sitter, lower anti De Sitter space) on which one can use conformal coordinates by introducing

$$(14d) \quad x^k = \frac{2X^k}{R + X^5}, \qquad x^5 = R, \qquad (k = 1 \ldots 4),$$

and choosing $R = R_0 = \text{const}$. The metric on the invariant variety is then:

$$(14e) \quad g^{ik} = \eta^{ik}\frac{(1 \pm \sigma^2/4)^2}{R^2}, \qquad \sigma^2 = \eta_{lm}x^lx^m, \qquad (i, k, l, m = 1 \ldots 4)$$

and the base vectors are:

$$(14f) \quad \xi^i_{[\alpha,\beta]} = x^k(\eta_{k\beta}\delta^i_\alpha - \eta_{k\alpha}\delta^i_\beta) \qquad (\alpha, \beta = 1 \ldots 4)$$

and

$$(14g) \quad \xi^i_{[\alpha,5]} = \pm\delta^i_\alpha(1 \pm \sigma^2/4) + \tfrac{1}{2}x^ix^k\eta_{k\alpha}.$$

We consider now the generalization discussed in Section 5. with the adjoint group as gauge group. Transforming at every point to a special system we give the four non-vanishing base vectors the indices $[\alpha, 5]$ whereas all $\xi_{[\alpha,\beta]} = 0$ $(\alpha, \beta = 1 \ldots 4)$. Remember the structure constants do not alter their form and values (14a) by the transformation, but the base vectors everywhere assume the components as at $x^i = 0$ of Equations (14f, g) expressed in general coordinates. We are able to solve Equations (13a, b) for the potentials and find:

$$\text{(15a)} \quad A_m^{[\alpha,5]} = -h_{Am}\eta^{A\alpha} \; (A, \alpha = 1 \ldots 4),$$

$$\text{(15b)} \quad A_E^{[\alpha,\beta]} = A_k^{[\alpha\beta]} h_E^k = \gamma_{ABE}\eta^{A\alpha}\eta^{B\beta},$$

with

$$\text{(16)} \quad \gamma_{ABE} = h_{A;k}^i h_{Bi} h_E^k:$$

the coefficients of rotation [11]. All the 40 components of the potential are thus expressed in terms of the tetrads and their derivatives. We express now also the fields $F_{ik}^{[\alpha,\beta]}$ of Equation (10d) by the tetrads:

$$\text{(15c)} \quad F_{ik}^{[\alpha,5]} = 0,$$

and

$$\text{(15d)} \quad F_{ik}^{[\alpha,\beta]} = \eta^{\alpha A}\eta^{\beta B}\left[\frac{\partial}{\partial x^i}(\gamma_{ABE}h_k^E) - \frac{\partial}{\partial x^k}(\gamma_{ABE}h_i^E) + \eta^{EF}(\gamma_{EAI}\gamma_{FBK} - \gamma_{EBI}\gamma_{FAK})h_i^I h_k^K\right] + C_{[I,5][K,5]}^{[\alpha,\beta]} h_i^I h_k^K$$
$$= (R_{pqki} - \mathring{R}_{pqki})h^{\alpha p}h^{\beta q}$$

(all capital and Greek indices, $1 \ldots 4$) with R_{pqki} the Riemann tensor of the metric related to the tetrads and

$$\text{(15e)} \quad \mathring{R}_{pqki} = \frac{1}{R_0^2}(g_{pk}g_{qi} - g_{pi}g_{qk}).$$

In case of the unperturbed space V_4 $R_{pqki} = \mathring{R}_{pqki}$ and all F_{ik} vanish.

We consider the following invariants formed out of F_{ik} for a Lagrangian:

$$\text{(17)} \quad F_{ik}^R \gamma_{RS} F^{Sik} = R_{pqki}R^{pqki} + \frac{4}{R_0^2}R + \frac{24}{R_0^4}$$

which is a well known quadratic Lagrangian with an admixture of an

Einstein term [10, 12] (necessary to avoid singularities in the solutions) and a cosmological constant.

Besides the Lagrangian of Equation (17) which is of the Maxwell type, as it occurs in all gauge theories, there exists here another Lagrangian linear in the fields F^M_{ik}

(17a) $F^{[\alpha,\beta]}_{ik} C^{[\varepsilon,5]}_{[\gamma,5][\alpha,\beta]} h^{\gamma i} h^k_\varepsilon.$

which is equivalent to the Einstein Lagrangian with a cosmologic member. One obtains thus essentially Einstein's theory if one adds the conventional matter Lagrangians e.g. Equation (5) with a constant and varies the total Lagrangian with respect to the tetrad fields $h^i_A(x)$ and the matter fields.

The law of motion in this theoretical framework has, no doubt, a deeper fundament than just the choice of a Lagrangian. We have given the condition for which in unperturbed space every geodesic is a group trajectory; The condition (6) is obviously fulfilled for the De Sitter group. The author has shown [10] that a principle of equivalence exists even for the De Sitter background that means in an analogy to Pauli's definition: Along the points of a geodesic an arbitrary metric can always be transformed into the metric of De Sitter space such that all its first derivatives vanish there.

This theorem contemplated from the point of view of the present paper leads to the following features:

(1) The potentials A^R_i can be transformed away on all points of a given line segment because the system of ordinary differential equations for the parameters of the adjoint group which achieves this has in general solutions.

(2) Killing's equations and the commutation relations of the $\xi^i_R(x)$ are then formally fulfilled along the points of the line and our considerations of Section 3 apply. A geodesic in the general case is then always a trajectory of a linear superposition of the transformed $\xi^i_R(x)$.

The system of differential equations which has as solutions all trajectories of the group is very complicated even in the unperturbed V_4. It is nonlinear of the 5th order in the De Sitter case. One obtains them by writing down the path of a general generator:

(18) $\dot{x}^i = \xi^i_{(x)} \equiv C^R \xi^i_R(x).$

Assume a special system of generators at an initial point P_0.

THEOREM. *A special system of generators can be introduced at any other point p of the trajectory such that the generator of the trajectory expressed in the transformed system is of the same form* (18) *as at* P_0.

The proof follows the considerations and the theorem of Section 3: For an infinitesimal displacement along the trajectory the parameters of the adjoint group which achieves the transformation $\delta\sigma^R$, are proportional to the C^R and thus

(19) $\delta C^R = C^R_{ST} C^S \delta\sigma^T = 0.$

One can thus express the C^R at every point in terms of $\dot{x}, \ddot{x}$, and higher derivatives up to the fifth from the equations of the trajectory:

(20) $\xi^i{}_{;k}\xi^k = C^B_{AR} h^i_B C^A C^R, \qquad (A,B,D = 1 \ldots 4; R,S = 1 \ldots r),$

(20a) $(\xi^i{}_{;k}\xi^k);m\xi^m = C^D_{AR} C^B_{DS} C^A C^R C^S$

and its higher covariant derivatives in the ξ-direction, the right side of which consists of a chain of structure constants of the same form with one more member for each derivative. More details of these equations and their possible relation to physics are postponed to a subsequent work. The equations of motion of higher order pertain no doubt–if they really play a role in physics–to the nonlinear Lagrangian (15d).

We finally mention a generalization of the above theory in which the gauge group consists of the direct product of the adjoint group and the group of scale transformations. The solution of the generalized commutation relations for the potentials A^R_k in terms of the tetrade fields and a simple vector gauge field can even be performed in such a case but the analogous results are not gauge covariant. We give here no details because in our opinion this approach will not result in a truly unified theory of gravitation and electromagnetism. Wider gauge groups result either in additional constraints between the potentials, or they introduce torsion. The methods introduced here can be applied to a multitude of situations the physical content of which should be investigated.

Dept. of Physics, Florida State University, Tallahassee

REFERENCES

[1] Dirac, P. A. M., *Ann. Math* **30**, 657 (1935).
[2] Eisenhart, L. P., *Continuous Groups of Transformations.*
[3] Halpern, L., SLAC-PUB-2166, July 1978 (T), FSU Preprint. See also Article to appear in Proc. Austin Symposium on Mathem. Physics 1978.

[4] Halpern, L., SLAC-PUB-FSU Preprint to appear shortly in JGRG.
[5] Halpern, L., in AIP Conference Proc. Nr. 48, Part. and Fields Sups., No. 15 Symposium in Honour of P. A. M. Dirac April 6, 1978; FSU Preprint HEP781011.
[6] Halpern, L., Article scheduled to appear in *Brazilian J. Phys.*
[7] See Refs. 4–6. A Lie derivative of spinors which is apparently related to the authors has also been suggested by W. Unruh (personal communication).
[8] DeWitt, B., in C. and B. DeWitt (eds.), 'Relativity, Groups and Topology', Proc. Les Houches, 1963 Summer School Gordon and Breach, New York.
[9] Gürsey, F., *Istambul Summer School on Theoretical Physics*, New York, 1962, Gordon and Breach.
[10] Halpern, L., *J. Gen. Relat. and Gravit.* **8**, No. 8, 623 (1977).
[11] Eisenhart, L. P., *Riemannian Geometry*, Princeton Univ. Press, 1964.
[12] Halpern, L., FSU-HEP-751230.
Halpern, L., FSU-HEP-761116, *Springer Lecture Notes in Mathematics*, 570 (1977), Differential Geometrical Methods in Mathematical Physics. Proc. of Bonn Symposium, July 1–4, 1975.
[13] Pauli, W., *Encycl. d. Mathem., Wissenschaften II*, p. 53g, Teubner, Leipzig, 1921.
[14] Utiyama, R., *Phys. Rev.* **101**, 1537 (1958).
[15] Yang, C. N., and Mills, R. L., *Phys. Rev.* **96**, 191 (1954).
[16] Lubkin, E., *Ann. Phys. (New York)* **23**, 233 (1963); and D. Finkelstein, personal communication.
[17] Yang, C. N., *Phys. Rev. Letters* **33**, No. 7, 445 (1974).
[18] Halpern, L., FSU-HEP-751230, FSU-HEP-761116.
[19] Lubkin, E., in C. Kuper and A. Peres (eds.), *Relativity and Gravitation Symposium Haifa (1969)*, Gordon and Breach, New York, 1971.
[20] Møller, C., *Mat. Fys. Skr., Canshe Vidensk. Selsh I.*, No. 10, (1969).
[21] Halpern, L., and Miketinač, M., *Can. J. Phys.* **48**, No. 2 (1970).

J. C. HOUARD and M. IRAC-ASTAUD

WARD–TAKAHASHI IDENTITIES AND GAUGE INVARIANCE

Cet article est dédié à la mémoire de Bernard Jouvet. C'est lui qui a guidé nos débuts dans la recherche. Il était notre ami, et nous lui témoignons ici notre reconnaissance.

ABSTRACT. The Ward–Takahashi identities are generally understood as expressing the gauge invariance. After a short critical review, we give a necessary and sufficient condition connecting the existence of these identities with the gauge invariance properties of the Lagrangian in the Abelian and non-Abelian cases.

1. A SHORT REVIEW

The common notions of gauge and of gauge transformations in classical electrodynamics give rise in quantum theory to many different ideas the connections of which are often intricate. After defining *a gauge* as a particular way to quantize the electromagnetic field, one has to speak of gauge transformations, change of gauge, gauge invariance and gauge independence [1–4], the main purpose being the construction of a well-behaved and unique S-matrix. We want to discuss here the criterion of the gauge invariance of the Lagrangian constituted by the Ward–Takahashi identities.

In the classical field theory of a vector field A_μ interacting with some charged fields ψ_k, the gauge transformations are the transformations of the form [5]

$$A_\mu \to A_\mu + \frac{1}{e}\partial_\mu \phi, \qquad \psi_k \to e^{-iQ_k\phi}\psi_k, \tag{1.1}$$

where ϕ is an arbitrary function, the charge associated with ψ_k being denoted by Q_k. Gauge invariance here means the invariance of the equations of motion, or more particularly of the Lagrangian, under the transformations (1.1). As is known, among the Lagrangians invariant under the gauge transformations of the first kind ($\phi = \text{const.}$ in (1.1)),

E. Tirapegui (Ed.), Field Theory, Quantization and Statistical Physics, 161–182.

the gauge invariant ones are those of the general form [6]

$$(1.2)\quad \mathscr{L} = \mathscr{L}(F_{\mu\nu}, \psi_k, \psi_k^+, (\partial_\mu + iQ_k eA_\mu)\psi_k, (\partial_\mu - iQ_k eA_\mu)\psi_k^+),$$

where $F_{\mu\nu} = \partial_\mu A_\nu - \partial_\nu A_\mu$. The particular case of the *minimal coupling* is obtained from the free Lagrangian of the ψ's by making the replacement $\partial_\mu \psi_k \to (\partial_\mu + iQ_k eA_\mu)\psi_k$, for any k, then adding the classical $-\frac{1}{4}F_{\mu\nu}F^{\mu\nu}$ as free Lagrangian of A_μ [5]. The most important example is of course the spinor electrodynamics in which a Dirac field ψ, associated with the electron, interacts with A_μ through the coupling $e\bar{\psi}\gamma^\mu\psi A_\mu$.

However, a system having such a kind of invariance is necessarily singular with regard to the determination of the motion from Cauchy data since, the function ϕ in (1.1) being arbitrary, it can be chosen so as to transform the solutions without changing the initial values [7]. In the Hamiltonian formalism the singularity takes the form of a constraint between canonical variables, the free Lagrangian $-\frac{1}{4}F_{\mu\nu}F^{\mu\nu}$ giving, for example, a vanishing conjugate momentum for A_0. This difficulty is usually settled by the choice of a gauge that is by prescribing a condition which restricts the potential A_μ. The most commonly used gauges are the Coulomb gauge defined by the condition $\partial_k A^k = 0$, $k = 1,2,3$, and the Lorentz gauge defined by $\partial_\mu A^\mu = 0$, $\mu = 0,1,2,3$. They are described in many textbooks [3,4,8]. The latter condition has the advantage to be a covariant one; it is preserved by the gauge transformations (1.1) in which the function ϕ is restricted to be a solution of the equation $\Box\,\phi = 0$ (restricted gauge transformations).

The difficulty just mentioned creates a problem for quantization since the usual method needs a canonical theory*. To satisfy this requirement, the Gupta–Bleuler solution for quantum electrodynamics consists in modifying the classical Lagrangian by the addition of the term $-\frac{1}{2}(\partial_\mu A^\mu)^2$; the corresponding equations of motion are however equivalent to the Maxwell equations only for the expectation values on the physical states $|\Psi\rangle$ defined by the subsidiary condition $\partial_\mu A^{\mu(+)}|\Psi\rangle = 0$ [3,4]. The equations of motion, the canonical commutation rules and the subsidiary condition are now invariant under the restricted gauge transformations only. The latter constitute a group of canonical transformations (gauge transformations), and the existence of this group is often considered as expressing the gauge invariance of the theory [3,4,9]. However, from a general point of view, the restricted transformations cannot be understood

* Non-canonical quantization procedures, in particular the quantization in the Coulomb gauge [8], will not be considered here.

as the fundamental invariance transformations of the Lagrangian, since otherwise the class of the allowed Lagrangians would be much larger than it is wanted (a possible term in the Lagrangian being, for example, $\bar{\psi}\psi\partial_\mu A^\mu$). It is in fact tacitly assumed that the Lagrangians relevant to the gauge invariance in the quantum case have to present the 'smallest deviation' from exact invariance under the general transformations (1.1).

It is in that context that the Ward–Takahashi identities for quantum electrodynamics were established. Let us briefly summarize the main known demonstrations of them. The original work of Ward [10] (see also [11, 12]) is exclusively concerned in proving the equality of the renormalization constants Z_1 and Z_2, as a consequence of the following relation

$$(1.3)\quad \Gamma_\mu(p,p) - \gamma_\mu = -e_0^2 \frac{\partial}{\partial p^\mu}\Sigma(p)$$

where Σ is the self-energy function of the electron, and Γ_μ the vertex function. The key of the proof is the identity

$$(1.4)\quad \frac{\partial}{\partial p^\mu}\frac{1}{\gamma p - m} = -\frac{1}{\gamma p - m}\gamma_\mu \frac{1}{\gamma p - m}.$$

It expresses that differentiating the electron propagator amounts to attach a photon of vanishing momentum on the corresponding line in the diagrams. Formula (1.3) follows by the way of the perturbation theory. A generalization of (1.3), namely [8, 13]

$$(1.5)\quad k^\mu[\Gamma_\mu(p+k,p) - \gamma_\mu] = -e_0^2[\Sigma(p+k) - \Sigma(p)]$$

can be analogously obtained by using, instead of (1.4), the relation [14]

$$(1.6)\quad \frac{1}{\gamma p - m} - \frac{1}{\gamma(p+k) - m} = \frac{1}{\gamma(p+k) - m}\gamma k \frac{1}{\gamma p - m}.$$

A more synthetical proof of the generalized relation (1.5), in configuration space, is due to Takahashi [15] (see also [2]). It is primarily based on the conservation of the Noether current J^μ and on the property of J^0 to generate the gauge transformations of the first kind. This allows to write

$$(1.7)\quad \frac{\partial}{\partial z^\mu}\langle T(J^\mu(z)\psi(x)\bar{\psi}(y))\rangle = (\delta(y-z) - \delta(x-z))\langle T(\psi(x)\bar{\psi}(y))\rangle.$$

Next, by relying on the form $J^\mu A_\mu$ of the interaction, the vacuum expectation value in the left member of (1.7) is interpreted as the improper vertex part. In fact only the proper part, with respect to the photon line, does contribute because of the identity

$$(1.8)\quad \partial^\mu D'_{\mu\nu} = \partial^\mu D_{\mu\nu},$$

where $D_{\mu\nu}$ and $D'_{\mu\nu}$ are, respectively, the bare and dressed propagators of the photon. Relation (1.8) was imputed by Takahashi to the gauge invariance of the second kind. However, it must be stressed that (1.8) can be established under the same assumptions as needed for (1.7) by considering $\langle T(J^\mu(z)A^\nu(x))\rangle$ instead of $\langle T(J^\mu(z)\psi(x)\bar{\psi}(y))\rangle$. In the same manner more general relations are obtained by replacing the product $\psi\bar{\psi}$ in (1.7) by any product of the form $A \ldots \psi \ldots \bar{\psi} \ldots$.

A more physically significant line of reasoning, really appealing to the gauge invariance of the second kind, founds upon the introduction of an external field, the gauge transformations of the quantized field being transferred on this latter. Such a kind of argument was often used to prove the relation $k^\mu \pi_{\mu\nu}(k) = 0$ (equivalent to (1.8)), for the self-energy function of the photon, by exploiting the formal analogy between $\pi_{\mu\nu}$ and the tensor defining the current induced in the vacuum by the external field [4, 16, 17]. More generally, a systematic way to get the Ward–Takahashi identities for all the gauge invariant theories, consists in introducing an external electromagnetic gradient field coupled in the same way as the quantized field A_μ (that is appearing in the combination $A_\mu + \partial_\mu \phi$ in the interaction Lagrangian). Since the couplings with this external field can be removed by a change in the phase of the matter fields, the identities result from expressing in two different ways the Green functions of the theory [4, 18]. A somewhat related argument, within the functional formalism, uses the fact that the generating functional of the Green functions, when expressed as a functional integral, is left unaltered by a change of the integration variables [19]. Making a gauge transformation then furnishes the Ward–Takahashi identities. Unlike the preceding argument the present one seems surprisingly to appeal, besides gauge invariance, to something else since here the identities are precisely given by the non-invariant terms of the action in the functional integral, namely the source term and the extra-term needed for quantization.

Although the original proofs do not refer to the gauge invariance, the Ward–Takahashi identities were rapidly considered as characterising it, probably because of the various examples already known to which they apply. However a precise proof of this property seems lacking.* What is more, many authors have stressed the fact that, for quantum electrodyna-

* The sufficiency of the Ward–Takahashi identities for the gauge invariance is asserted in [20], but the proof does not reveal clearly what it is meant by an invariant Lagrangian.

mics, the identities follow from the conservation of the current alone [2,8,15,21]. On the other hand, the Ward–Takahashi identities may be proved for massive vector meson theories, when it has been traditionally considered that gauge invariance implies the vanishing of the photon mass [16,22]. But, as to this last assertion, if it is true in classical theory, it is not the same in quantum theory since in this case the Lagrangian cannot be strictly invariant. In fact, gauge invariance in the sense of the relation $k^{\mu}\pi_{\mu\nu}=0$ only implies that the bare and the physical mass simultaneously vanish. Thus the question arises to know what kind of invariance characterize exactly the Ward–Takahashi identities. Of course, such a question takes its sense only by considering a wide class of theories, a priori not necessarily invariant, and not by only reasoning upon a single model as electrodynamics.

In what follows, we examine in general the derivation of the Ward–Takahashi identities, starting from the first principles. In Section 2, we establish a formula giving the variation of the Green functions under any transformation of the field variables, in terms of the variation of the Lagrangian. This formula is used in Section 3 to deduce a necessary and sufficient condition connecting the invariance of the Lagrangian and the Ward–Takahashi identities. In Section 4, an analogous theorem is obtained for the non-Abelian gauge theories. The difficulties of unitarity present in these last models are only touched. The relations of the Ward–Takahashi identities with the renormalization are not discussed in this paper, the interest being centered on their connection with the invariance properties.

2. The basic identity associated with a local transformation

We consider the theory defined by a Lagrangian $\mathscr{L}(\phi,\partial\phi)$, where $\phi=(\phi_i)$ is a given set of fields. Let $\phi\to\phi'=\phi+\delta\phi$ be an arbitrary infinitesimal local transformation (that is such that the variation $\delta\phi$ at the point x only depends on x and $\phi(x)$). With such a transformation is associated the current

$$\delta j^{\mu}=\frac{\partial\mathscr{L}}{\partial(\partial_{\mu}\phi)}\delta\phi \tag{2.1}$$

together with the variation of the Lagrangian

$$(2.2)\quad \delta\mathscr{L} = \frac{\partial\mathscr{L}}{\partial\phi}\delta\phi + \frac{\partial\mathscr{L}}{\partial(\partial_\mu\phi)}\partial_\mu\delta\phi$$

$$= \left[\frac{\partial\mathscr{L}}{\partial\phi} - \partial_\mu\frac{\partial\mathscr{L}}{\partial(\partial_\mu\phi)}\right]\delta\phi + \partial_\mu\delta j^\mu .$$

When the equations of motion are satisfied, the corresponding variation of the action may therefore be written as

$$(2.3)\quad \delta\mathscr{A} = \int_{\sigma_1}^{\sigma_2} \mathrm{d}x\,\delta\mathscr{L} = \delta Q(\sigma_2) - \delta Q(\sigma_1)$$

where

$$(2.4)\quad \delta Q(\sigma) = \int_\sigma \mathrm{d}\sigma_\mu\,\delta j^\mu .$$

The Schwinger Action Principle then reads

$$(2.5)\quad [\delta Q(\sigma), \phi(x)] = -i\delta\phi(x), \qquad x\in\sigma .$$

Let us now consider the integral

$$(2.6)\quad -i\int \mathrm{d}x\, T(\delta\mathscr{L}(x)\phi_{i_1}(x_1)\dots\phi_{i_n}(x_n)).$$

We always can assume that the x_k's are in the chronological order so that (2.6) is:

$$(2.7)\quad -i\int_{-\infty}^{t_n} \mathrm{d}x\,\phi_{i_1}(x_1)\dots\phi_{i_n}(x_n)\delta\mathscr{L}(x)$$

$$-i\sum_{k=1}^{n-1}\int_{t_{k+1}}^{t_k} \mathrm{d}x\,\phi_{i_1}(x_1)\dots\phi_{i_k}(x_k)\delta\mathscr{L}(x)\phi_{i_{k+1}}(x_{k+1})\dots\phi_{i_n}(x_n)$$

$$-i\int_{t_1}^{+\infty} \mathrm{d}x\,\delta\mathscr{L}(x)\phi_{i_1}(x_1)\dots\dots\phi_{i_n}(x_n).$$

By introducing the expression (2.3) of $\delta\mathscr{A}$ and by using the Schwinger Action Principle (2.5) we get

$$(2.8)\quad -i\int dx\, T(\delta\mathscr{L}(x)\phi_{i_1}(x_1)\ldots.\phi_{i_n}(x_n))$$
$$=\delta(\phi_{i_1}(x_1)\ldots\phi_{i_n}(x_n))-i[\delta Q(+\infty)\phi_{i_1}(x_1)\ldots\phi_{i_n}(x_n)$$
$$-\phi_{i_1}(x_1)\ldots\phi_{i_n}(x_n)\delta Q(-\infty)]$$

in which the existence of $\delta Q(\pm\infty)$ is assumed.

From now we require that the variations $\delta\phi$ are such that the following relations hold, for some real constant λ,*

$$(2.9)\quad \delta Q(\mp\infty)|0,{}^{\text{in}}_{\text{out}}\rangle=\lambda|0,{}^{\text{in}}_{\text{out}}\rangle.$$

In particular, these conditions are automatically satisfied when the variations vanish outside a compact subset of the space-time, for we then have $\delta Q(\pm\infty)=0$.

By introducing the notation

$$(2.10)\quad \langle T(\phi_{i_1}(x_1)\ldots\phi_{i_n}(x_n))\rangle=\frac{\langle 0,\text{out}|T(\phi_{i_1}(x_1)\ldots\phi_{i_n}(x_n))|0,\text{in}\rangle}{\langle 0,\text{out}|0,\text{in}\rangle}$$

we then obtain from (2.8), for any time ordering of the x_k's,

$$(2.11)\quad -i\int dx\langle T(\delta\mathscr{L}(x)\phi_{i_1}(x_1)\ldots\phi_{i_n}(x_n))\rangle$$
$$=\langle\delta\,(T(\phi_{i_1}(x_1)\ldots\phi_{i_n}(x_n)))\rangle$$

Graphically, this relation may be written as

(2.12) 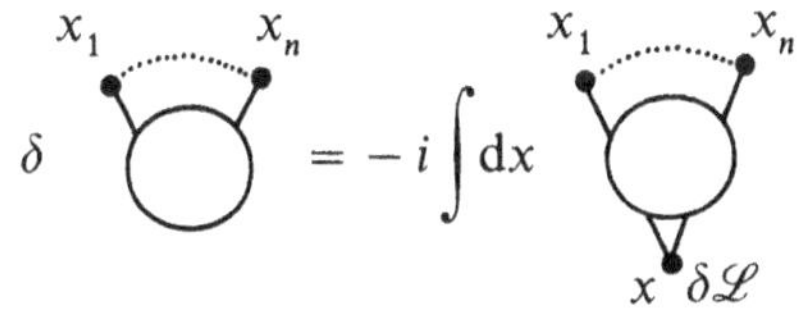

This is the basic relation which will be used in the next section.

Before proceeding any farther, let us indicate that a local form of (2.12)

---

* To reserve the possibility to consider interactions with external fields we do not necessarily assume the equality $|0,\text{in}\rangle=|0,\text{out}\rangle$.

can be obtained by replacing the variations previously considered by the following ones

(2.13) $\bar{\delta}\phi(x) = u(x)\delta\phi(x)$,

where $u(x)$ is a given function with compact support. Now we have

(2.14) $\bar{\delta} j^{\mu}(x) = \dfrac{\partial \mathscr{L}}{\partial(\partial_{\mu}\phi)} \bar{\delta}\phi(x) = u(x)\delta j^{\mu}(x)$

and

(2.15) $\bar{\delta}\mathscr{L}(x) = u(x)\delta\mathscr{L}(x) + \partial_{\mu}u(x)\delta j^{\mu}(x)$.

Relation (2.12) is then replaced by

(2.16)

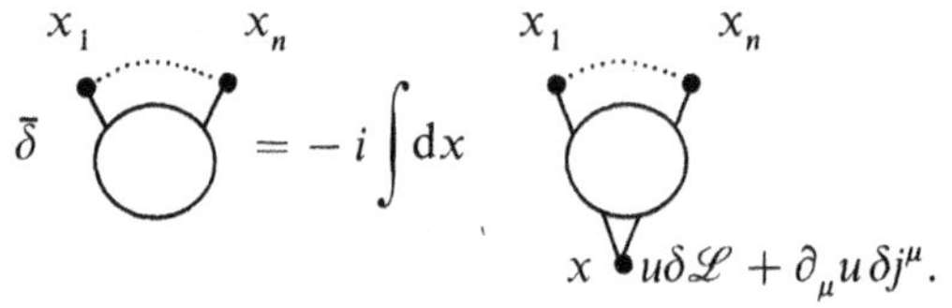

By putting

(2.17) $\bar{\delta}\phi(y) = \int dx\, u(x)\delta_x\phi(y)$

that is

(2.18) $\delta_x\phi(y) = \delta\phi(y)\delta(x-y)$

the relation (2.16) becomes

(2.19)

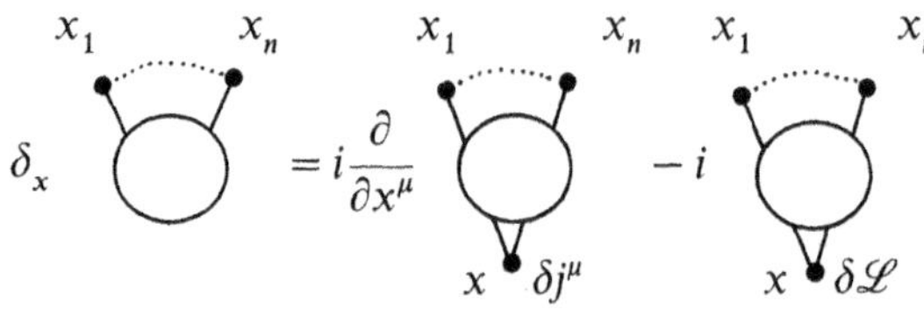

Let us note that the vacuum expectation value of (2.8) follows from (2.19) by integration on x. The last formula can also be seen as an immediate consequence of the local form of the Schwinger Principle: disregarding the eventual Schwinger terms, we in fact have from (2.5)

(2.20) $[\delta j^{0}(x), \phi(y)]_{x_0 = y_0} = -i\delta\phi(y)\delta(\vec{x} - \vec{y})$

then the calculation of $\partial_\mu \langle T\delta j^\mu \phi \ldots . \phi \rangle$ (first term in the right member of (2.19)) directly leads to (2.19) if we remark that $\partial_\mu \delta j^\mu = \delta\mathscr{L}$ when the equations of motion are satisfied.

Remark. The expression of $\delta\mathscr{L}$ in (2.12) can eventually be simplified by adding a divergence $\partial_\mu \Omega^\mu$, provided that Ω^0 commutes with all the ϕ's. This condition in fact insures that the differentiation can be permuted with the T-product, the resulting divergence then disappearing by integration.

To conclude this section, we consider the special case of a boson field transformation of the form

$$(2.21)\quad \phi \to \phi + \delta\alpha$$

where $\delta\alpha = (\delta\alpha_i)$ is a set of arbitrary functions with compact supports. It is easy to see that (2.12) here is equivalent to the equations of motion combined with the canonical commutation relations. To get a more convenient form of that result let us write the Lagrangian under the form $\mathscr{L} = \mathscr{L}_{\text{quad}} + l$, where $\mathscr{L}_{\text{quad}}$ is the quadratic part of $\mathscr{L}$. The right member of (2.12) is accordingly decomposed into two terms, the first of which is

(2.22) $-i\int dx$ 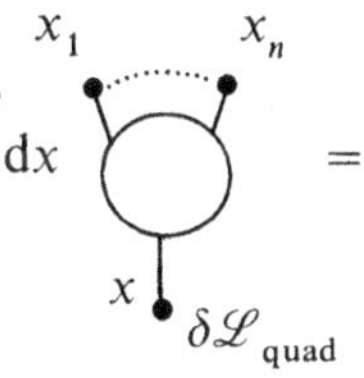

$= -i\int dx$ 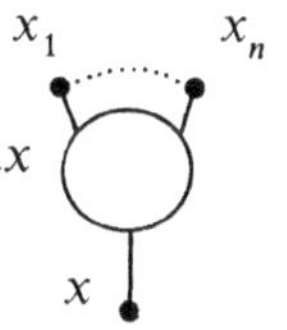

$$\frac{\partial\mathscr{L}_{\text{quad}}}{\partial\phi}\delta\alpha + \frac{\partial\mathscr{L}_{\text{quad}}}{\partial(\partial_\mu\phi)}\partial_\mu\delta\alpha$$

Let us now proceed to calculate this term. Integrating by parts in the right member of (2.22) we have to transform the expression

$$(2.23)\quad \left\langle T\frac{\partial\mathscr{L}_{\text{quad}}}{\partial\phi_i}(x)\phi_{i_1}(x_1)\ldots\phi_{i_n}(x_n)\right\rangle - \partial_\mu\left\langle T\frac{\partial\mathscr{L}_{\text{quad}}}{\partial(\partial_\mu\phi_i)}(x)\phi_{i_1}(x_1)\ldots\phi_{i_n}(x_n)\right\rangle.$$

Let $\mathscr{D}_{ij}$ be the second-order differential operator appearing in the free field equations

$$(2.24)\quad \partial_\mu\frac{\partial\mathscr{L}_{\text{quad}}}{\partial(\partial_\mu\phi_i)} - \frac{\partial\mathscr{L}_{\text{quad}}}{\partial\phi_i} \equiv \mathscr{D}_{ij}\phi_j = 0.$$

This operator may also be exhibited in the free Lagrangian written as follows

(2.25) $\mathscr{L}_{\text{quad}}(\phi,\partial\phi) = -\frac{1}{2}\phi_i\mathscr{D}_{ij}\phi_j + \text{(divergence)}$

and the corresponding propagator $\Delta_{ij}(x-y) = \langle T\phi_i(x)\phi_j(y)\rangle$ satisfies the equation

(2.26) $\mathscr{D}_{ij}\Delta_{jk}(x) = -i\delta_{ik}\delta(x).$

Since in (2.23) the terms $\partial\mathscr{L}_{\text{quad}}/\partial\phi_i$ and $\partial\mathscr{L}_{\text{quad}}/\partial(\partial_\mu\phi_i)$ are linear with respect to the derivatives of the fields, the corresponding differentiations can be taken out of the T-products, that leads to

$$-\mathscr{D}_{ij}\langle T\phi_j(x)\phi_{i_1}(x_1)\ldots\phi_{i_n}(x_n)\rangle.$$

Thus (2.22) becomes, by using (2.26),

(2.27) $-i\int \mathrm{d}x$ 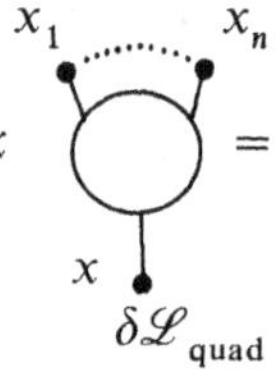

$= i\int \mathrm{d}x\,\delta\alpha^i(x)\mathscr{D}_{ij}$ 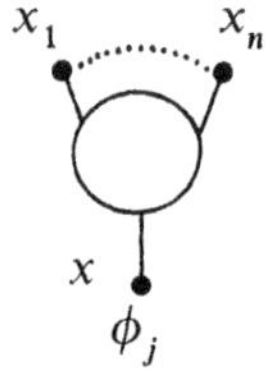

$= \int \mathrm{d}x\,\delta\alpha^k(x)$ 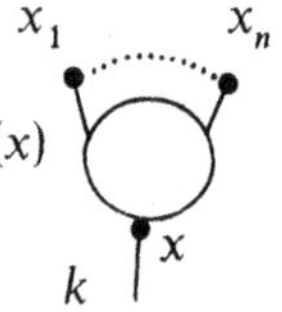

In the last diagram, the contribution of the direct terms, that is those for which x coincides with one of the x_k's, is equal to

(2.28) $\sum_{k=1}^{n} \delta\alpha^{i_k}(x_k)$ 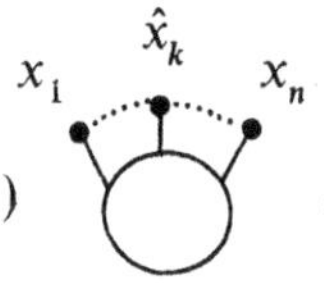

$= \delta$ 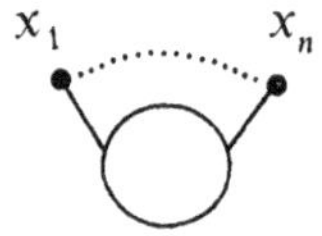

Therefore (2.12) furnishes the relation

(2.29) $\int \mathrm{d}x$ 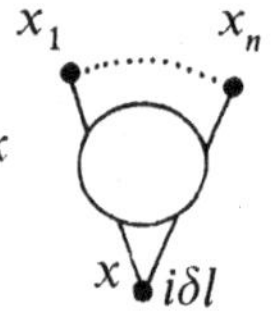

$= \int \mathrm{d}x\,\delta\alpha^k(x)$ 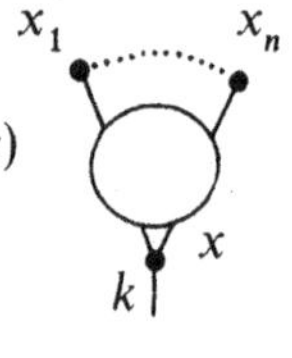

where, in the second member, the k-line enters at the point x by the full interaction.

3. Abelian Gauge Transformations

We will here apply the preceding results to the case of the fields $\phi = (A_\mu, \psi)$, where A_μ is a vector field and ψ a Dirac field, subject to the transformations

$$(3.1)\qquad \begin{aligned} \delta\psi &= ie_0\,\delta\omega\,\alpha_1\psi, \\ \delta A_\mu &= \delta\omega\,\partial_\mu\alpha_2, \end{aligned}$$

where α_1 and α_2 are arbitrary real functions. The current is

$$(3.2)\qquad \delta j^\mu = \delta\omega\left\{\alpha_1 J^\mu + \frac{\partial\mathscr{L}}{\partial(\partial_\mu A_\nu)}\partial_\nu\alpha_2\right\}$$

with

$$(3.3)\qquad J^\mu = ie_0\left[\frac{\partial\mathscr{L}}{\partial(\partial_\mu\psi)}\psi - \bar\psi\frac{\partial\mathscr{L}}{\partial(\partial_\mu\bar\psi)}\right].$$

We only consider Lagrangians invariant with respect to the gauge transformations of the first kind, namely the transformations (3.1) with $\alpha_1 = \alpha_2 = 1$. Under this hypothesis the current J^μ is the Noether conserved current associated with these particular transformations, while the variation of the Lagrangian under the general transformation (3.1) is

$$(3.4)\qquad \delta\mathscr{L} = \delta\omega\left\{\partial_\mu\alpha_1 J^\mu + \partial_\nu\alpha_2\frac{\partial\mathscr{L}}{\partial A_\nu} + \partial_{\mu\nu}\alpha_2\frac{\partial\mathscr{L}}{\partial(\partial_\mu A_\nu)}\right\}.$$

The asymptotic condition (2.9) is satisfied if α_1 and α_2 are constant outside a compact support: in this case we in fact have

$$(3.5)\qquad \delta Q(\mp\infty) = \alpha_1(\infty)\delta\omega Q$$

where Q is the conserved charge. In addition, from Relation (2.5) written for the positive frequency part of the asymptotic fields, we deduce

$$(3.6)\qquad \phi^{\text{in}\,(+)}_{\text{out}}(x)(\delta Q(\mp\infty)|0, {}^{\text{in}}_{\text{out}}\rangle) = 0$$

and therefore, due to the uniqueness of the vacuum,

$$(3.7)\qquad \begin{aligned} \alpha_1(\infty)\delta\omega Q|0, {}^{\text{in}}_{\text{out}}\rangle &= \delta Q(\mp\infty)|0, {}^{\text{in}}_{\text{out}}\rangle \\ &= \lambda_{\substack{\text{in}\\\text{out}}}|0, {}^{\text{in}}_{\text{out}}\rangle. \end{aligned}$$

Since Q is Hermitian, the constants λ_{in} and λ_{out} are necessarily equal provided that the vacuum amplitude $\langle 0, \text{out} | 0, \text{in} \rangle$ is nonvanishing.

To proceed with (2.12), we again separate the total Lagrangian into $\mathscr{L} = \mathscr{L}_{\text{quad}} + l$, where $\mathscr{L}_{\text{quad}}$ is the quadratic part with respect to A and ∂A only (not depending on ψ). Moreover the independence of the functions α_1 and α_2 allows the decomposition of the total variation δ into $\delta = \delta_1 + \delta_2$, and accordingly the separation of Formula (2.12) into two relations. We thus get*

(3.8) $\dfrac{1}{\delta\omega}\delta_1$ 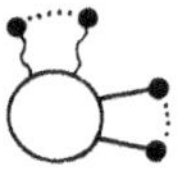 $= A(\alpha_1),$

(3.9) $\dfrac{1}{\delta\omega}\delta_2$ 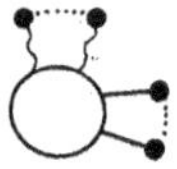 $= -i\displaystyle\int \mathrm{d}x$ 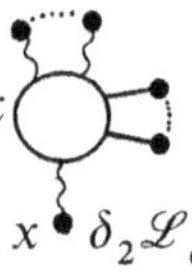 $+ B(\alpha_2),$

where A and B are defined by

(3.10) $A(\alpha_1) = -i\displaystyle\int \mathrm{d}x\,\partial_\mu\alpha_1(x)$ 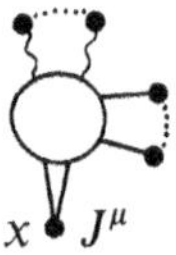 ,

(3.11) $B(\alpha_2) = -i\displaystyle\int \mathrm{d}x$ 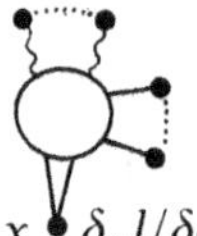 .

Let us now introduce the quantity

(3.12) $C(\alpha_2) = \displaystyle\int \mathrm{d}x\,\partial_\mu\alpha_2(x)$ 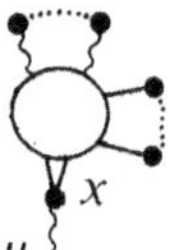 .

---

*Wavy or straight lines, respectively, represent A-or ψ-lines.

Since δ_2 is a translation of the fields, the argument developed at the end of the preceding section applied to (3.9) furnishes a relation similar to (2.29), namely

(3.13) $B(\alpha_2) + C(\alpha_2) = 0.$

Relation (3.8) is identical to the relation demonstrated by Takahashi [15], and rests on the gauge invariance of the first kind only, while (3.9), or equivalently (3.13), is merely an expression of the equation of motion of A_μ.

A necessary and sufficient condition for the gauge invariance of the second kind is now contained in the following theorem.

THEOREM. *The part $l = \mathscr{L} - \mathscr{L}_{\text{quad}}$ of the Lagrangian is gauge invariant, up to a divergence, if and only if the following condition is satisfied*

(3.14) ∂_μ 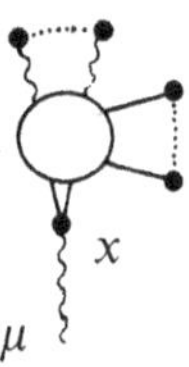

$= -i\partial_\mu$ 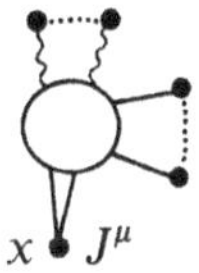

Proof. By adding (3.8) and (3.9), and taking into account (3.13) we have in general

(3.15) $\frac{1}{\delta\omega}\delta$ 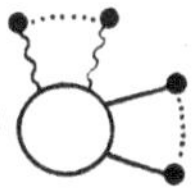 $= -i\int \mathrm{d}x$ 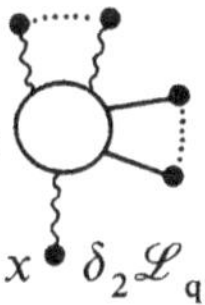

$+ A(\alpha_1) - C(\alpha_2).$

The gauge transformations of the second kind are defined by (3.1) with $\alpha_1 = \alpha_2 = \alpha$. Condition (3.14) is clearly equivalent to $A(\alpha) = C(\alpha)$ when α has a compact support. Then, by using the general relation (2.12), and remarking that $\delta_2\mathscr{L}_{\text{quad}} = \delta\mathscr{L}_{\text{quad}}$ we obtain from (3.15) for a gauge transformation

(3.16) $\int \mathrm{d}x$ 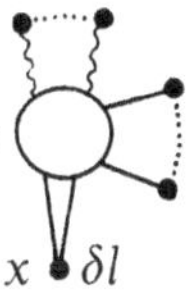

$= 0.$

This condition is therefore equivalent to (3.14). It also may be written as the functional equation

$$\int \mathrm{d}x\, \delta l(\phi(x), \partial\phi(x)) = 0, \qquad \forall\phi. \tag{3.17}$$

In fact, (3.16) means that the vacuum expectation values corresponding to $\mathscr{L}$ and $\mathscr{L} + \delta l$ are equal; since the generating functionals of these vacuum expectation values may be represented by the Feynman path integral, the equality of the action integrals for $\mathscr{L}$ and $\mathscr{L} + \delta l$ follows. Finally, (3.17) holds if and only if the equations of motion deduced from δl identically vanish, that is if δl reduces to a divergence. □

Condition (3.14), or $A = C$, may be written with the help of (3.8) under the equivalent form

(3.18) $\dfrac{1}{\delta\omega}\delta_1$ 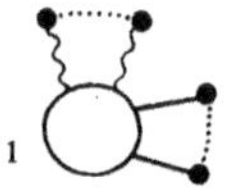 $= \displaystyle\int \mathrm{d}x\, \partial_\mu \alpha(x)$ 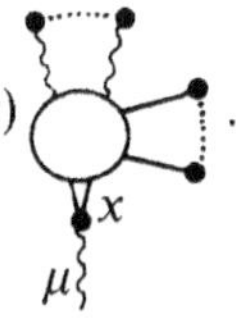 .

These relations are the *Ward identities*. According to the preceding theorem they characterize the gauge invariance (up to a divergence) of the Lagrangian, *except for the quadratic part* $\mathscr{L}_{\text{quad}}$.

Let us add that the relations (3.14) and (3.18) may be replaced by the relations having the same form but containing only proper diagrams with respect to the additional photon line (the line labelled by μ). In fact, (3.18) for the diagrams having one photon line only, reads

(3.19) $\displaystyle\int \mathrm{d}x\, \partial_\mu \alpha(x)$  $= 0.$

This can be written*

*Hatched blobs represent proper diagrams with respect to the μ-line.

(3.20) 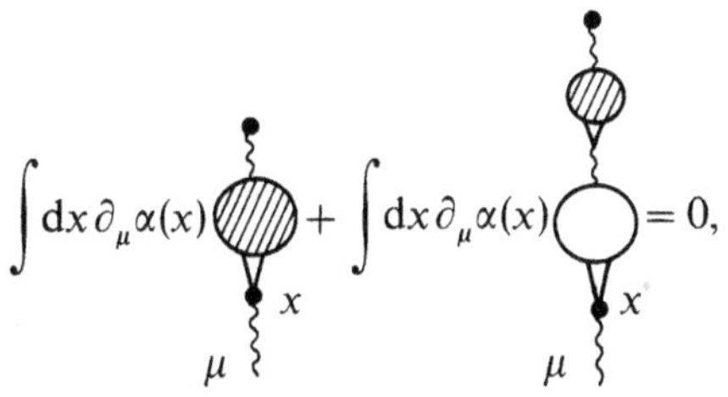

where the second term vanishes according to (3.19), so that it remains

(3.21) 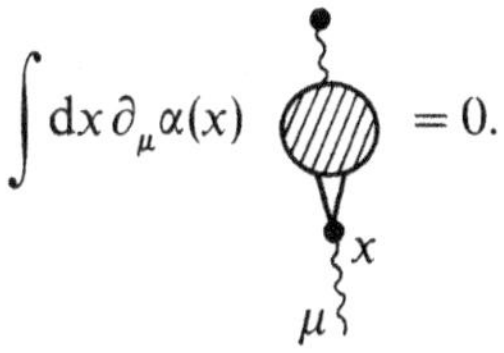

On account of that relation, (3.18) becomes for any diagram

(3.22) 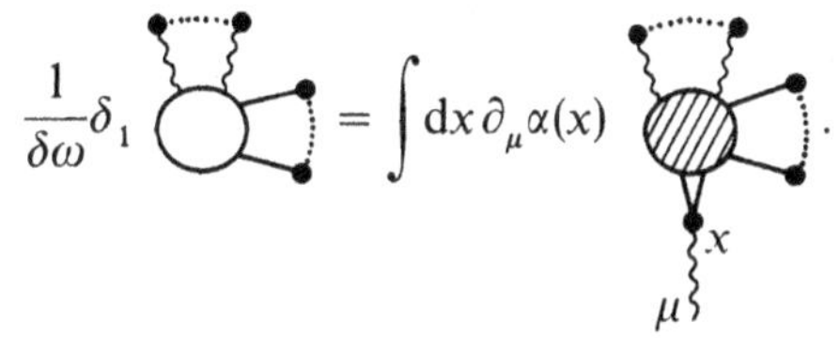

Conversely (3.22) implies (3.21) and therefore (3.18). An analogous reasoning applies to (3.14). Let us also note that (3.19) is nothing but (1.8).

Of course these results apply to the particular case of electrodynamics, since the relation $A = C$ is there an immediate consequence of the form $J_\mu A^\mu$ of the coupling term. It comes to say that, as soon as the coupling is so fixed, the Ward relations (3.18) are trivially identified with (3.8). This explains why the Ward identities were often attributed to the gauge invariance of the first kind only [2, 8, 15, 23], because the demonstrations were made, explicitly or implicitly, in quantum electrodynamics.

The theorem demonstrated above allows to characterize the Lagrangians offering the 'smallest deviation' from exact invariance by the property to give rise to the identities (3.18).* In particular the form of these identities does not depend on the specific choice made for $\mathscr{L}_{\text{quad}}$ (provided that the resulting theory is canonical). Such is the case because

*This property was called 'invariance pratique de jauge' by B. Jouvet [18].

the diagram

$$(3.23)\quad \int dxvy \sim\!\sim\!\sim\!\sim x\delta\mathscr{L}_{\text{quad}} = i\delta\omega\,\partial_\nu\alpha(y)$$

is independent of $\mathscr{L}_{\text{quad}}$. Let us also note that the necessary non-invariance of the total Lagrangian $\mathscr{L}$ is a direct consequence of the relation (2.12), for if $\delta\mathscr{L}=0$ this relation gives the contradiction

$$(3.24)\quad \delta\omega\,\partial_\mu\alpha(x) = \delta$$  $$= 0.$$

The general form of $\mathscr{L}_{\text{quad}}$ is, with the usual normalization of A_μ,

$$(3.25)\quad \mathscr{L}_{\text{quad}} = -\tfrac{1}{2}\partial_\mu A_\nu \partial^\mu A^\nu + \tfrac{1}{2}\varkappa(\partial_\mu A^\mu)^2 + \tfrac{1}{2}\mu_0^2 A_\mu A^\mu.$$

Except for the notations it is identical to that of Ref. 24. A non-vanishing mass is possible. The term factor of $\varkappa$ is known as the gauge fixing term, leading to a mass $\mu_0^2/(1-\varkappa)$ for the longitudinal part of A_μ. As shown in [24] the physical amplitudes do not depend on $\varkappa$, that expresses the gauge independence of the S-matrix. In fact, by letting the fermion lines in (3.18) go to infinity, the left member goes to zero since α has a compact support, that gives the relation

$$(3.26)\quad k^\mu \times$$ 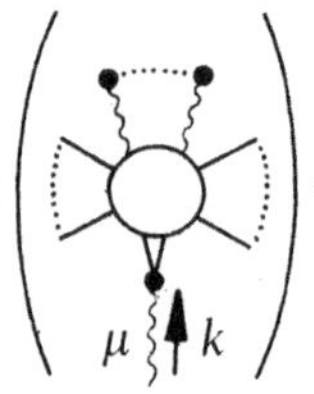 $$= 0.$$

It follows that, for any amplitude, the term factor of $k_\mu k_\nu$ in the A-propagator $\Delta_{\mu\nu}$, the only depending on $\varkappa$ [24], cancels out, leaving the classical $g_{\mu\nu}$-term. From the same relation (3.26) one can derive the unitarity property for the S-matrix between transverse A-states, since the projection operator for these states differs from the total projection operator $-g_{\mu\nu}$ by terms containing k_μ or k_ν.

Remark. The part l of the Lagrangian which appears in the foregoing theorem is only restricted to be gauge invariant up to a divergence. One can show that such a Lagrangian has the general form

$$(3.27)\quad l = l_{\text{inv}} + \xi^\mu A_\mu + \zeta^{\mu\nu\rho} F_{\mu\nu} A_\rho$$

where l_{inv} is strictly invariant and where ξ^μ and $\zeta^{\mu\nu\rho}$ are constant sources, with ζ antisymmetrical. Of course, the source terms are ruled out by the relativistic invariance.

4. Non-abelian gauge transformations

We shall now parallel the non-Abelian case with the Abelian one. The transformation law (3.1) is replaced by [7, 25]

$$\begin{aligned} \delta\psi &= -i\delta\omega_1\cdot\tau\psi, \\ \delta b_\mu &= \delta\omega_1 \times b_\mu - \frac{1}{\varepsilon}\partial_\mu\delta\omega_2. \end{aligned} \tag{4.1}$$

Here, by denoting $\delta\omega = (\delta\omega^a)$ the infinitesimal parameters of the 'isospin' group G, we have introduced two arbitrary functions $\delta\omega_1$ and $\delta\omega_2$, while $b_\mu = (b_\mu^a)$ is the Yang–Mills field and $\tau = (\tau_a)$ are the generators of the representation to which the fermion field ψ belongs; we also use the notations $u\cdot v = u^a v^a$ and $u \times v$ defined by $(u \times v)^a = C^a_{bc} u^b v^c$, the coefficients C^a_{bc} being the totally antisymmetrical structure constants of G displayed in the relations $[\tau_a, \tau_b] = iC^c_{ab}\tau_c$.

The current associated with (4.1) is

$$\delta j^\mu = \delta\omega_1\cdot J^\mu - \frac{1}{\varepsilon}\frac{\partial\mathscr{L}}{\partial(\partial_\mu b_\nu)}\cdot\partial_\gamma\delta\omega_2 \tag{4.2}$$

where

$$J^\mu = -i\left[\frac{\partial\mathscr{L}}{\partial(\partial_\mu\psi)}\tau\psi - \bar\psi\tau\frac{\partial\mathscr{L}}{\partial(\partial_\mu\bar\psi)}\right] + b_\nu \times \frac{\partial\mathscr{L}}{\partial(\partial_\mu b_\nu)}. \tag{4.3}$$

Once more, the Lagrangian is assumed to be invariant with respect to the gauge transformations of the first kind ($\delta\omega_1 = \delta\omega_2 = \text{const.}$), so giving

$$\delta\mathscr{L} = \partial_\mu\delta\omega_1\cdot J^\mu - \frac{1}{\varepsilon}\frac{\partial\mathscr{L}}{\partial b_\nu}\cdot\partial_\nu\delta\omega_2 - \frac{1}{\varepsilon}\frac{\partial\mathscr{L}}{\partial(\partial_\mu b_\nu)}\cdot\partial_{\mu\nu}\delta\omega_2 \tag{4.4}$$

where J^μ is the isospin current.

The same method as in Section 3 now gives analogous results. Formulas (3.8) and (3.9) respectively become

(4.5) δ_1 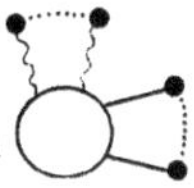 $= A(\delta\omega_1)$

with

(4.6) $A(\delta\omega_1) = -i \int \mathrm{d}x\, \partial_\mu \delta\omega_1(x) \cdot$ 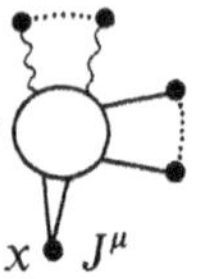

,

and

(4.7) δ_2 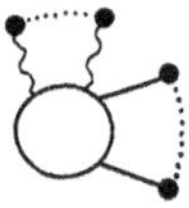 $= -i \int \mathrm{d}x$ 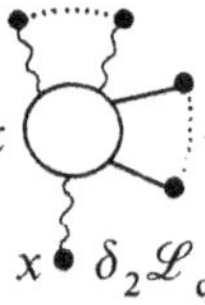

$+ B(\delta\omega_2)$,

where B is defined similarly to (3.11). Relation (3.13) remains obviously valid with

(4.8) $C(\delta\omega_2) = -\frac{1}{\varepsilon} \int \mathrm{d}x\, \partial_\mu \delta\omega_2(x) \cdot$ 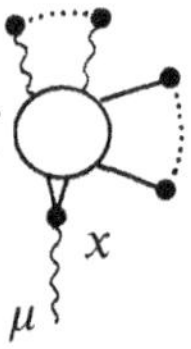

.

We thus get, instead of (3.15),

(4.9) δ 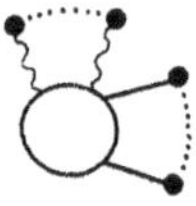 $= -i \int \mathrm{d}x$ 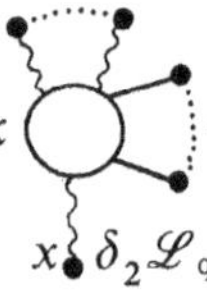

$+ A(\delta\omega_1) - C(\delta\omega_2)$

Before generalizing the theorem of Section 3, it is important to realize that any term of the form $(\partial_\mu b_\nu - \partial_\nu b_\mu)\cdot(\partial^\mu b^\nu - \partial^\nu b^\mu)$ contained in $\mathscr{L}_{\text{quad}}$ does not contribute to $\delta_2 \mathscr{L}_{\text{quad}}$. Therefore if, for an arbitrary constant α, we put

$$(4.10) \quad \mathscr{L}_{\text{quad}} = \bar{\mathscr{L}}_{\text{quad}} + \alpha\left(-\tfrac{1}{4}(\partial_\mu b_\nu - \partial_\nu b_\mu)\cdot(\partial^\mu b^\nu - \partial^\nu b^\mu)\right)$$

we have, in addition to (4.9), the same relation in which $\delta_2 \bar{\mathscr{L}}_{\text{quad}}$ replaces $\delta_2 \mathscr{L}_{\text{quad}}$. Then, by expressing the left member of this last relation with the help of (2.12), we get the equality:

(4.11) $-i\int \mathrm{d}x$ 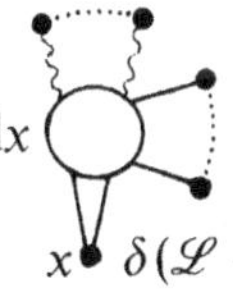 $= i\int \mathrm{d}x$ 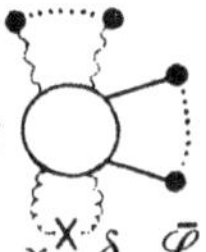 $+ A(\delta\omega_1) - C(\delta\omega_2).$

$x\quad \delta(\mathscr{L} - \bar{\mathscr{L}}_{\text{quad}})\qquad\qquad x\quad \delta_1\bar{\mathscr{L}}_{\text{quad}}$

We observe that the sum of the first two terms in the right-hand side of (4.11) is represented by the same formula (4.6) as A with J^μ replaced by $\bar{J}^\mu$, the part of the isospin current coming from $\mathscr{L} - \bar{\mathscr{L}}_{\text{quad}}$. This enables the following theorem to be stated:

THEOREM. *A necessary and sufficient condition for the gauge invariance of the second kind (up to a divergence) of* $\bar{l} = \mathscr{L} - \bar{\mathscr{L}}_{\text{quad}}$ *is*

(4.12) ∂_μ 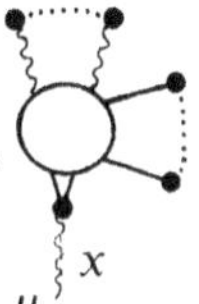 $= i\varepsilon\partial_\mu$ 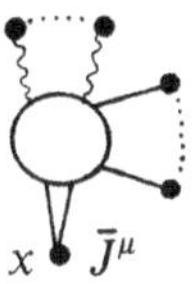 .

$\mu\quad x\qquad\qquad x\quad \bar{J}^\mu$

The corresponding Ward identities are deduced as in Section 3 by taking (4.12) into (4.5) that gives

(4.13) δ_1 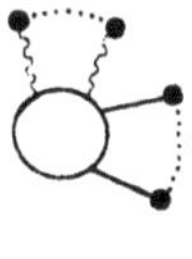 $= -\frac{1}{\varepsilon}\int \mathrm{d}x\,\partial_\mu\delta\omega(x).$ 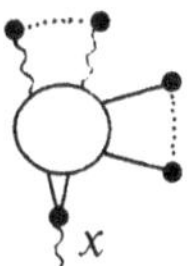 $- i\int \mathrm{d}x$ 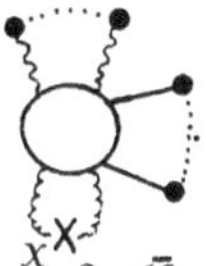 .

$x\quad \mu\qquad\qquad x\quad \delta_1\bar{\mathscr{L}}_{\text{quad}}$

From the general expression of $\bar{\mathscr{L}}_{\text{quad}}$

(4.14) $$\bar{\mathscr{L}}_{\text{quad}} = \beta(-\tfrac{1}{4}(\partial_\mu b_\nu - \partial_\nu b_\mu)\cdot(\partial^\mu b^\nu - \partial^\nu b^\mu)) + \tfrac{1}{2}(\varkappa - 1)\partial_\mu b^\mu\cdot\partial_\nu b^\nu + \tfrac{1}{2}\mu_0^2 b_\mu\cdot b^\mu,$$

we derive

(4.15) $$\delta_1\bar{\mathscr{L}}_{\text{quad}} = \partial_\mu\delta\omega_1\cdot\left(b_\nu \times \frac{\partial\bar{\mathscr{L}}_{\text{quad}}}{\partial(\partial_\mu b_\nu)}\right)$$
$$= \partial_\mu\delta\omega_1\cdot\{-\beta b_\nu \times (\partial^\mu b^\nu - \partial^\nu b^\mu) + (\varkappa - 1)b^\mu \times \partial_\nu b^\nu\}.$$

Actually, we easily verify that the term $b_\nu \times \partial^\nu b^\mu$ may be replaced by $b^\mu \times \partial_\nu b^\nu$ in the last diagram in Formula (4.13), so that we may write

$$(4.16)\quad \delta_1 \bar{\mathscr{L}}_{\text{quad}} = -\beta\partial_\mu\delta\omega_1\cdot(b_\nu \times \partial^\mu b^\nu) + (\varkappa - 1 + \beta)\partial_\mu\delta\omega_1\cdot(b^\mu \times \partial_\nu b^\nu).$$

Contrarily to the Abelian case, the form of the Ward relations (4.13) depends on the choice of $\bar{\mathscr{L}}_{\text{quad}}$, because here $\delta_1\, \bar{\mathscr{L}}_{\text{quad}}$ does not identically vanish. The reason for this fact obviously is that the field b_μ, now carries isospin. Conversely, giving the additional term depending on $\delta_1 \bar{\mathscr{L}}_{\text{quad}}$ in (4.13) determines $\bar{\mathscr{L}}_{\text{quad}}$, except for the mass term, as can be seen from (4.14) and (4.16).

The standard normalization of b_μ implies that the constants α and β defined by (4.10) and (4.14) are such that $\alpha + \beta = 1$. One must observe that the choice of these constants affects the self-couplings of the meson field, since a term of the form $(\partial_\mu b_\nu - \partial_\nu b_\mu)\cdot(\partial^\mu b^\nu - \partial^\nu b^\mu)$ contained in the invariant part $\bar{l}$ of $\mathscr{L}$ necessarily appears in the combination $f_{\mu\nu}\cdot f^{\mu\nu}$, where $f_{\mu\nu} = \partial_\mu b_\nu - \partial_\nu b_\mu + \varepsilon b_\mu \times b_\nu$. The Lagrangian usually considered belongs to the case $\alpha = 1$ so that $\bar{l}$ contains the Yang–Mills Lagrangian

$$(4.17)\quad \mathscr{L}_{\text{YM}} = -\tfrac{1}{4} f_{\mu\nu}\cdot f^{\mu\nu}.$$

It is precisely to include this case that we had to distinguish in (4.10) between $\mathscr{L}_{\text{quad}}$ and $\bar{\mathscr{L}}_{\text{quad}}$. Really, the self-interaction terms contained in $\mathscr{L}_{\text{YM}}$ cannot appear in the opposite case defined by $\alpha = 0$. However, if the idea of 'smallest deviation' from exact invariance is interpreted to mean that $\bar{\mathscr{L}}_{\text{quad}}$ (or $\delta_1 \bar{\mathscr{L}}_{\text{quad}}$) is as simple as possible, it is not sufficient for choosing from the instances $\alpha = 1, \varkappa \neq 1$ and $\alpha = \varkappa = 0$, for which $\delta_1 \bar{\mathscr{L}}_{\text{quad}}$ is reduced to a single term. To select the first case, one has to impose that $\bar{l}$ contains the classical Lagrangian.

Till now the non-Abelian case is fairly similar to the Abelian one. However, the physical requirement of unitarity of the S-matrix between transverse meson states now gives rise to some difficulties [26]. To end this paper we will restrict ourselves to make a few remarks on this subject, in connection with the Ward identities (4.13).

The only case for which the additional term in (4.13) vanishes for any diagram is obtained for $\beta = 0$, $\mu_0 \neq 0$, by taking the limit $\varkappa \to 1$. The b-propagator then goes to the usual Proca propagator for a massive vector boson, and the Ward identities take the same form as in the Abelian case. In order to derive from these relations an equation analogous to (3.26), it is here necessary to take the asymptotic limit for the boson lines also, since b_μ is transformed under δ_1, that gives instead of (3.26)

(4.18) $k^\mu \times$ 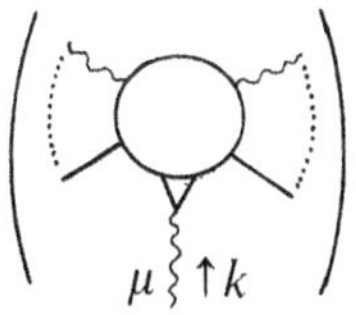 $= 0.$

This allows the unitarity to be demonstrated of the S-matrix between transverse b-states as in the Abelian case.

The last term of (4.13) also vanishes, when $\beta = 0$ and $\varkappa \neq 1$, for the tree diagrams restricted on the mass shell and involving only transverse external mesons, that leads to the unitarity property for these diagrams [26]. In all the other cases, the last term of (4.13) subsists so that the unitarity condition is not fulfilled in general. The solution of this problem for $\beta = 0$ is provided by the introduction of the Faddeev–Popov ghosts [26–29]. This brings another non-invariant term $\mathscr{L}_{\mathrm{FP}}$ to the Lagrangian, namely

$$(4.19) \quad \mathscr{L}_{\mathrm{FP}} = \partial^\mu \xi \cdot \partial_\mu \eta - \mathrm{i}\varepsilon b_\mu \cdot \partial^\mu \xi T \eta$$

where ξ and η are the ghost fields, and where $T = (T_a)$ are the generators of the adjoint representation of the isospin group. Correspondingly, the Ward identities (4.13) are modified by replacing $\delta_1 \bar{\mathscr{L}}_{\mathrm{quad}}$ by $\delta_1 \bar{\mathscr{L}}_{\mathrm{quad}} + \delta \mathscr{L}_{\mathrm{FP}}$, where ξ and η are not transformed, and one can check that they are still characteristic of the form of the Lagrangian. Finally let us mention that these identities differ from those introduced by Slavnov and Taylor [19, 30], which are derived with the help of transformations different from (4.1), and relate other diagrams than those appearing in (4.13).

Laboratoire de Physique Théorique et Mathématique, Université Paris VII

References

[1] Strocchi, F., and Wightman, A. S., *J. Math. Phys.* **15**, 2198 (1974).
[2] Symanzik, K., DESY T-71/1, (1971).
[3] Jauch, J. M., and Rohrlich, F., *The Theory of Photons and Electrons*, Cambridge (Mass.), Addison-Wesley, 1955.
[4] Schweber, S. S., *An Introduction to Relativistic Quantum Field Theory*, Harper and Row, New York, 1961.
[5] Glashow, S. L., Gell-Mann, M., *Ann. Phys.* **15**, 437 (1961).
[6] Weinberg, S., in *Lectures on Elementary Particles and Quantum Field Theory*, Vol. 1, Brandeis summer institute, M.I.T. Press, 1970.

[7] Lee, B. W., in *Les Houches, Session XXVIII*, 1975, North-Holland, 1976.
[8] Bjorken, J. D., and Drell, S. D., *Relativistic Quantum Fields*, McGraw-Hill, 1965.
[9] Brandt, R. A., *Ann. Phys.* **52**, 122 (1969).
[10] Ward, J. C., *Phys. Rev.* **78**, 182 (1950).
[11] Källen, G., *Helv. Phys. Acta* **26**, 755 (1953).
[12] Bogoliubov, N. N., and Shirkov, D. V., *Introduction to the Theory of the Quantized Fields*, Interscience, New-York, 1959.
[13] Kazes, E., *Nuovo Cimento* **13**, 1226 (1959).
[14] Feynman, R. P., *Phys. Rev.* **76**, 769 (1949).
[15] Takahashi, Y., *Nuovo Cimento* **6**, 371 (1957).
[16] Schwinger, J., *Phys. Rev.* **75**, 651 (1949).
[17] Pauli, W., and Villars, F., *Rev. Mod. Phys.* **21**, 434 (1949).
[18] Jouvet, B., *Nuovo Cimento* **20**, 28 (1961).
[19] Slavnov, A. A., *Theor. Math. Phys.* **10**, 152 (1972) [English translation, **10**, 99 (1972)].
[20] Nishijima, K., *Phys. Rev.* **119**, 485 (1960).
[21] Boyer, T. H., *Ann. Phys.* **44**, 1 (1967).
[22] Wentzel, G., *Phys. Rev.* **74**, 1070 (1948).
[23] Johnson, K., in *Acta Physica Austriaca*, Suppl. II, 109 (1965).
[24] Feldman, G., and Matthews, P. T., *Phys. Rev.* **130**, 1633 (1963).
[25] Yang, C. N., Mills, R. L., *Phys. Rev.* **96**, 191 (1954).
[26] Feynman, R. P., *Acta Physica Polonica* **24**, 697 (1963).
[27] Faddeev, L. D., and Popov, V. N., *Phys. Letters* **25B**, 29 (1967).
[28] Popov, V. N., and Faddeev, L. D., Kiev Report, I.T.P. 67–36, (1967).
[29] 't Hooft, G., *Nucl. Phys.* **B33**, 173 (1971).
[30] Taylor, J. C., *Nucl. Phys.* **B33**, 436 (1971).

A. PETERMAN

TOWARDS THE EARLY STAGES OF THE UNIVERSE

ABSTRACT. The physical significance of coupling parameters in quantum field theories is questioned. As a result of the renormalization such couplings depend on parameters defining the renormalization procedure. On the other hand, the variation of these couplings when a renormalization parameter, like the scale of the renormalization mass for instance varies, makes physical sense when comparing physical strengths at different energies. This renormalization group machinery provides the possibility to go backwards in time in a Universe sprung from a big-bang for different theories of grand unification. It is this travel in the past which is explored in the present paper.

1. Generalities on Coupling Parameters

In 1953, a fundamental paper [1] questioned the notion of coupling parameters, used to expand in power series physical quantities in quantum field theories. This was somewhat revolutionary at a time when the fine structure constant in QED in the limit of vanishing energy momentum (from $2P_{3/2} - 2P_{1/2}$ separation in deuterium or from the Thomson limit etc.) seemed to be the only possible definition and was moreover consistent with the classical definition of the electron charge via the correspondence principle.

In the summary, at the beginning of this 1953 paper, one can read indeed the following:

... A coupling parameter, however, can only be specified in terms of a chosen development of a function $S(xy \ldots \varkappa; c_1 \ldots)$ of physical significance.* However, the terms of the actual correspondence development (in terms of e^2) $S = S_2 + S_4 + \ldots$ have no physical meaning. Therefore the coefficient e^2 in S_2 has only a mathematical significance. It requires that the functions of $xy \ldots; S_4, S_6, \ldots, S_n$ have all been specified. As this specification involves the c_i's,** we must expect that a group of infinitesimal operations $P_i = (\partial/\partial c_i)_{c=0}$ exists,[1] satisfying

$$P_i S = h_{ie}(\varkappa,\mu,e)\frac{\partial}{\partial e}S(\varkappa,\mu,e;0,0\ldots), \qquad \text{(I)}$$

admitting thus a renormalization of e.***

* $\varkappa$ stands for the electron mass.

** c_i are parameters originating in *finite* renormalizations.

*** S, appearing on the left is of course the $S(\varkappa,\mu,e;c_1,c_2\ldots)$.

E. Tirapegui (Ed.), Field Theory, Quantization and Statistical Physics, 183–200.

The above equation is the general form of the renormalization group equations for QED. Its restriction to the trivial one-parameter group for overall scale transformations of the subtraction point μ

$$\mu' = \mu \exp(-c) \tag{1}$$

is obvious. It gives the usual renormalization group equation for observable physical quantities

$$-\mu\frac{\partial}{\partial\mu}S_0(e,\varkappa,\mu) = \beta\frac{\partial}{\partial e}S_0(e,\varkappa,\mu) \tag{2}$$

(with $\beta = +h_e$ and $S_0(e,\varkappa,\mu) \equiv S(e,\varkappa,\mu;c=0)$) since

$$\frac{\partial}{\partial c}S_0(e,\varkappa,\mu\exp(-c))\Big|_{\substack{c=0\\ e,\mu,\varkappa\,\mathrm{fixed}}} = -\mu\frac{\partial}{\partial\mu}S_0(e,\varkappa,\mu)\Big|_{\substack{e,x\\ \mathrm{fixed}}}$$

Therefore, Equation (2) is, as announced, exactly equivalent to

$$\frac{\partial}{\partial c}S(e,\varkappa,\mu;c)\Big|_{\substack{c=0\\ e,\varkappa,\mu\,\mathrm{fixed}}} = h_e\frac{\partial}{\partial e}S(e,\varkappa,\mu;0), \tag{4}$$

the Equation (I) of the summary quoted above. Indeed, for the special subgroup considered (scale transformation of μ)

$$S(e,\varkappa,\mu;c) = S_0(e,\varkappa,\mu\exp(-c)). \tag{5}$$

This kind of equation* has played various roles since then in several topics of physics. For the first time, one had an equation with a derivative with respect to a coupling parameter, allowing thus to relate different orders in the perturbation expansion amongst themselves[2]. This feature has been explored and exploited to determine the asymptotic behaviour of all terms in perturbation theory. But the most exciting aspect is that it questions seriously the concept of coupling constant. Indeed, since physical quantities must satisfy relations like

$$S_0(e,\mu\exp(-c)) = S_0(e(c),\mu) \tag{6)**$$

---

*Equation (I) extracted from the summary of Ref. 1. For multiplicatively renormalized quantities (Green functions, 1PI amplitudes, etc.):

$$S(e,\varkappa,\mu;c) = z^K(c)S_0(e,\varkappa,\mu\exp(-c))$$

there is an additional term in (I): $K(\partial z(c)/\partial c)|_{c=0}$ (the so-called anomalous dimensions γ). $z^K(c)$ stands for a product of various powers of z's with $z(0)=1$.

** $\varkappa$ has from now on been set equal to zero.

a change in $\mu' = \mu e^{-c}$ corresponds to a change in $e(c)$. More concretely, if

$$\mu \to \mu'' = \mu \exp(c), \quad \text{then } e \to e(c) \tag{7}$$

with $e(0) = e$ and the equation[3]

$$\left.\frac{\partial e(c)}{\partial c}\right|_{c=0} = \beta(e) = h_e(e) \tag{8}$$

is easily derived from the preceding and also the RG equation

$$\frac{\partial e(c)}{\partial c} = \beta(e(c)). \tag{9}$$

Equation (7) allows us to write

$$c = \operatorname{Log} \mu''/\mu$$

$e(c)$, the solution of Equation (9), is most often written $\bar{e}(t)$ in the current literature and called the 'running' or 'effective' coupling constant.

At this point one sees why the preceding examination of the real meaning of a coupling parameter in quantum field theory was relevant. It led to a coupling which, if fixed at $\mu'' = \mu$ is equal to e, yields, at $\mu'' = \mu'' : \bar{e}(\log \mu''/\mu)$ which is a solution of Equation (9) and can be written implicitly as

$$\int_{e}^{\bar{e}(t)} \frac{\mathrm{d}x}{\beta(x)} = t. \tag{10}$$

One knows in perturbation theory what $\beta(x)$ is. This is usually done by computing the relevant 1PI amplitudes and retaining their dimensionless part $S^{(i)}(p^2/\mu^2, e)$ evaluated at $p^2/\mu^2 = e^{2c}$ and then using the definition (8) of β (for details, see Ref. 2). In field theories with one single coupling g, β is of the general form:

$$\beta(g) = b_0 g^K + b_1 g^m + \ldots, \qquad K < m, \tag{11}$$

and the non-written terms being of order bigger than m in g. For example, in a $g\Phi^4$ theory, $K = 2, m = 3$; in QED, $K = 3, m = 5$, etc. In the so-called gauge theories with a single gauge coupling β has the same form as in QED.

2. Several couplings

The situation becomes more involved if more than two fields are in interaction and several couplings occur. This situation has been examined in the fourth section of the paper [1]. More comprehensive formulations have been made since then and the reader can refer to these standard reviews [3]. In general if one has a coupling constant space with vectors $\boldsymbol{g}$:

$$\boldsymbol{g} = (g_1, g_2, \dots g_n) \tag{12}$$

then the term we had in Equation (2) $\beta(e)\partial/\partial e$ becomes

$$\sum_{i=1}^{n} \beta^{(i)} \frac{\partial}{\partial g_i}. \tag{13}$$

There is a β function $\beta^{(i)}$ for each g_i which depends on all these g_i, i.e.,

$$\boldsymbol{\beta} = (\beta^{(1)}, \beta^{(2)}, \dots \beta^{(n)}, \tag{14}$$

and (13) is written

$$\boldsymbol{\beta} \cdot \frac{\partial}{\partial \boldsymbol{g}} \equiv \sum_{i=1}^{n} \beta^{(i)} \frac{\partial}{\partial g_i}.$$

Similarly, the 'anomalous dimensions' that we encountered in multiplicatively renormalizable quantities like Green functions of fields, 1PI amplitudes, Green functions of an operator, also depend on all the g_i (see the footnote on p. 184):

$$\gamma = \gamma(\boldsymbol{g}).$$

However, even for very complicated situations with several couplings, the evolution of the effective couplings $\bar{g}_i(t)$ can be determined, at least by numerical methods. The only essential assumption that must be fulfilled is that perturbation theory makes sense for each of these constants.

Working at the one-loop level, and considering the renormalization group equations for gauge coupling constants only a simplification arises in that the various equations decouple [4] and one has equations of the type

$$\frac{\mathrm{d}\bar{\alpha}_i}{\mathrm{d}t} = -2\frac{\beta_0^{(i)}}{4\pi}\bar{\alpha}_i^2; \qquad i = 1, 2, \dots n, \tag{15}$$

with $\bar{\alpha}_i = \bar{g}_i^2(t)/4\pi$, and β_0 being pure numbers (chosen with this definition without geometrical factors $(4\pi)^k$).

3. Sign of β_0

The sign of the one-loop term in the expansion of the β function is of decisive importance. Indeed the solution of (15) is

$$2t = -4\pi \int_{\alpha_i}^{\bar{\alpha}_i} \frac{dx}{\beta_0^i x^2} = \frac{4\pi}{\beta_0^i} \frac{1}{x}\bigg|_{\alpha_i}^{\bar{\alpha}_i} = \frac{4\pi}{\beta_0^i}(\bar{\alpha}_i^{-1} - \alpha_i^{-1}),$$

$$\bar{\alpha}_i(0) = \alpha_i,$$

so that

$$\bar{\alpha}_i(t) = \frac{\alpha_i}{1 + (\beta_0^i/2\pi)\alpha_i t}. \qquad (16)^*$$

When t increases, $\bar{\alpha}$ decreases logarithmically with the ratio of the energy M at which $\bar{\alpha}$ is computed and the energy μ at which α is defined so that (1.6) is also often written

$$\alpha_i(M) = \alpha_i(\mu)\left[1 + \frac{\beta_0^i}{2\pi}\alpha_i(\mu)\,\mathrm{Log}\, M/\mu\right]^{-1} \qquad (17)$$

This phenomenon ($\beta_0 > 0$) is proper exclusively to non-Abelian gauge theories in four dimensions. It has received the name of asymptotic freedom [5] and occurs spectacularly in QCD, the $SU(3)$ colour non-Abelian gauge theory of strong interactions. But, for instance in QED, it is well-known that $\beta_0 < 0$ and, when log $M/\mu \to \infty, \alpha(M)$ has a ghost pole when

$$\left[1 + \frac{\beta_0}{2\pi}\alpha(\mu)\,\mathrm{Log}\, M/\mu\right]$$

is zero. This, however, is not a genuine difficulty as for these values of $t = \log M/\mu$, the two-loop terms are more important than the simple one-loop approximation term. It simply indicates that the perturbation expansion of $\beta(\alpha)$ breaks down. This is a good example of how important is the examination of the second term in the expansion of β. For instance, in QCD, even if the precision obtained at present in the current experiments does not require the knowledge of this two-loop term, it is important to know its sign and magnitude. This in order to make sure that at present Q^2 leading logarithms are indeed dominant (which is not the case near the Landau pole in QED) and that the results of the leading log approximation are not altered by the next-to-leading ones. It can be readily shown that they are not, and this is very satisfactory from the purely theoretical point of view.

---

*Remember that $t = \log M/\mu$.

4. Back in time for $U(1)$, $SU(2)$ and $SU(3)$ couplings

In the present view of the origin of the Universe in a big bang, at very early times the temperature was very high and all particles were very energetic. 'Upward in energy' or 'upward in temperature' can therefore be expressed in a fancy way as 'backward in time'. Indeed knowing the value of gauge couplings at present energies* allows, through the machinery of the renormalization group developed in the previous sections, to calculate what their value will be at much higher energies like those of the order of the Planck mass $M_P = 1.2 \times 10^{19}$ GeV, provided, of course, that one has verified that perturbational techniques are still valid in this range of energy. The only things to know are the β_0^i for i = 1,2,3 corresponding respectively to the $U(1)$,$SU(2)$ and $SU(3)$ β_0 functions. We have seen that in the one-loop approximation the equations decouple so that it becomes elementary to perform this calculation. One has [6], neglecting possible mass terms

$$\begin{aligned} \beta_0^1 &= -\tfrac{2}{3}f, \\ \beta_0^2 &= (22/3) + \beta_0^1, \\ \beta_0^3 &= 11 + \beta_0^1. \end{aligned} \tag{18}$$

At first one remarks that g_1,g_2 and g_3 evolve quite differently, g_1 increasing with energy whereas g_2 and g_3 decrease, the latter faster than g_2. It is therefore to be expected that for some large value of M, these couplings, very different at $\mu \simeq 10$ GeV, will be of the same order of magnitude, and this suggests the possibility of a unification of these three different interactions with quantization of electromagnetic charge. For instance, one sees that α_2 and α_3 will satisfy

$$\alpha_2(M) = \alpha_3(M)$$

for a value of M defined by

$$\alpha_2^{-1}(\mu) - \alpha_3^{-1}(\mu) = \frac{11}{6\pi} \operatorname{Log} M/\mu. \tag{19}$$

Equation (19) cannot be used at once to fix the value of M in terms of μ. If $\alpha_3(\mu)$ is a somewhat well-known quantity that can be derived from QCD

---

*The gauge group at low energies seems to be QCD $SU(3)$ and $SU(2) \times U(1)$ for the weak and electromagnetic interactions.

this is not immediately true of $\alpha_2(\mu)$. At present moderately low energies ($\sim 0(10$ GeV$)$), the only exact symmetries are the $SU(3)$ colour for strong interaction and the $U(1)$ for electromagnetic phenomena. However, it is strongly believed that around 100 GeV, weak and electromagnetic interactions get unified with an $SU(2) \times U(1)$ symmetry. A popular model for this symmetry is the Weinberg–Salam [7] model (WS) and it is so far supported by all experimental data on this topic, so that it does not take special courage to say that the WS model is the 'right' one. Thus one can express α_2 in terms of $\alpha_{em} = e^2/4\pi$ and the weak mixing angle θ_W. One has

$$\alpha_2^{-1} = \sin^2\theta_w/\alpha_{em} \tag{20}$$

$\sin^2\theta_W$ is of course also μ dependent and it has been measured in neutral current experiments at an energy of about $\mu \ll 100$ GeV.

Its value is not yet firmly established and an average of all available results gives [8]

$$\sin^2\theta_w(100\,\text{GeV}) = 0.23 \pm 0.01. \tag{21}$$

On the other hand, knowing $\alpha_{em}^{-1}(0) = 137.0$ one can, using the β function for QED with sharp thresholds of the θ function type, calculate the value of $\alpha_{em}^{-1}(100$ GeV$)$. One starts at $Q^2 = 4m_e^2$ and one considers the contributions of the different steps to α_{em}^{-1} (Table I).

TABLE I
The contributions to the finite renormalization of α_{em}^{-1} between $4m_e^2$ and $4m_W^2$ (with step functions).

Q^2 range	Contributions to α_{em}^{-1}
$4m_e^2 - 4m\mu^2$	-1.132
$4m_\mu^2 - 0.5$ GeV^2	-0.513
0.5 GeV$^2 - 12$ GeV^2	-1.349
12 GeV$^2 - 10^2$ GeV^2	-1.425
10^2 GeV$^2 - 10^3$ GeV^2	-1.629
10^3 GeV$^2 - 4m_W^2$	-2.835
Total for $4m_e^2 - 4m_W^2$	-8.883

Taking $\alpha_{em}^{-1}(2m^e) \cong 137$, one has

$$\alpha_{em}^{-1}(2m_w) \cong 137.0 - 8.9 = 128.1.$$

This estimation is very crude but agrees rather well with more refined analyses done recently [9]. We know therefore the value $\alpha_2^{-1}(2m_W)$. If we choose $\sin^2\theta_W = 0.2$, then Equation (19) gives $M = 7.25 \times 10^{15}$ GeV. For $\sin^2\theta_W = 0.21$, $M = 9.2 \times 10^{16}$ GeV, and for $\sin^2\theta_W = 0.22$, $M = 8.2 \times 10^{17}$ GeV. The sensitivity of M to the value of $\sin^2\theta_W$ is enormous. More refined estimations taking into account vector boson thresholds and two-loop β functions reduce appreciably the above crude evaluation and give $M \cong 4.8 \times 10^{14}$ GeV, but for $\sin^2\theta_W = 0.19$ [9]. The exact value of M will be of interest later. But now, more interesting is the common value α_{GUM} taken both by $\alpha_2(M)$ and $\alpha_3(M)$ which is

$$\alpha_{GUM} = 0.024 = 1/41.7$$

By considering the order of magnitude of the energy at which the two coupling constants become equal, one sees that 10^{15} GeV is an extremely high temperature which has only been reached in the far past of the Universe, that is to say a very short time after its birth; at least if we believe in the scenario of a big bang.

5. The Grand Unification

Up to now we have not considered the constant $\alpha_1(\mu)$ and its evolution. Enlarging the frame of our assumptions (WS model) α_1 gets related to α_{em} and α_2 if we assume that the $SU(3)$, $SU(2)$ and $U(1)$ groups are subgroups of a larger group G [10]

$$\alpha_{em}^{-1}(\mu) = \alpha_2^{-1}(\mu) + C^2\alpha_1^{-1}(\mu) \tag{22}$$

where C is a parameter depending on the grand group chosen for unification (in fact $C = (T_3 - Q)/T_0$ where $\vec{T}$ and T_0 are, respectively, the $SU(2)$ and $U(1)$ generators and Q the charge). Having $\sin^2\theta_W(\mu)$, $\alpha_{em}(\mu)$ and $\alpha_3(\mu)$ as input, one can eliminate $\log M/\mu$ of Equation (19) with the help of

$$\alpha_1^{-1}(\mu) - \alpha_2^{-1}(\mu) = \frac{11}{3\pi}\,\mathrm{Log}\, M/\mu \tag{23}$$

and writing $\alpha_1^{-1} = (\cos^2\theta/\alpha_{em} C^2)$, using (22), and have an equation for C^2. Taking as before $\alpha_{em}^{-1}(100) = 127$, $\alpha_3(100) = 0.139$ and $\sin^2\theta_W(100) = 0.20$, C^2 turns out to be equal to 1.65 which is very close to 5/3. But if

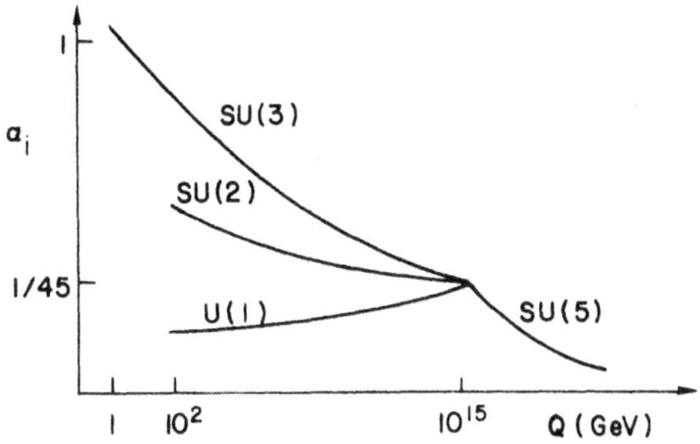

Fig. 1. Qualitative picture of the evolution of the $SU(3), SU(2)$ and $U(1)$ couplings in a one-step grand unified theory such as $SU(5)$.

$\sin^2 \theta_W(100) = 0.21$, then $C^2 = 1.52$ and for $\sin^2 \theta_W(100) = 0.22$, $C^2 = 1.42$. So either G is $SU(5)$ and the present measurements of $\sin^2 \theta_W(100)$ suffer from some systematic error, or the measurements can be trusted at face value and one has to deal with another group which, perhaps, contains $SU(5)$ as a subgroup.

A lot has been written on $SU(5)$ chosen as the grand group [11] and very interesting results have come out which suggest that it has some role to play in the unification. But on the other hand there are also some conflicting results which could be better solved, though not destroying its nice features, in larger groups and especially by proceeding to the grand unification by steps like, for instance [12]:

$$\begin{array}{c} SU(3) \times SU(2) \times U(1) \xleftarrow[m]{} SU(4) \times SU(2) \times U(1) \\ \uparrow M_x \\ O(10) \end{array} \tag{24}$$

or [13]

$$\begin{array}{c} SU(3) \times SU(2)_L \times U(1) \xleftarrow[M_R]{} SU(3) \times SU(2)_L \times SU(2)_R \times U(1). \\ \uparrow M_X \\ E_6 \end{array} \tag{25}$$

In these particular cases, it is possible to accommodate a theoretical value of $\sin^2 \theta_W$ (Figure 2) which, through renormalization group equations, decreases from 3/8 at M_x (the grand unification mass) down to a $\sin^2 \theta(100)$ which agrees with present experimental data (Table II).

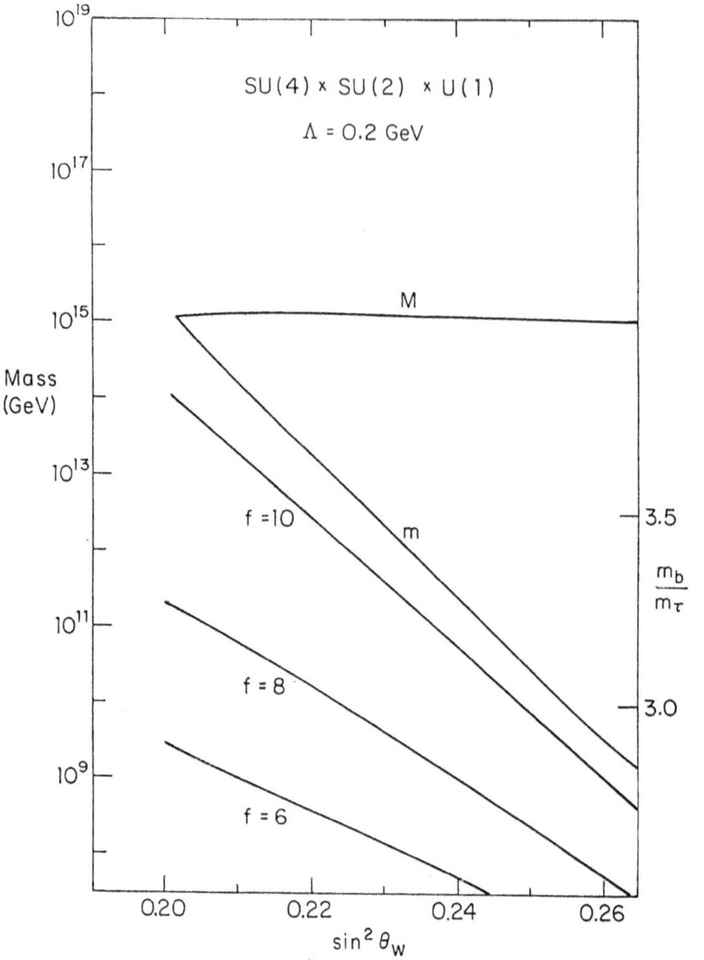

Fig. 2. M_x and m (in GeV) and the ratio m_b/m_τ (at 10 GeV) for $f = 6$ and 8, plotted versus $\sin^2\theta_W$ for the subgroup $SU(4) \times SU(2) \times U(1)$ of $O(10)$. $\Lambda = 0.2$ GeV.

One sees that the procedure of unification is perhaps not as simple as going directly from $SU(3) \times SU(2) \times U(1)$ to $SU(5)$. But whatever this procedure is, it will have to pass all the tests shown above. Namely going from the simple grand group G values of $\sin^2\theta_W$ down to the observed $\sin^2\theta_W(100)$ through the renormalization group equations. Moreover the

TABLE II
Plots of M_R/M_L versus $\sin^2\theta_W$ and of $\sin^2\theta_W$ versus $\alpha_s(10\text{ GeV})$. (M_x is taken to be 1.2×10^{19} GeV and $M_L = 80$ GeV) [13].

M_R/M_L	$\sin^2\theta_W$	$\alpha_s(10\,\text{GeV})$
10	0.274	0.093
10^2	0.266	0.102
10^3	0.259	0.113
10^4	0.251	0.128
10^5	0.244	0.144
10^6	0.236	0.169
10^7	0.229	0.199
10^8	0.221	0.251
10^9	0.214	0.328
10^{10}	0.206	0.487
10^{11}	0.199	0.869
10^{12}	0.191	
10^{13}	0.184	

β^i functions for the couplings of $SU(3)$, $SU(2)$ and $U(1)$ must be such that they provide an evolution which brings the three coupling parameters to a common value α_{GUM}, the unique coupling parameter of the simple group G. And this at a value of $M \gtrsim (2\text{–}4) \times 10^{14}$ GeV. This is necessary if one wishes that the proton lifetime does not become smaller than the present experimental lower bound of $\sim 2 \times 10^{30}$ years [14]. This is because once grand unification arises, quarks and leptons are to be put in the same multiplets and there will be interactions changing quarks into leptons via the baryon number violating forces gauged by the bosons of mass M_x. Therefore the proton becomes instable and one can show that its lifetime τ_p is proportional to M_x^4. A theoretical estimation of τ_p is:

$$\tau_p \sim K \times 10^3 M_x^4/m_p^5. \tag{26}$$

K being of the order of unity.

With a value of $M_x \cong (2\text{–}4) \times 10^{14}$ GeV on gets $\tau_p = 10^{31 \pm 2}$ years.

We will soon end our travel back to the very early time of the Universe. But we shall, before concluding, have a look on the evolution of the masses.

6. Fermion masses

Masses in perturbative quantum field theory can be considered on the same footing as the coupling parameters [15]. One is led to consider 'effective' or 'running' masses $\bar{m}_i(t)$ which evolve according to an equation of the type

$$(\partial/\partial t)\,\mathrm{Log}\,\bar{m}(t) = \gamma_m(\bar{g}(t)) \tag{27}$$

where γ_m is a mass anomalous dimension similar to the γ encountered for multiplicatively renormalizable quantities. Its definition comes from the fact that masses can get renormalized by finite renormalization z_m such that $m_R = z_m m_{R'}$ (R and R' are two different renormalization prescriptions). In our rescaling of $\mu \to \mu' = \mu e^{-c}$, z_m is a function of the parameter c and

$$\gamma_m = (\partial/\partial_c) z_m(c)\big|_{c=0}, \qquad z_m(0) = 1. \tag{28}$$

When one has several interactions like in our cases, fermion masses receive contributions from the $SU(3)$, $SU(2)$ and $U(1)$ types of interaction. They have been computed and one can write [16]

$$\gamma_{m_f} \cong \gamma^{(3)}_{m_f} + \gamma^{(2)}_{m_f} + \gamma^{(1)}_{m_f} \tag{29}$$

with, at the one-loop approximation and neglecting mass corrections

$$\gamma^{(3)}_{m_f} = \begin{cases} -\dfrac{2}{\pi}\alpha_3 & \text{for quarks} \\ 0 & \text{for leptons} \end{cases}$$

$$\gamma^{(1)}_{m_f} = \begin{cases} -\dfrac{1}{16\pi}\alpha_1 & \text{for } u, c, t \text{ quarks} \\ \dfrac{7}{80\pi}\alpha_1 & \text{for } d, s, b \text{ quarks} \\ -\dfrac{33}{80\pi}\alpha_1 & \text{for } e, \mu, \tau \text{ leptons*} \end{cases} \tag{30}$$

Using the values of the one-loop β functions for α_1, α_2 and α_3, one has the

* $\gamma^{(1)}_{m_f} = 3\alpha_1 (80\pi)^{-1}$ for neutrinos.

following solutions [16] for the various m_{if} (μ)

$$\frac{m_{d,s,b}(\mu)}{m_{e,\mu,\tau}(\mu)} = \frac{m_{d,s,b}(M_x)}{m_{e,\mu,\tau}(M_x)}\left[\frac{\alpha_3(\mu)}{\alpha_3(M_x)}\right]^{12/(33-2f)}\left[\frac{\alpha_1(\mu)}{\alpha_1(M_x)}\right]^{3/2f}$$

$$\frac{m_{u,c,t}(\mu)}{m_{d,s,b}(\mu)} = \frac{m_{u,c,t}(M_x)}{m_{d,s,b}(M_x)}\left[\frac{\alpha_1(\mu)}{\alpha_1(M_x)}\right]^{-9/20f} \tag{31}$$

If the grand group G gives simple relations between quarks and fermion masses, then one can use them as an input in the ratios (31) at M_x and get their values at present energies, i.e., those values which are experimentally determined nowadays. For instance, if $m_b(M_x) = m_\tau(M_x)$ one gets $m_b(10 \text{ GeV}) = 5.2$ GeV.* Depending from the models of symmetry breaking and the grand group chosen, zeroth-order mass ratios (i.e., at M_x) can be chosen in different ways.** As an example, within the $O(10)$ model [12] one can account for several quark to lepton mass ratios (m_t is predicted to be 13–14 GeV, for instance).

7. Symmetry breaking

Having gone back in time close to the Planck time, $t_p = \hbar/KT_p \cong 10^{-43}$ sec after the birth of the Universe, we have to come back now at our starting point, namely at energies of the order 10–100 GeV. During this return journey, we shall observe the most noteworthy features that we encounter. At the Planck temperature T_p, we suppose that grand unification has taken place, there is one single coupling (we neglect gravitation which has not entered the game yet). The particles are classified in generations, i.e., they have been put in well-defined representations of the grand group G in $\bar{5} + 10$ for the ($SU(5)$ case). The coupling parameter increases as we decrease T by several orders of magnitude. In this region, aside the fermions, there are also scalar particles present, the so-called Higgs particles. We then attend to the breaking of the G symmetry which will take place by same mechanism. As discussed in Ref. 17 for $SU(5)$, the breaking can arise through radiative corrections. The quartic Higgs couplings, which are assumed to be of order g^2 at $M_p = 1.2 \times 10^{19}$ GeV, decrease and

* $m_q(\mu)$ is always computed in such *a* way that $\mu_{\text{th}} = 2m_q(\mu_{\text{th}})$, μ_{th} being the approximate position of the $q\bar{q}$ threshold.

** Depending the way we choose to distribute the fermions in the various representations of G.

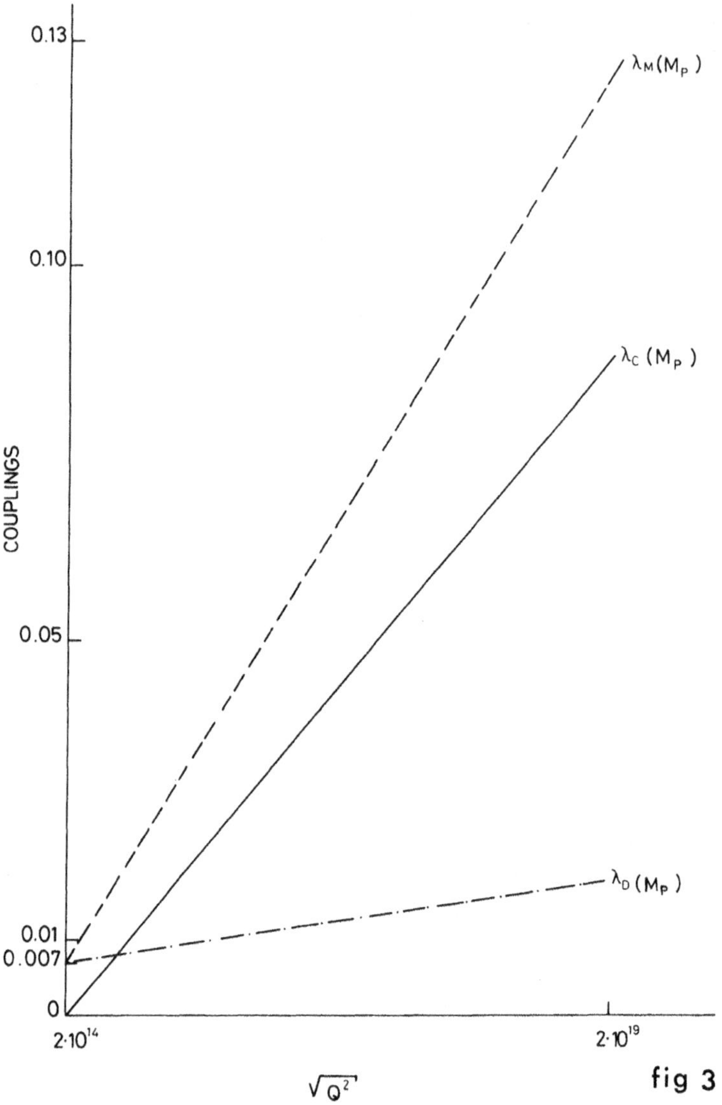

Fig. 3. The two adjoint (λ_M and λ_c) and the fundamental (λ_D) couplings. Their average slope between $Q = 2 \times 10^{14}$ GeV and $Q = 2 \times 10^{19}$ GeV are plotted. These are solutions of a system of coupled renormalization group equations. λ_M and λ_D have been normalized at the same value for $Q = 2 \times 10^{14}$ in order to show the difference in average slope [$SU(5)$ case].

when one of these couplings (more precisely one of the two adjoint Higgs couplings) has a zero, the symmetry breaks. This will occur rather rapidly in the $SU(5)$ model, when Q will be of the order 10^{-4}–10^{-5} times the Planck mass, i.e., $\lambda_{\text{adj}}\ (Q_c) = 0$ when Q_c is around 10^{15} GeV. In $SU(5)$, according to which of the two adjoint couplings vanishes first, then the grand group will break into $SU(4) \times U(1)$ or into $SU(3) \times SU(2) \times U(1)$. The first way of breaking is undesired but nothing compelling forces $SU(5)$ to break into the other way. Concretely,

$$SU(5) \xrightarrow{\lambda_M(Q_c)=0} SU(4) \times U(1)$$

$$SU(5) \xrightarrow[\lambda_c(Q_c)=0]{} SU(3) \times SU(2) \times U(1)$$

Meanwhile, the vector Higgs couplings ($\bar{5}$ in $SU(5)$) evolve much more slowly than the adjoint which have a large non-Abelian charge (see Figure 3).

After passing $Q_c = 10^{15}$ GeV, nothing occurs in $SU(5)$, because these

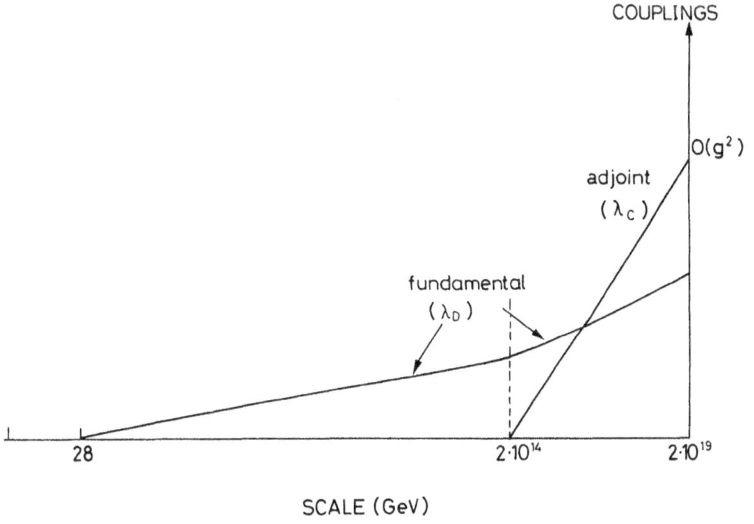

Fig. 4. Qualitative picture of the Q evolution of the effective Higgs couplings of fundamental and adjoint representations in the $SU(5)$ model.

fundamental couplings evolve even more slowly below Q_c, in part because large masses have been generated for some particles which therefore do not contribute to the renormalization group equations below Q_c. $\lambda_{\text{vect}}(\mu)$ can be arranged to vanish for $\mu = O(10)^2$ GeV [18]. Therefore between the time corresponding to $T_c = 10^{15}$ GeV and that corresponding to $T_0 = 10^2$ GeV we have crossed a real desert. At T_0, $\lambda_{\text{vect}}(T_0) = 0$ (see Figure 4) and

$$SU(3) \times SU(2) \times U(1) \longrightarrow SU(3) \times U(1).$$

This is the second symmetry breaking generating the W and Z masses and exhibiting finally the symmetry which we left at the beginning of our trip, namely the $SU(3) \times U(1)$ characteristic of our current energies of a few GeV. Of course, as Equations (24) and (25) show, in other schemes, something may occur between 10^{15} and 10^2 GeV. While the first symmetry breaking (which is not necessarily due to radiative corrections) occurs somewhere in between the Planck mass and 10^{15} GeV, it is followed by a second step of breaking, somewhere between 10^{15} GeV and 10^2 GeV.* And then breaks finally to $SU(3) \times U(1)$ at $T_0 = O(10^2)$ GeV. This makes the return journey a little less boring than it was for $SU(5)$ above.

As a conclusion, one may say that we have discovered a strange world ($M_x < Q < M_p$) where all gauge particles are governed by a single coupling parameter; where scalar particles are in great abundance and where no stable baryonic matter exists. Moreover, the interactions of the scalar particles with the fermions create forces which violate also a C and CP symmetry. They might very well be a source of the residual baryons which are the constituents of our present Universe; this Universe where we have returned and where we live again now.

* In (24), $M_x = O(M_p)$ and $M_R > 3M_L$, with $M_L = m_w$. In (25), $M_x = O(10^{15})$ GeV and 10^9 GeV $< m < M_x$.

Acknowledgements

The author wishes to thank John Ellis for the numerous and enlightening discussions.

Notes

[1] The fact that one restricts oneself to continuous groups results from the fact that non-parametrized renormalization prescriptions $R, R', R'', R''' \ldots$ do not form a group [19].

This is due to the absence of a rule for multiplying $R \to R'$ and $R'' \to R'''$, which, however, can be obtained for special subsets of prescriptions which can be explicitly parametrized.

[2] A well-known application has been made in the calculations of the muon anomalous magnetic moment for the coefficients of the powers of log (m_μ^2/m_e^2). See Ref. 20).

[3] Gell-Mann and Low [21] have been the first to give explicitly the first two terms of h_e (ψ in their notation) in an expansion in e^2, in the limit of vanishing electron mass. They have shown that this limit exists in all orders of perturbation expansion.

References

[1] Stueckelberg, E. C. G., and Peterman, A., *Helv. Phys. Acta* **26**, 499 (1953).

[2] Peterman, A.,—Physics Reports **53**, 157 (1979).

[3] (a) Bogoliubov, N. N., and Shirkov, D., *Introduction to the Theory of Quantized Fields*, Interscience, New York, 1959.
(b) Crewther, R. J., 'Weak and Electromagnetic Interactions at High Energies', *Proceedings Cargèse Summer School 1975*, Plenum Press.
(c) Politzer, H. D., *Phys. Rep.* **14C**, 129 (1974).
(d) Gross, D. J., 'Application of the Renormalization Group'. In R. Balian and J. Zinn-Justin (eds.), *Les Houches (1975) Methods in Field Theory*, North Holland, 1976.

[4] See, for example: Cheng, T., Eichten, E., and Li, L. *Phys. Rev.* **D9**, 2259 (1974).

[5] Gross, D. J., and Wilczek, F., *Phys. Rev. Letters* **30**, 1343 (1973); Politzer, H. D., *Phys. Rev. Letters* **30**, 1346 (1973).

[6] Gross, D. J., and Wilczek, F., *Phys. Rev.* **D8**, 3633 (1973) and Ref. 3(c).

[7] Weinberg, S., *Phys. Rev. Letters* **19**, 1264 (1967); Salam, A., in N. Svartholm (ed.), Proceedings 8th Nobel Symposium 1968 Almqvist and Wiksell, Stockholm, 1968.

[8] Dydak, F., 'Neutral Currents', Talk at the EPS International Conference on High Energy Physics, Geneva, 1979.

[9] Goldman, T., and Ross, D. A., *Phys. Letters* **84B**, 208 (1979); Marciano, W., *Phys. Rev.* **D20**, 274 (1979).

[10] Georgi, H., Quinn, H., and Weinberg, S., *Phys. Rev. Letters* **33**, 451 (1979).

[11] Georgi, H., and Glashow, S. L., *Phys. Rev. Letters* **32**, 438 (1974); Fritzsch, H., and Minkowski, P., *Ann. Phys. (New York)* **93**, 193 (1975). For a review, see the talk by J. Ellis at the EPS International Conference on High Energy Physics, Geneva (1979), CERN Preprint TH 2723 (1979); See also: Nanopoulos, D., Harvard Preprint HUTP-78/A062 (1978).

[12] Georgi, H., and Nanopoulos, D., Harvard Preprint HUTP-79/A039 (1979).

[13] Shafi, Q., and Wetterich, C., CERN Preprint TH. 2667 (1979).

[14] Reines, F., and Crouch, M., *Phys. Rev. Letters* **32**, 493 (1974).

[15] Eriksson, K. E., *Nuovo Cimento* **19**, 1010, 1044 (1961); Weinberg, S., *Phys. Rev.* **D8**, 3497 (1973).

[16] Buras, A. J., Ellis, J., Gaillard, M. K., and Nanopoulos, D., *Nuclear Phys.* **B135**, 66 (1978).

[17] Ellis, J., Gaillard, M. K., Peterman, A., and Sachrajda, C., CERN Preprint TH. 2696 (1979).

[18] Weinberg, S., *Phys. Letters* **82B**, 387 (1979).
[19] Astaud, M., and Jouvet, B., *Comptes Rendus* **264**, 1433 (1967); *Nuovo Cimento* **53A**, 841 (1968) and **63A**, 5 (1969).
[20] Kinoshita, T., *Nuovo Cimento* **51B**, 140 (1967).
Lautrup, B., and de Rafael, E., *Nuclear Phys.* **B70**, 317 (1974).
[21] Gell-Mann, M., and Low, F., *Phys. Rev.* **95**, 1300 (1954). See also Ref. 3(a).

G. RIDEAU

COVARIANT QUANTIZATIONS OF THE MAXWELL FIELD

Cet article ne saurait commencer sans que j'évoque la mémoire de Bernard Jouvet, à qui il est dédié. Parmi ceux qui ont collaboré à ce volume, je me trouve sans doute être un des plus anciens à l'avoir connu. Nos premières rencontres se firent au cours des années cinquante dans la fréquentation commune du séminaire Proca, dont le rôle fut si déterminant pour le renouveau de la physique théorique française. Nous y découvrions, non sans échanger parfois nos perplexités,les travaux alors nouveaux de Feynman, de Schwinger, de Tomonaga et de bien d'autres. Nous y primes l'un et l'autre ce qui correspondait à nos goûts respectifs, plus pragmatiques, plus instrumentaux sans doute chez Bernard Jouvet que chez moi, plus attiré par l'élucidation du jeu mathématique. Ce fut l'objet de discussions nombreuses et passionnées, tout au long de ces presque trente années qui nous procurèrent souvent l'occasion de confronter nos options et nos résultats dans des conversations dont les détours ultimes nous entrainaient invariablement vers des domaines vastes et variés, bien éloignés en tout cas de l'initiale motivation scientifique.

Bernard Jouvet n'aura rien connu du présent travail, complété, pour sa plus grande part, depuis sa disparition. Il n'aurait sans doute pas manqué de le disséquer, d'en critiquer les aspects qui lui auraient paru incompatibles avec le sens profound qu'il avait de ce qui était physique ou non physique. Nous aurions noué à nouveau le fil de nos amicales querelles, dans la fougue et l'élan qui etaient siens. C'est avec beaucoup d'émotion et de tristesse que j'évoque ici son absence et que je salue une dernière fois celui qui fut, pour moi, un chercheur exemplaire et un ami.

ABSTRACT. The basic axioms of a covariant quantization of the Maxwell field are reviewed, and the general form of the resulting two-points function for the vector potential field is given. Cohomological considerations are decisive for this derivation and allow a significant weakness of the axioms. An application is made in the particular case of linear conformal invariant gauge. Explicit constructions are obtained. The main tool is the theory of extensions of mass-zero representations of the Poincaré group. Under a supplementary condition concerning the indecomposability of the mono-particle states, it is shown that the linear conformal invariant gauges are identical with the so-called generalized Lorentz gauges. If this condition is removed, one is faced with a field theory where the spectral condition is no longer valid except for the physical states.

E. Tirapegui (Ed.), Field Theory, Quantization and Statistical Physics, 201–226.

Introduction

It is well known that there arise special problems in the quantization of the Maxwell field. Two main ways are open in order to get a consistent answer. In the first one, the redundant variables are systematically eliminated and the remaining transverse components are quantized. Then, we are concerned with physically meaningful quantities only, but the vector potential field does not transform like a true vector under the Lorentz group. This becomes cumbersome when this quantization is used in quantum electrodynamics, for then the renormalization programme exhibits ambiguities which can be overcome only if we have learnt from covariant renormalization which diagramms must be put together.

Therefore, a second way of quantization has been developed where the vector potential field is actually a four-vector in Minkowski space. This covariant quantization has been initiated in the 1950s by Gupta and Bleuler and has been applied with complete success to quantum electrodynamics. But, the necessary use of redundant variables has three consequences: the Maxwell equations are no longer valid as field equations, there exists a subsidiary condition in order to eliminate the physically meaningless state vectors and, finally, the space of state vectors is provided with an invariant Hermitian form which is not a true scalar product.

Obviously, the usual axiomatic formulation of the quantum field theory in term of vacuum expectation functions (Wightman functions) must be accordingly modified. This has been done in [1] where there is provided a general framework well adjusted to the situation above. As a consequence, many other quantization processes as the Gupta–Bleuler one are a priori possible and can be consistently described. In the present paper, we shall study mainly in this framework the particular case of covariant quantization. Then it will appear that some basic hypotheses can be significantly weakened in a way which agrees at the best with the minimal physical needs. We shall show it in our first part which brings together some personal recent works yet unpublished or published only in short form. In the second part, we shall be concerned with the important case of quantization in a linear conformal invariant gauge and we shall be able to give a complete description of the quasi-free field situation. As we shall see, group theoretical considerations concerning the extension of mass zero representations of Poincaré group will play a crucial and illuminating role. Finally, in the appendix, we give the proof of the chomological theorem which is basic for the statements of the first part, but which has a too technical character for being incorporated easily in the main text.

PART I: THE TWO-POINTS FUNCTION OF THE VECTOR POTENTIAL FIELD

I.1. THE BASIC AXIOMS (SEE [1])

Let $\mathscr{H}$ be some Hilbert space provided with a non-degenerate sesquilinear Hermitian form $\langle\, , \rangle$, non-negative on a subspace $\mathscr{H}'$, the so-called subspace of the physical states.

Let $U(a,\Lambda)$ be a representation of the Poincaré group $\mathscr{P}$ leaving invariant the form $\langle\, , \rangle$ (and the subspace $\mathscr{H}'$).

Let the vector potential field $A_\mu(x)$ be a vector distribution on $\mathscr{S}(\mathbb{R}^4)$ with values in the set of unbounded operators of $\mathscr{H}$ which are symmetric with respect to $\langle\, , \rangle$.

The electromagnetic field $F_{\mu\nu}(x)$ is given by:

$$F_{\mu\nu}(x) = \partial_\mu A_\nu(x) - \partial_\nu A_\mu(x)$$

and we require that $\mathscr{H}'$ is globally invariant under the $F_{\mu\nu}$'s.

We shall say we get a covariant quantization of the Maxwell field if the following conditions are fulfilled:

I.1.1. *Existence and Unicity of the Vacuum State*

There exists $\psi_0 \in \mathscr{H}'$ such that:

$$U(a,\Lambda)\psi_0 = \psi_0, \qquad (a,\Lambda) \in \mathscr{P},$$

and ψ_0 is the unique vector in $\mathscr{H}$ invariant under the translation subgroup of $\mathscr{P}$.

I.1.2. *Covariance*

The representation $U(a,\Lambda)$ and the field $A_\mu(x)$ are related by:

$$U(a,\Lambda)A_\mu(x)U(a,\Lambda)^{-1} = \Lambda^\nu_\mu A_\nu(\Lambda x + a)$$

I.1.3. *Spectral Condition*

For $\phi \in \mathscr{H}'$, the distribution:

$$\langle \phi, F_{\mu\nu}(x)\psi_0 \rangle$$

has Fourier transform with support in $\bar{V}_+$, the closed future cone.

I.1.4. *Maxwell Equation*

The remaining Maxwell equation is satisfied in mean on the physical subspace $\mathcal{H}'$. In other words, for $\phi,\psi \in \mathcal{H}'$:

$$\langle \phi, \partial^{\mu} F_{\mu\nu}(x)\psi \rangle = 0.$$

I.1.5. *CPT Invariance*

There exists an antilinear operator Θ, antiunitary with respect to $\langle\ ,\ \rangle$, leaving invariant ψ_0 and such that, for $\phi,\psi \in \mathcal{H}'$:

$$\langle \phi, \Theta \mathscr{P}(F_{\mu\nu}(x))\Theta^{-1}\psi \rangle = \langle \phi, \mathscr{P}(F_{\mu\nu}(-x))\psi \rangle$$

where $\mathscr{P}(F_{\mu\nu}(x))$ denotes a polynomial in the $F_{\mu\nu}$'s with real coefficients.

Remark. For the sake of rigor, it would be necessary to assume there exists a dense domain in $\mathcal{H}$ on which would be defined all the unbounded smeared out operators $A(f)$ and also the $U(a,\Lambda)$ which are not necessarily bounded operators on $\mathcal{H}$ (see Ref. 2, for the Gupta–Bleuler case). It would be not difficult to introduce the corresponding restrictions in the statements above but we neglect to do it in so far as this does not play a role in the following.

I.2. The two-points function of the electromagnetic field

Let us write:

(I.2.1) $G_{\alpha\beta\lambda\mu}(x-y) = \langle \psi_0, F_{\alpha\beta}(x)F_{\lambda\mu}(y)\psi_0 \rangle.$

Since the electromagnetic field is a covariant antisymmetrical tensor, we have:

(I.2.2) $G_{\alpha\beta\lambda\mu}(x) = -G_{\beta\alpha\lambda\mu}(x), \qquad G_{\alpha\beta\lambda\mu}(x) = -G_{\alpha\beta\mu\lambda}(x),$

and for any Λ in the Lorentz group:

(I.2.3) $\Lambda^{\alpha'}_{\alpha}\Lambda^{\beta'}_{\beta}\Lambda^{\lambda'}_{\lambda}\Lambda^{\mu'}_{\mu}G_{\alpha'\beta'\lambda'\mu'}(x) = G_{\alpha\beta\lambda\mu}(\Lambda x).$

Furthermore, from the global invariance of $\mathcal{H}'$ under the $F_{\mu\nu}$'s and the spectral condition, we deduce that $G_{\alpha\beta\lambda\mu}(x)$ is the boundary value as $\eta \to 0$ of a function $G_{\alpha\beta\lambda\mu}(z)$ holomorphic in $z = x - i\eta$ for $\eta \in V_+$. By analytic continuation, $G_{\alpha\beta\lambda\mu}(z)$ verifies also (I.2.2) and (I.2.3).

Therefore, from ref. 3 and with the same standard trick as in ref. 1, we can write:

$$(\text{I.2.4}) \quad \begin{aligned} G_{\alpha\beta\lambda\mu}(x) = {} & (g_{\alpha\lambda}\partial_\beta\partial_\mu - g_{\beta\lambda}\partial_\alpha\partial_\mu - g_{\alpha\mu}\partial_\beta\partial_\lambda + g_{\beta\mu}\partial_\alpha\partial_\lambda)G_1(x) \\ & + (\partial_\alpha\varepsilon_{\beta\lambda\mu\rho}\partial^\rho - \partial_\beta\varepsilon_{\alpha\lambda\mu\rho}\partial^\rho)G_2(x) \\ & + (\partial_\lambda\varepsilon_{\mu\alpha\beta\rho}\partial^\rho - \partial_\mu\varepsilon_{\lambda\alpha\beta\rho}\partial^\rho)G_3(x) \\ & + (g_{\alpha\lambda}g_{\beta\mu} - g_{\alpha\mu}g_{\beta\lambda})G_4(x) + \varepsilon_{\alpha\beta\lambda\mu}G_5(x) \end{aligned}$$

where the $G_i(x)$, $i = 1,2,3,4,5$ are boundary values of holomorphic functions of z^2 for z in the domain above.

Lemma I. 2.1 *From the CPT invariance, we have:*

$$G_2(x) = G_3(x).$$

Proof. The CPT invariance implies:

$$(\partial_\alpha\varepsilon_{\beta\lambda\mu\rho}\partial^\rho - \partial_\beta\varepsilon_{\alpha\lambda\mu\rho}\partial^\rho - \partial_\lambda\varepsilon_{\mu\alpha\beta\rho}\partial^\rho + \partial_\mu\varepsilon_{\lambda\alpha\beta\rho}\partial^\rho)(G_2(x) - G_3(x)) = 0.$$

This is equivalent to:

$$(\partial_0^2 + \partial_i^2)(G_2(x) - G_3(x)) = 0, \qquad i = 1,2,3.$$

Therefore, using the Lorentz invariance, we can write:

$$(\text{I.2.5}) \quad G_2(x) = G_3(x) + a + \tfrac{1}{2}bx^2.$$

But, the constant a does not contribute to the right-hand side of (I.2.4) and the contribution from $\frac{1}{2}bx^2$ is absorbed in the last term if we add to $G_5(x)$ a suitable constant. Thus we can assume $a = b = 0$ in (I.2.5), hence $G_2(x) = G_3(x)$.

Lemma I. 2.2 Let us express that the field $F_{\mu\nu}(x)$ is a rotational field. Then we get:

$$(\text{I.2.6}) \quad G_{\alpha\beta\lambda\mu}(x) = (g_{\alpha\lambda}\partial_\beta\partial_\mu - g_{\beta\lambda}\partial_\alpha\partial_\mu - g_{\alpha\mu}\partial_\beta\partial_\lambda + g_{\beta\mu}\partial_\alpha\partial_\lambda)G(x)$$

with $G(x) = G_1(x) + \frac{1}{2}bx^2$, where b is some constant.
Proof. Since the field $F_{\mu\nu}(x)$ is a rotational, we have:

$$\partial^\alpha\varepsilon_{\alpha\beta\rho\sigma}G^{\rho\sigma}{}_{\lambda\mu}(x) = 0.$$

Applied to (I.2.4) where $G_2(x) = G_3(x)$, this gives:

$$(g_{\beta\mu}\partial_\lambda - g_{\beta\lambda}\partial_\mu)(\Box G_2(x) - G_5(x)) + \partial^\alpha\varepsilon_{\alpha\beta\lambda\mu}G_4(x) = 0$$

with the general solution (c_4 and c_5 are some constants):

$$G_4(x) = c_4, \qquad \Box G_2(x) - G_5(x) = c_5$$

If we take into account the following identity:

$$(\partial_\alpha \varepsilon_{\beta\lambda\mu\rho}\partial^\rho - \partial_\beta \varepsilon_{\alpha\lambda\mu\rho}\partial^\rho + \partial_\lambda \varepsilon_{\mu\alpha\beta\rho}\partial^\rho - \partial_\mu \varepsilon_{\lambda\alpha\beta\rho}\partial^\rho) G_2(x) + \varepsilon_{\alpha\beta\lambda\mu} \Box G_2(x) = 0,$$

we get:

$$G_{\alpha\beta\lambda\mu}(x) = (g_{\alpha\lambda}\partial_\beta\partial_\mu - g_{\beta\lambda}\partial_\alpha\partial_\mu - g_{\alpha\mu}\partial_\beta\partial_\lambda + g_{\beta\mu}\partial_\alpha\partial_\lambda) G(x) + \varepsilon_{\alpha\beta\lambda\mu} c_5$$

with $G(x) = G_1(x) + \frac{1}{4}c_4 x^2$.

Let us consider now the following distribution:

$$G_{\alpha\lambda\mu}(x - y) = \langle \psi_0, A_\alpha(x) F_{\lambda\mu}(y) \psi_0 \rangle.$$

We can set down:

$$G_{\alpha\lambda\mu}(x) = (g_{\alpha\lambda}\partial_\mu - g_{\alpha\mu}\partial_\lambda) + \tfrac{1}{2}c_5 \varepsilon_{\alpha\rho\lambda\mu} x^\rho + H_{\alpha\lambda\mu}(x)$$

where the distribution $H_{\alpha\lambda\mu}(x)$ verifies:

$$\partial_\alpha H_{\beta\lambda\mu}(x) - \partial_\beta H_{\alpha\lambda\mu}(x) = 0.$$

Therefore, there exists a distribution $H_{\lambda\mu}(x)$ such that:

$$H_{\alpha\lambda\mu}(x) = \partial_\alpha H_{\lambda\mu}(x).$$

Let us express again that $F_{\mu\nu}(x)$ is a rotational. We obtain:

$$3c_5 g_{\mu\alpha} + \partial_\alpha \partial^\lambda \varepsilon_{\lambda\mu\rho\sigma} H^{\rho\sigma}(x) = 0$$

which is equivalent to:

$$\partial^\lambda \varepsilon_{\lambda\mu\rho\sigma} H^{\rho\sigma}(x) + 3c_5 x_\mu = c_\mu$$

where the c_μ are some constants. If we contract this last equation with ∂^μ, we get:

$$12 c_5 = 0.$$

Therefore $c_5 = 0$, and $G_{\alpha\beta\lambda\mu}(x)$ has well the form (I.2.6).

THEOREM I.2.1 *Under the hypothesis given in the previous section, we have:*

$$\text{(I.2.7)} \quad G_{\alpha\beta\lambda\mu}(x) = \gamma(g_{\alpha\lambda}\partial_\beta\partial_\mu - g_{\beta\lambda}\partial_\alpha\partial_\mu - g_{\alpha\mu}\partial_\beta\partial_\lambda + g_{\beta\mu}\partial_\alpha\partial_\lambda)\Delta^+(x)$$

where $\gamma \geqslant 0$ and the distribution $\Delta^+(x)$ is given by:

$$\Delta^+(x) = \int_{k_o = |k|} \frac{d^3k}{|\vec{k}|} \exp(-ikx).$$

Proof. From the validity in mean on $\mathscr{H}'$ of the equation $\partial^\mu F_{\mu\nu} = 0$ it results:

$$\partial^\alpha G_{\alpha\beta\lambda\mu}(x) = 0.$$

With (I.2.6), this gives:

(I.2.8) $\Box G = c.$

But G is an invariant distribution and the limit as $\eta \to 0$ of a holomorphic function of $z = x - i\eta$, $\eta \in V_+$.
The general solution of (I.2.8) which fits these conditions can be written:

$$G(x) = \gamma\Delta^+(x) + \tfrac{1}{8}cx^2.$$

After substitution into (I.2.6), we obtain:

(I.2.9) $$G_{\alpha\beta\lambda\mu}(x) = \gamma(g_{\alpha\lambda}\partial_\beta\partial_\mu - g_{\beta\lambda}\partial_\alpha\partial_\mu - g_{\alpha\mu}\partial_\beta\partial_\lambda + g_{\beta\mu}\partial_\alpha\partial_\lambda)\Delta^+(x) + \tfrac{1}{2}c(g_{\alpha\lambda}g_{\beta\mu} - g_{\beta\lambda}g_{\alpha\mu}).$$

Now, from the positivity of $\langle\ ,\ \rangle$ on $\mathscr{H}'$, we must have for any antisymmetrical tensor f, with components in $\mathscr{S}(\mathbb{R}^4)$:

$$\int d^4x d^4y f^{\alpha\beta}(x) G_{\alpha\beta\lambda\mu}(x-y) f^{\lambda\mu}(y) \geqslant 0.$$

Let us specify f to be a tensor with only its 01 (or 23) component different from zero and equal to a function $f(x) \in \mathscr{S}(\mathbb{R}^4)$ such that $\int f(x) d^4x = \mathbf{1}$. Then, we get from (I.2.9) the following inequality:

(I.2.10) $$-\gamma \int \frac{d^3k}{|\vec{k}|}(k_2^2 + k_3^2)|\hat{f}(|\vec{k}|,\vec{k})|^2 \leqslant c/2 \leqslant \gamma \int \frac{d^3k}{|\vec{k}|}(k_2^2 + k_3^2)|\hat{f}(|\vec{k}|,\vec{k})|^2$$

where $\hat{f}(k)$ is the Fourier transform of $f(x)$. This implies readily $\gamma \geqslant 0$. Let us replace now in (I.2.10) $\hat{f}(k)$ by $\hat{f}(k/\varepsilon)$. The only change in (I.2.10) is to multiply the left- and right-hand sides by ε^4. Therefore, they go to zero with ε, so that necessarily $c = 0$, and (I.2.9) becomes (I.2.7).

Remark. The Maxwell equation $\partial^\mu F_{\mu\nu} = 0$ has been used only in the proof of the last theorem. As a consequence, we conclude that the form (I.2.6) for the two-points function of the electromagnetic field is also valid when the field interacts with charged matter.

I.3. THE FUNCTION $G_{\alpha\lambda\mu}(x)$

We have already introduced $G_{\alpha\lambda\mu}(x)$ and have found for it the following form:

$$(\text{I.3.1})\quad G_{\alpha\lambda\mu}(x) = \gamma(g_{\alpha\lambda}\partial_\mu - g_{\alpha\mu}\partial_\lambda)\Delta^+(x) + \partial_\alpha H_{\lambda\mu}(x)$$

where $H_{\lambda\mu}(x)$ is a tempered distribution such that:

$$(\text{I.3.2})\quad \partial^\lambda \varepsilon_{\lambda\mu\rho\sigma} H^{\rho\sigma} = c_\mu$$

LEMMA I.3. *The constants c_μ in (I.3.2) are necessarily equal to zero.*
Proof. The general solution of (I.3.2) is given by:

$$H_{\lambda\mu} = \tfrac{1}{6}\varepsilon_{\lambda\mu\rho\sigma} c^\rho x^\sigma + \partial_\lambda H_\mu(x) - \partial_\mu H_\lambda(x)$$

where the $H_\lambda(x)$ are tempered distributions. Now, the covariance of the vector-potential field implies:

$$(\text{I.3.3})\quad \partial_\lambda(\Lambda^{\mu'}_\mu H_{\mu'}(\Lambda x) - H_\mu(x)) - \partial_\mu(\Lambda^{\lambda'}_\lambda H_{\lambda'}(\Lambda x) - H_\lambda(x)) - \tfrac{1}{6}\varepsilon_{\lambda\mu\rho\sigma}(\Lambda^\rho_{\rho'} c^{\rho'} - c^\rho)x^\sigma = \gamma_{\lambda\mu}(\Lambda)$$

where the $\gamma(\Lambda)$ are indefinitely differentiable functions on $SL(2,\mathbb{C})$ (see Ref. 1, p. 2208 for the argument). Let us express now that the first two terms in the left-hand side are the component of a rotational. We obtain, after some calculation:

$$\Lambda_{\mu\beta'} c^{\beta'} - c_\mu = 0$$

i.e., $c_\mu = 0$, since the $D(\frac{1}{2},\frac{1}{2})$ representation of $SL(2,\mathbb{C})$ is irreducible.

THEOREM I.3.1. *Under the hypotheses of the first section, we have necessarily*:

$$(\text{I.3.4})\quad G_{\alpha\lambda\mu}(x) = \gamma(g_{\alpha\lambda}\partial_\mu - g_{\alpha\mu}\partial_\lambda)\Delta^+(x).$$

Proof. From (I.3.3), with $c_\mu = 0$, we deduce easily:

$$\gamma_{\lambda\mu}(\Lambda_1\Lambda_2) = \gamma_{\lambda\mu}(\Lambda_2) + (\Lambda_2)^{\lambda'}_\lambda(\Lambda_2)^{\mu'}_\mu \gamma_{\lambda'\mu'}(\Lambda_1)$$

so that $\gamma_{\lambda\mu}(\Lambda^{-1})$ is a 1-cocycle defined on $SL(2,\mathbb{C})$ with respect to the antisymmetrical tensorial product of $D(\frac{1}{2},\frac{1}{2})$ by itself. Since $SL(2,\mathbb{C})$ is semi-simple and simply connected and the representation is finite-dimensional, such a 1-cocycle is trivial. Therefore, we can find an antisymmetrical tensor $\gamma_{\lambda\mu}$ such that:

$$\gamma_{\mu\lambda}(\Lambda) = \gamma_{\lambda\mu} - \Lambda_\lambda^{\lambda'}\Lambda_\mu^{\mu'}\gamma_{\lambda'\mu'}.$$

Let us write:

$$(I.3.5) \quad H'_\lambda(x) = H_\lambda(x) + \tfrac{1}{2}x^\rho\gamma_{\rho\lambda}.$$

There (I.3.3) becomes:

$$\partial_\lambda(\Lambda_\mu^{\mu'}H'_{\mu'}(\Lambda x) - H'_\mu(x)) - \partial_\mu(\Lambda_\lambda^{\lambda'}H'_{\lambda'}(\Lambda x) - H'_\lambda(x)) = 0$$

Therefore, there exists a distribution $H(\Lambda,x)$, which is a C^∞-function on $SL(2,\mathbb{C})$ and such that:

$$(I.3.6) \quad \Lambda_\mu^{\mu'}H'_{\mu'}(\Lambda x) - H'_\mu(x) = \partial_\mu H(\Lambda,x).$$

From this, we deduce easily the following equation:

$$H(\Lambda_2^{-1}\Lambda_1^{-1},x) - H(\Lambda_1^{-1},x) - H(\Lambda_2^{-1},\Lambda_1 x) = \gamma(\Lambda_1,\Lambda_2),$$
$$\Lambda_1,\Lambda_2 \in SL(2,\mathbb{C})$$

where the C^∞-function $\gamma(\Lambda_1,\Lambda_2)$ is a 2-cocycle with respect to the trivial representation of $SL(2,\mathbb{C})$. Therefore it is trivial and we can find a C^∞-function $d(\Lambda)$ such that:

$$\gamma(\Lambda_1,\Lambda_2) = d(\Lambda_1) + d(\Lambda_2) - d(\Lambda_1\Lambda_2)$$

But then $H(\Lambda^{-1},x) + d(\Lambda)$ is finally a 1-cocycle with value in $\mathscr{S}'(\mathbb{R}^4)$ with respect to the following representation $V(\Lambda)$ of $SL(2,\mathbb{C})$:

$$T(x) \to V(\Lambda)T(x) = T(\Lambda^{-1}x).$$

It will be proved in the appendix to the present paper that such a 1-cocycle is indeed trivial, so that there exists a distribution $H(x)$ with the property:

$$H(\Lambda,x) + \mathrm{d}(\Lambda^{-1}) = H(x) - H(\Lambda x)$$

After substitution in (I.3.6) we have:

$$\Lambda_\mu^{\mu'}(H'_{\mu'}(\Lambda x) + \partial_{\mu'}H(\Lambda x)) = H'_\mu(x) + \partial_\mu H(x).$$

But since an invariant vectorial distribution is a gradient, we can find a distribution $T(x)$ such that:

$$H_\lambda(x) = -\tfrac{1}{2}x^\rho\gamma_{\rho\lambda} + \partial_\mu T(x)$$

This implies first that:

$$H_{\lambda\mu}(x) = \gamma_{\lambda\mu}$$

and consequently we have $H_{\alpha\lambda\mu}(x) = 0$ in (I.3.1).

I.4. The two-points function of the vector potential field

Let us set down:

$$G_{\alpha\beta}(x - y) = \langle \psi_0, A_\alpha(x) A_\beta(y) \psi_0 \rangle.$$

We can state the following theorem:

THEOREM I.4.1 *Under the hypotheses of the first section, we have:*

$$(I.4.1) \quad G_{\alpha\beta}(x) = -\gamma g_{\alpha\beta} \Delta^+(x) + \partial_\alpha \partial_\beta H(x)$$

where $H(x)$ is a Lorentz invariant tempered distribution.
Proof. Let us write:

$$(I.4.2) \quad G_{\alpha\beta}(x) = -\gamma g_{\alpha\beta} \Delta^+(x) + h_{\alpha\beta}(x).$$

The distribution $h_{\alpha\beta}(x)$ is such that:

$$\partial_\lambda h_{\alpha\mu}(x) - \partial_\mu h_{\alpha\lambda}(x) = 0$$

In other words, there exists $h_\alpha(x)$ such that:

$$(I.4.3) \quad h_{\alpha\beta}(x) = \partial_\beta h_\alpha(x)$$

From the covariance of the vector-potential field, we get:

$$(I.4.4) \quad \Lambda_\alpha^{\alpha'} h_{\alpha'}(\Lambda x) - h_\alpha(x) = c_\alpha(\Lambda^{-1})$$

where the $c_\alpha(\Lambda)$ are C^∞-function on $SL(2, \mathbb{C})$. It results from (I.4.4) that they are the components of an indefinitely differentiable 1-cocycle with respect to the $D(\frac{1}{2}, \frac{1}{2})$ representation: therefore it is trivial and we can find constants c_α such that:

$$c_\alpha(\Lambda^{-1}) = \Lambda_\alpha^{\alpha'} c_{\alpha'} - c_\alpha$$

After substitution in (I.4.2), we find that $h_\alpha(x) - c_\alpha$ is a vectorial invariant distribution; therefore it is a gradient and there exists a Lorentz invariant distribution $H(x)$ such that:

$$h_\alpha(x) = c_\alpha + \partial_\alpha H(x).$$

But, from (I.4.3) and (I.4.2), this implies (I.4.1).

Remark. In the proof of Theorem I.3.1, we have used only the covariance of the vector-potential field and a general result on the triviality of some 1-cocycle. In other words, if we started with the form (I.2.6) for the two-points function of the electromagnetic field, we should be able to prove with the same methods that:

$$G_{\alpha\lambda\mu}(x) = (g_{\alpha\lambda}\partial_\mu - g_{\alpha\mu}\partial_\lambda)G(x).$$

Then, proceeding as above, we get readily:

$$G_{\alpha\beta}(x) = -g_{\alpha\beta}G(x) + \partial_\alpha\partial_\beta H(x).$$

This appears to be the general expression of the two-points function for the vector potential. Here $G(x)$ is the boundary value of a holomorphic function of z^2 when $z = x - i\eta, \eta \in V_+$, such that the associated kernel $G_{\alpha\beta\lambda\mu}(x-y)$ is the kernel of a non-negative sesquilinear Hermitian form.

PART II: THE LINEAR CONFORMAL INVARIANT GAUGES

II. 1. GENERALITIES

In all this part, we add to the hypotheses of Section I.1, the following ones:

II.1.1. *Gauge Condition*

We select among the various covariant quantizations described above those which verify the following equation:

(II.1.1) $\Box\partial^\mu A_\mu(x) = 0.$

This equation characterizes the linear conformal invariant gauges (cf. Ref. 4) and it appears particularly suitable for quantization since it allows to speak unambiguously of the positive or negative frequency parts of $\partial^\mu A_\mu(x)$ (see Ref. 5)

II.1.2. *Quasi-free Field Condition*

We assume now that the n-points truncated functions of the quantum field $A_\mu(x)$ are identically zero for $n > 2$.

Moreover, we require that the finite linear complex combinations of the vectors:

$$A(f_1)\dots A(f_n)\psi_0, \qquad n=0,1,2,\dots$$

from a total set in $\mathscr{H}$.

From the first condition, we deduce readily the following equation:

(II.1.2) $\partial_\mu \square^2 H(x)=0.$

Its general Lorentz invariant solution is given by:

(II.1.3) $$H(x)=d_+G^+(x)+d_-G^-(x)+c_1+c_2x^2+c_3(x^2)^2 + M_+\Delta^+(x)+M_-\Delta^-(x)$$

where $G^\pm(x)$ are those particular solutions of the equations:

$$\square G^\pm(x)=\Delta^\pm(x)$$

which are defined by:

$$G^\pm(x)=\int d^4k e^{+ikx}\theta(k_0)\delta'(k^2).$$

If we remark that:

$$k_\beta\theta(k_0)\delta'(k^2)=\frac{1}{2}\frac{\partial}{\partial k^\beta}\theta(k_0)\delta(k^2)$$

and if we take into account the following identity for any $F(k)\in\mathscr{S}(\mathbb{R}^4)$:

$$\frac{\partial F}{\partial k^i}(\pm|\vec{k}|,\pm\vec{k})=\frac{k_i}{|\vec{k}|}\frac{\partial F}{\partial k_0}(\pm|\vec{k}|,\pm\vec{k})\pm\frac{\partial}{\partial k^i}F(\pm|\vec{k}|,\pm\vec{k})$$

we get after a suitable integration by parts:

(II.1.4) $$\int d^4x d^4y f^\alpha(x) f'^\beta(y)\partial_\alpha\partial_\beta G_\pm(x-y) = \frac{1}{2}\int\frac{d^3k}{|\vec{k}|}\left[\frac{\partial}{\partial k_0}\frac{\Omega(\pm k)\overline{\Omega'(\pm k)}}{k_0}\right]_{k_0=|\vec{k}|}$$

with $\Omega(k)=k^\alpha\hat{f}_\alpha(k), \Omega'(k)=k^\alpha\hat{f}'_\alpha(k)$ where $\hat{f}_\alpha(k)$ and $\hat{f}'_\alpha(k)$ are the Fourier transforms of the test-functions in $\mathscr{S}(\mathbb{R}^4), f_\alpha(x)$ and $f'_\alpha(x)$. Obviously, the hermicity of $\langle\ ,\ \rangle$ implies that the constants in (II.1.3) are real constants.

II.2. Indecomposable representations of the Poincaré group and explicit constructions

In this section, we introduce some non-unitary representations of the Poincaré group $\mathscr{P}$ which belong to the rather large family of indecomposable representations. As we shall see, they play a basic role in the realization of a quantum theoretical model which fits the two-points function corresponding to the value (II.1.3) of $H(x)$. We delay up to the next section the detailed investigation of their mathematical properties. At the present time, we give only the basic definitions and the ensuing applications.

II.2.1. Let H_+ be the Hilbert space of four-components functions $\varphi_\mu(\vec{k})$, $\mu = 0,1,2,3$, such that:

$$\int \frac{d^3k}{|\vec{k}|}\left(\sum_0^3 |\varphi_\mu(\vec{k})|^2 + |\omega(\vec{k})|^2 + \sum_1^3 \left|\frac{\partial\omega}{\partial k_l}(\vec{k})\right|^2\right) < \infty \tag{II.2.1}$$

where $\omega(\vec{k}) = k^\mu \varphi_\mu(k), k^0 = |\vec{k}|$.

For real d_+, we define in H_+ the following indecomposable representation, $U_{d_+}(a,\Lambda)$, of the group $\mathscr{P}$:

$$U_{d_+}(a,\Lambda)\varphi_\mu(\vec{k}) = e^{ia\cdot k}(\Lambda_\mu^\nu \varphi_\nu(\Lambda^{-1}k) - \tfrac{1}{2}d_+ \frac{k_\mu}{|\vec{k}|}\left(\Lambda_l^0 \frac{\partial\omega}{\partial k_l}(\vec{\Lambda}^{-1}k) + ia_0\omega(\vec{\Lambda}^{-1}k))\right). \tag{II.2.2}$$

It can be shown that the representations indexed by different value of d_+ are inequivalent representations and there exists for each d_+ a one-parameter family of invariant non-degenerate Hermitian forms given by:

$$B_{d_+}^{M_+}(\varphi,\varphi') = -\int \frac{d^3k}{|\vec{k}|}(\varphi_\mu(\vec{k})\overline{\varphi'^\mu(\vec{k})} + M_+\omega(\vec{k})\overline{\omega'(\vec{k})}) - \tfrac{1}{2}d_+ \int \frac{d^3k}{|\vec{k}|}\frac{\omega(\vec{k})\overline{\omega'(\vec{k})}}{|\vec{k}|^2} \tag{II.2.3}$$

where M_+ is a real constant.

The proofs are given in Ref. 6 for functions $\varphi_\mu(\vec{k})$ equal to zero in a neighbourhood of the origin, but they can be extended easily to H_+ by a density argument.

II.2.2 Let H_- be the Hilbert space of two-components functions $\omega_1(\vec{k})$, $\omega_2(\vec{k})$, such that:

$$\int \frac{d^2k}{|\vec{k}|}\left(\frac{|\vec{k}|^2}{1+|\vec{k}|^2}|\omega_1(\vec{k})|^2 + \frac{1+|\vec{k}|^2}{|\vec{k}|^2}|\omega_2(\vec{k})|^2 + \sum_{i=1}^3 \left|\frac{\partial\omega_2}{\partial k_i}(\vec{k})\right|^2\right) < \infty. \tag{II.2.4}$$

We define in H_- the following indecomposable representation $U_-(a,\Lambda)$

of the group $\mathscr{P}$:

$$U_-(a,\Lambda)\omega_1(\vec{k}) = \mathrm{e}^{-iak}(\omega_1(\overrightarrow{\Lambda^{-1}k})$$
$$+\frac{1}{|\vec{k}|}\left(\Lambda^0_p\frac{\partial\omega_2}{\partial k_l}(\overrightarrow{\Lambda^{-1}k}) - ia_0\omega_2(\overrightarrow{\Lambda^{-1}k}))\right)$$

$$\text{(II.2.5)} \quad U_-(a,\Lambda)\omega_i(\vec{k}) = \mathrm{e}^{-iak}\omega_p(\overrightarrow{\Lambda^{-1}k}).$$

It can be shown that there exists a one-parameter family of invariant non-degenerate Hermitian forms given by:

$$\text{(II.2.6)} \quad B^{m-}_-(\omega,\omega') = \int\frac{\mathrm{d}^3k}{|\vec{k}|}\left(\omega_1(\vec{k})\overline{\omega_2'(F)} + \omega_2(\vec{k})\overline{\omega_1'(\vec{k})} = \frac{\omega_2(\vec{k})\overline{\omega_2'(\vec{k})}}{|\vec{k}|^2}\right)$$
$$- m_-\int\frac{\mathrm{d}^3k}{|\vec{k}|}\omega_2(\vec{k})\omega_2'(\vec{k})$$

where m_- is a real constant.

II.2.3 Let us denote by $V(a,\Lambda)$ the representation of $\mathscr{P}$ defined in the test-function space by:

$$\text{(II.2.7)} \quad V(a,\Lambda)f_\mu(2) = \Lambda^\nu_\mu f_\nu(\Lambda^{-1}(x-1)).$$

Then we can state the following propositions, the proofs of which will be omitted owing to their purely technical character:

PROPOSITION II.2.1. *There exists an intertwining operator Π_{d_+} between $V(a,\Lambda)$ and $U_{d_+}(a,\Lambda)$. If $\hat{f}_\mu(k)$, $\mu = 0,1,2,3$, are the Fourier transforms of the components of the vector test-function $f_\mu(x)$, we have*:

$$\text{(II.2.8)} \quad (\Pi_{d_+}f)_\mu(\vec{k}) = \hat{f}_\mu(|\vec{k}|,\vec{k}) - \tfrac{1}{2}d_+\frac{k_\mu}{|\vec{k}|}\frac{\partial\Omega}{\partial k_0}(|\vec{k}|,\vec{k})$$

where $\Omega(k) = k^\mu\hat{f}_\mu(k), k\in\mathbb{R}.^4$

PROPOSITION II.2.2 *There exists an intertwining operator Π_- between $V(a,\Lambda)$ and $U_-(a,\Lambda)$. With the same notations as in the previous proposition, we have*:

$$\text{(II.2.9)} \quad \begin{aligned} (\Pi_-f)_1(\vec{k}) &= -\frac{1}{|\vec{k}|}\frac{\partial\Omega}{\partial k_0}(-|\vec{k}|,-\vec{k}),\\ (\Pi_-f)_2(\vec{k}) &= \Omega(-|\vec{k}|,-\vec{k}). \end{aligned}$$

Let us consider now the Hilbertian sum $H = H_+ \oplus H_-$ the elements of which will be denoted by $F = (\varphi, \omega)$. In H, is defined a representation of $\mathscr{P}, U(a,\Lambda)$, direct sum of $U_{d_+}(a,\Lambda)$ and $U_-(a,\Lambda)$. The operator $\Pi = \Pi_+ \oplus \Pi_-$ is obviously an intertwining operator between $V(a,\Lambda)$ and $U(a,\Lambda)$. Among the set of invariant non-degenerate Hermitian forms, let us choose the following:

(II.2.10) $\quad B(F,F^1) = B_{d_+}^{M_+}(\varphi,\varphi') + d_- B_-^{m_-}(\omega,\omega')$

where M_+, d_+, d_- are as in (II. 1.3) and $m_- = M_-/d_-$.

Suppose, for the moment $c_1 = c_2 = c_3 = 0$ in (II.1.3). Then it is readily seen that:

(II.2.11) $\quad B(\Pi f, \Pi f') = \int G_{\alpha\beta}(x-y) f^{\alpha}(x) f'^{\beta}(y) \mathrm{d}^4x \mathrm{d}^4y$

for test functions f and f'.

Let now $\mathscr{H}$ be the symmetrical Fock space built on H. Let $\mathscr{U}(a,\Lambda)$ be the representation of $\mathscr{P}$ induced in $\mathscr{H}$ by $U(a,\Lambda)$ and let $\mathscr{B}(\phi,\phi')$ be the invariant non-degenerate Hermitian form induced by $B(F,F')$. We introduce the creation operator $a^+(F)$ in the usual way and we obtain the annihilation operator $a(F)$ by adjunction with respect to $\mathscr{B}(\phi,\phi')$:

(II.2.12) $\quad \mathscr{B}(a(F)\phi,\phi') = \mathscr{B}(\phi, a^+(\bar{F})\phi')$.

We define now the vector potential field by:

(II.2.13) $\quad A(f) = a^+(\Pi f) + a(\overline{\Pi f})$

PROPOSITION II.2.3. *The field defined by* (II.2.13) *is a covariant field.*

Proof. First we have, since Π is an intertwining operator:

$$A(V(a,\Lambda)f) = a^+(U(a,\Lambda)\Pi f) + a(\overline{U(a,\Lambda)\Pi f}).$$

But, by construction:

$$\mathscr{U}(a,\Lambda)a^+(F)\mathscr{U}(a,\Lambda)^{-1} = a^+(U(a,\Lambda)F).$$

Therefore, from (II.2.12):

$$\mathscr{U}(a,\Lambda)a(\bar{F})\mathscr{U}(a,\Lambda)^{-1} = a(\overline{U(a,\Lambda)F}).$$

Consequently:

$$A(V(a,\Lambda)f) = \mathscr{U}(a,\Lambda)A(f)\mathscr{U}(a,\Lambda)^{-1}.$$

This is just the covariance property.

Moreover, if we choose on $\mathscr{H}$ the sesquilinear form $\langle\ ,\ \rangle$ to be identical with $\mathscr{B}(\phi,\phi')$, then it results from (II.2.11) that $A(f)$ has just the wanted two-points function. In other words, at least for $c_1 = c_2 = c_3 = 0$, we have given an explicit construction for all cases a priori possible.

Assume now the constants c_i different from zero. Our description can be easily adapted: it is sufficient to start with a Hilbert space H' which is direct sum of H and of a finite-dimensional Hilbert space. The representation $U(a,\Lambda)$ is changed in a representation $U'(a,\Lambda)$ which is the direct sum of $U(a,\Lambda)$ and of a reducible but indecomposable finite dimensional representation of $\mathscr{P}$. As a matter of fact, there exists always a vector of H' invariant with respect to the action of the translation subgroup of $\mathscr{P}$. But this is inconsistent with the unicity property which we have explicitly required from the vacuum state. Therefore, we have necessarily $c_1 = c_2 = c_3 = 0$ and the construction given above covers finally all cases.

As usual in quantum field theory we introduce the annihilation (creation) operators at the point $\vec{k}$ by the formal writing

$$\text{(II.2.14)}\qquad \begin{aligned} a(F) &= \int \mathrm{d}^3\vec{k}(a^{\mu}(\vec{k})\varphi_{\mu}(\vec{k}) + b^{i}(\vec{k})\omega_i(\vec{k})), \\ a^{+}(F) &= \int \mathrm{d}^3\vec{k}(a^{+\mu}(\vec{k})\varphi_{\mu}(\vec{k}) + b^{+i}(\vec{k})\omega_i(\vec{k})). \end{aligned}$$

Then, we deduce easily the following commutation relations:

$$\text{(II.2.15)}\qquad \begin{aligned} [a^{\mu}(\vec{k}),a^{+\nu}(\vec{l})] &= -\frac{1}{|\vec{k}|}\delta(\vec{k}-\vec{l})\left(g^{\mu\nu} + \tfrac{1}{2}d_{+}\frac{k^{\mu}k^{\nu}}{|\vec{k}|^2} + M_{+}k^{\mu}k^{\nu}\right), \\ [b^{1}(\vec{k}),b^{+2}(\vec{k})] &= \frac{d_{-}}{|\vec{k}|}\delta(\vec{k}-\vec{l}), \\ [b^{2}(\vec{k}),b^{+1}(\vec{k})] &= \frac{d_{-}}{|\vec{k}|}\delta(\vec{k}-\vec{l}), \\ [b^{2}(\vec{k}),b^{+2}(\vec{k})] &= -\frac{1}{|\vec{k}|}\delta(\vec{k}-\vec{l})\left(M_{-} + d_{-}\frac{1}{|\vec{k}|^2}\right), \end{aligned}$$

all other commutators being equal to zero. As to the supplementary conditions which define the subspace $\mathscr{H}'$ of physical states, it can now be written in the following form:

$$\text{(II.2.16)}\qquad \begin{aligned} k'^{\mu}a_{\mu}(\vec{k})\phi &= 0, \\ b^{1}(\vec{k})\phi &= 0. \end{aligned}$$

Finally, we summarize this discussion in the following statement:

THEOREM II.2.1. *The most general two-points function of a vector potential field $A_\mu(x)$ satisfying to the gauge condition* (II.1.1) *and to the hypotheses of Part* I *is given by:*

$$\begin{aligned}\langle\psi_0, A_\mu(x)A_\nu(y)\psi_0\rangle = & \, (-g_{\mu\nu} + M_+\partial_\mu\partial_\nu)\Delta^+(x-y) \\ & + d_+\partial_\mu\partial_\nu G^+(x-y) + M_-\partial_\mu\partial_\nu\Delta^-(x-y) \\ & + d_-\partial_\mu\partial_\nu G^-(x-y),\end{aligned}$$

where the various constants are real and where the distribution $\Delta^\pm(x)$ and $G^\pm(x)$ are given by: .

$$\Delta^\pm(x) = \int d^4k e^{\mp ikx}\theta(k_0)\delta(k^2),$$

$$G^\pm(x) = \int d^4k e^{\mp ikx}\theta(k_0)\delta'(k^2).$$

An explicit construction of $A_\mu(x)$ as a quasi-free field corresponding to the two-points function above can be given for each set of real constants d_+, d_-, M_+, M_-. It is realized in a Fock space $\mathscr{H}$ where the six creation $(a^{+\mu}(\vec{k}), b^{+i}(\vec{k}))$ and the six annihilation operators $(a^\mu(\vec{k}), b^i(\vec{k}))$ verify the relations (II.2.15). The vector potential field itself is defined by:

$$A(f) = a^+(\Pi f) + a(\Pi f).$$

Here, we use the notation (II.2.14) and, for the test function $f_\mu(x)$ with Fourier transform $\hat{f}^\mu(k)$; the various components of Πf are given by:

$$(\Pi f)_\mu(\vec{k}) = \hat{f}_\mu(|\vec{k}|,\vec{k}) - \tfrac{1}{2}d_+\frac{k_\mu}{|\vec{k}|}\frac{\partial\Omega}{\partial k_0}(|\vec{k}|,\vec{k})$$

$$(\Pi f)_1(\vec{k}) = -\frac{1}{|\vec{k}|}\frac{\partial\Omega}{\partial k_0}(-|\vec{k}|,\vec{k})$$

$$(\Pi f)_2(\vec{k}) = \Omega(-|\vec{k}|,-\vec{k})$$

where $\Omega(k) = k^\mu\hat{f}_\mu(k)$.

Finally, the physical subspace $\mathscr{H}'$ is the set of elements in $\mathscr{H}$ which satisfy the supplementary condition (II.2.16). The mean values of the physical quantities have to be evaluated by means of a sesquilinear Poincare invariant hermitian form $\mathscr{B}(\phi,\phi')$ given in the one-particle space by:

$$B(F,F') = \int \frac{d^3k}{|\vec{k}|}\left(- \varphi^{\mu}(\vec{k})\overline{\varphi'_{\mu}(\vec{k})} + M_{+}\,\omega(\vec{k})\overline{\omega'(\vec{k})} - \tfrac{1}{2}d_{+}\frac{\omega(\vec{k})\overline{\omega'(\vec{k})}}{|\vec{k}|^2}\right)$$

$$+ d_{-}\int \frac{d^2k}{|\vec{k}|}\Big(\omega'_1(\vec{k})\overline{\omega_2(\vec{k})} + \omega_2(\vec{k})\overline{\omega'_1(\vec{k})}$$

$$- \frac{\omega_2(\vec{k})\overline{\omega'_2(\vec{k})}}{|\vec{k}|^2} - \frac{M_{-}}{d_{-}}\omega_2(\vec{k})\overline{\omega'_2(\vec{k})}\Big).$$

It remains now to deduce the field equations and to precise the actual meaning of the construction above. This will be done in the last section, after we have afforded a deeper knowledge of the representations $U_{d_+}(a,\Lambda)$ and $U_{-}(a,\Lambda)$.

II. 3. The mathematical structures of $U_{d_+}(a,\Lambda)$ and $U_{-}(a,\Lambda)$

The representations $U_{d_+}(a,\Lambda)$ and $U_{-}(a,\Lambda)$ belong to the set of representations of $\mathscr{P}$ which are extensions of unitary representations of $\mathscr{P}$, in the present case, of mass zero unitary representations.

Let us recall first some definitions. Let G be a group and let U_1, U_2 be continuous representations of G in topological spaces E_1, E_2. Then, in the topological sum $E = E_1 + E_2$ we call extension of U_2 by U_1 a representation U given by:

$$\text{(II.3.1)}\quad U(g) = \begin{vmatrix} U_1(g) & Z(g)U_2(g) \\ 0 & U_2(g) \end{vmatrix}$$

where $Z(g)$ is a continuous mapping from E_2 to E_1, verifying the following equation:

$$\text{(II.3.2)}\quad Z(g_1 g_2) = Z(g_1) + U_1(g_1)Z(g_2)U_2^{-1}(g_1).$$

$Z(g)$ is called a 1-cocycle of extension and it is trivial if there exists $Z \in \mathscr{L}(E_2, E_1)$ such that;

$$Z(g) = Z - U_1(g)ZU_2(g)^{-1}.$$

We are not actually interested by these 1-cocycles since the representation they define is equivalent to the direct sum $U_1 + U_2$. Similarly, when two 1-cocycles $Z_1(g), Z_2(g)$ differ by a trivial 1-cocycle, it is easily shown that they define equivalent representations. Therefore, we are interested with the solutions of (II.3.1) up to trivial ones. But it may appear that two inequivalent 1-cocycles correspond in fact to equivalent representations. In particular, it is the case when the classes of 1-cocycles form a one-dimensional vectorial space.

PROPOSITION II.3.1. *The representation $U_-(a,\Lambda)$ is a non-trivial extension by itself of a representation with mass zero, helicity zero, and negative energy.*

Proof. $U_-(a,\Lambda)$ has clearly the structure (II.3.1). The representations

$$\omega_i(\vec{k}) \to e^{-iak}\omega_i(\overrightarrow{\Lambda^{-1}k}), \quad k_0 = |\vec{k}|, \quad i = 1,2,$$

are well unitary representations of $\mathscr{P}$ with mass zero, helicity zero, and negative energy. As to the non-triviality of the extension, it proceeds from general results proved in Ref. 7.

The structure of $U_{d_+}(a,\Lambda)$ is more involved. For the sake of simplicity, we introduce on C_+ the coordinates $t\in\mathbb{R}, \zeta\in\mathbb{C}$ defined by:

$$\text{(II.3.3)} \quad t = -\tfrac{1}{2}\log 2|\vec{k}|, \quad \zeta = -\frac{k_2 + ik_1}{|\vec{k}| - k_3}.$$

Furthermore, at each point k on the cone C_+, we associate a set of four independent vectors: the vector k itself, the vector with components $\hat{k}_0 = \tfrac{1}{4}|\vec{k}|, \hat{k}_i = -k_i/4|\vec{k}|^2$, and the complex conjugate vectors $X_\mu(k), \bar{X}_\mu(k)$ given by:

$$X_0(k) - X_3(k) = \frac{\bar{\zeta}}{1+|\zeta|^2}, \qquad X_0(k) + X_3(k) = -\frac{\zeta}{1+|\zeta|^2},$$

$$X_2(k) + iX_1(k) = \frac{1}{1+|\zeta|^2}, \qquad X_2(k) - iX_1(k) = -\frac{\bar{\zeta}^2}{1+|\zeta|^2}.$$

Let us set down $(k\in C_+)$:

$$\text{(II.3.4)} \quad \begin{aligned} z(\Lambda,k) &= -e^{2t}\frac{\overline{(\alpha\zeta+\gamma)}(\alpha-\gamma\bar{\zeta}) + \overline{(\beta\zeta+\delta)}(\beta-\delta\bar{\zeta})}{|\alpha J+\gamma|^2 + |\beta\zeta+\delta|^2}, \\ \omega(\Lambda,k) &= \left(\frac{\beta\zeta+\delta}{|\beta\zeta+\delta|}\right)^{-2}, \end{aligned}$$

where the matrix $\begin{vmatrix}\alpha & \beta \\ \gamma & \delta\end{vmatrix} \in SL(2,\mathbb{C})$ is associated to the element Λ of the Lorentz group.

Then, by direct calculations, we can prove the following useful relations:

$$\text{(II.3.5)}\quad \begin{aligned} &\hat{k}_\mu - \Lambda_\mu^\nu(\widehat{\Lambda^{-1}k})_\nu = k_\mu|z(\Lambda,k)|^2 + z(\Lambda,k)X_\mu(k) + z(\Lambda,k)X_\mu(k),\\ &X_\mu(k) - k_\mu z(\Lambda,k) = \omega(\Lambda,k)\Lambda_\mu^\nu X_\nu(\Lambda^{-1}k). \end{aligned}$$

Let us write now:

$$\varphi_\mu(\vec{k}) = b_1(\vec{k})\hat{k}_\mu + b_2(\vec{k})X_\mu(k) + b_3(\vec{k})\overline{X_\mu(k)} + b_4(\vec{k})k_\mu.$$

From (II.2.2) and (II.3.5), we deduce readily the following law of transformation of the $b_i(\vec{k})$, $i = 1,2,3,4$:

$$\text{(II.3.6)}\quad \begin{aligned} &b_4(\vec{k}) \to \mathrm{e}^{iak}[b_4(\overrightarrow{\Lambda^{-1}k}) - \omega(\Lambda,k)\overline{z(\Lambda,k)}b_3(\overrightarrow{\Lambda^{-1}k})\\ &- \overline{\omega(\Lambda,k)}z(\Lambda,k)b_2(\overrightarrow{\Lambda^{-1}k}) + |z(\Lambda,k)|^2 b_1(\overrightarrow{\Lambda^{-1}k})\\ &- \tfrac{1}{4}d_+\frac{1}{|\vec{k}|}\left(\Lambda_l^0\frac{\partial b_1}{\partial k_l}(\Lambda^{-1}k) + ia_0 b_1(\overrightarrow{\Lambda^{-1}k})\right)\Bigg],\\ &b_3(\vec{k}) \to \mathrm{e}^{iak}(\omega(\Lambda,k)b_3(\overrightarrow{\Lambda^{-1}k}) - z(\Lambda,k)b_1(\overrightarrow{\Lambda^{-1}k})),\\ &b_2(\vec{k}) \to \mathrm{e}^{iak}(\overline{\omega(\Lambda,k)}b_2(\overrightarrow{\Lambda^{-1}k}) - \overline{z(\Lambda,k)}b_1(\overrightarrow{\Lambda^{-1}k})),\\ &b_1(\vec{k}) \to \mathrm{e}^{iak}b_1(\Lambda^{-1}k). \end{aligned}$$

On this reduced form the structure of $U_{d_+}(a,\Lambda)$ becomes manifest. Indeed it appears as the extension of four irreducible unitary mass zero representations of $\mathscr{P}$; two representations have their helicity equal to zero, and correspond to the components $b_1(\vec{k})$ and $b_4(\vec{k})$; the two last representations have their helicity equal to ± 1 and correspond to the components $b_3(\vec{k})$ and $b_2(\vec{k})$ respectively. We have studied elsewhere (see Ref. 8) these extensions and we have shown that their equivalence classes form a one-parameter family; as a matter of fact the representation (II.3.6) can be chosen as a representative in each class. It is worth noticing that the indexing parameter arises actually from the existence of a one-parameter family of non trivial 1-cocycles of extension by itself of the representation of $\mathscr{G}$ with mass zero and helicity zero. This gives the mathematical meaning of the constant d_+ in (II.2.2).

II. 4. Field equations

From (I.3.4), we deduce:

$$\langle\psi_0, A_\alpha(x)\Box F_{\lambda\mu}(y)\psi_0\rangle = 0$$

Therefore, by the completeness hypothesis of Section II.1, we must have:

$$\Box F_{\lambda\mu}(x) = 0$$

But this is equivalent to:

$$\Box A_{\mu}(x) = \partial_{\mu} B(x)$$

where $B(x)$ is a scalar field. To determine it, we apply the definition (II.2.13) with a test function of the form $\Box g_{\mu}(x), g_{\mu}(x) \in \mathscr{S}(\mathbb{R}^4)$. Let us set down:

(II.4.1) $$B_{+}(x) = \frac{1}{i} \int_{k_o = |\vec{k}|} d^3k (k^{\mu} a_{\mu}^{+}(\vec{k}) e^{ikx} - k^{\mu} a_{\mu}(\vec{k}) e^{-ikx}),$$

(II.4.2) $$B_{-}(x) = \frac{2}{i} \int_{k_o = |\vec{k}|} d^3k (b_1^{+}(\vec{k}) e^{-ikx} - b_1(\vec{k}) e^{ikx}).$$

Then we get:

(II.4.3) $$\Box A_{\mu}(x) = d_{+} \partial_{\mu} B_{+}(x) - \partial_{\mu} B_{-}(x).$$

Similarly if we apply (II.2.13) with a test function of the form $ik_{\mu}\hat{f}(k)$, $\hat{f} \in \mathscr{S}\mathbb{R}^4)$, we obtain:

(II.4.4) $$\partial^{\mu} A_{\mu}(x) = (d_{+} - 1) B_{+}(x) - B_{-}(x).$$

From this, we deduce the quantum Maxwell field equation in the following form:

(II.4.5) $$\partial^{\mu} F_{\mu\nu}(x) = \partial_{\nu} B_{+}(x)$$

Thus the field $B_{-}(x)$ does not appear in the operatorial equation which must replace the classical Maxwell field equation. In this respect the field $B_{-}(x)$ appears as a spurious field. It can be easily eliminated if we add to our set of postulates the following one:

INDECOMPOSABILITY CONDITION. The representation of $\mathscr{P}$ restricted to the one-particle space must be an indecomposable representation.

With this postulate, the representation $U_{-}(a,\Lambda)$ disappears, we must have $d_{-} = M_{-} = 0$ in (II.1.3) and we find a two-points function characteristic of the generalized Lorentz gauges (see Ref. 6 for a detailed study) Therefore we have proved the following theorem:

THEOREM II.4.1. *Under the hypotheses of Part I and of Section* II.1 *and with the indecomposability condition, the linear conformal covariant gauges are identical with the generalized Lorentz gauges. In this case, the spectral condition is satisfied by the vector potential field itself.*

But in the general case, there is no reason to drop out the pathological field $B_-(x)$ so that the two-points function cannot be the boundary value of some analytical function. Such a strange situation appears obviously as a direct consequence of the non-physical nature of the field $A_\mu(x)$ implied by the gauge invariance of the theory. Nevertheless, it is worth noticing that in any case the vector potential field is a local field.

APPENDIX: COMOHOLOGICAL RESULTS

Let us consider in $\mathscr{S}'(\mathbb{R}^4)$ the representation $V(\Lambda)$ of $SL(2,\mathbb{C})$ defined by:

$$T(x) \to T(\Lambda^{-1}x), \qquad T(x) \in \mathscr{S}'(\mathbb{R}^4), \qquad \Lambda \in SL(2,\mathbb{C}).$$

An indefinitely differentiable 1-cocycle with respect to V is an indefinitely differentiable mapping $H(x,\Lambda)$ from $SL(2,\mathbb{C})$ to $\mathscr{S}'(\mathbb{R}^4)$ such that:

$$\text{(A.1)} \quad H(x,\Lambda_1\Lambda_2) - H(x,\Lambda_1) - H(\Lambda_1^{-1}x,\Lambda_2) = 0, \qquad \Lambda_1,\Lambda_2 \in SL(2,\mathbb{C}).$$

We prove here the following theorem:

THEOREM A.1. *The general solution of* (A.1) is given by

$$\text{(A.2)} \quad H(x,\Lambda) = H(x) - H(\Lambda^{-1}x), \qquad \Lambda \in SL(2,\mathbb{C})$$

where $H(x) \in \mathscr{S}'(\mathbb{R}^4)$

In other words, each 1-cocycle with respect to V is a coboundary.

Proof. We can always assume that (A.2) is true when Λ is restricted to the compact group $SU(2)$. Then, following Ref. 9, for each $X \in sl(2,\mathbb{C})$, the Lie algebra of $SL(2,\mathbb{C})$, we introduce:

$$\text{(A.3)} \quad h(x,X) = \frac{\mathrm{d}}{\mathrm{d}t} H(x,\exp tX)\Big|_{t=0}.$$

Let $\mathrm{d}V$ denote the representation of $sl(2,\mathbb{C})$ derived from V. The linear functional $h(x,X)$ verifies:

$$\text{(A.4)} \quad h(x,[X,Y]) = \mathrm{d}V(X)h(x,Y) - \mathrm{d}V(Y)h(x,X)$$

with the supplementary condition:

(A.5) $h(x,X) = \mathrm{d}V(X)h(x), \qquad X \in su(2),$

for some $h(x) = \mathscr{S}'(\mathbb{R}^4)$.
The theorem will be proved if we can prove that $h(x,X)$ has the form (A.5) for all $X \in sl(2,\mathbb{C})$. It is obviously sufficient to prove it for a basis of $sl(2,\mathbb{C})$. We choose this basis as in Ref. 10. The commutation relations of $sl(2,\mathbb{C})$ and (A.4) give:

(A.6) $h(x,F_{\pm}) = \pm \mathrm{d}V(H_{\pm})h(x,F_3) \pm \mathrm{d}V(F_3)\mathrm{d}V(H_{\pm})h(x).$

Let us set down:

(A.7) $t(x) = h(x,F_3) - \mathrm{d}V(F_3)h(x)$

The distribution $t(x)$ verifies the following set of equations:

$$\text{(A.8)} \quad \begin{aligned} &\mathrm{d}V(H_+)^2 t(x) = 0, \\ &\mathrm{d}V(H_-)^2 t(x) = 0, \\ &\mathrm{d}V(H_3)t(x) = 0, \\ &(\mathrm{d}V(H_+)\mathrm{d}V(H_-) - 2)t(x) = 0, \end{aligned}$$

and:

$$\text{(A.9)} \quad \begin{aligned} &\mathrm{d}V(F_+) + \mathrm{d}V(F_3)\mathrm{d}V(H_+)t(x) = 0, \\ &\mathrm{d}V(F_-) - \mathrm{d}V(F_3)\mathrm{d}V(H_-)t(x) = 0. \end{aligned}$$

The first set means that:

$$\text{(A.10)} \quad t_0^1(x) = t(x), \qquad t_{\pm 1}^1(x) = \frac{1}{\sqrt{2}}\mathrm{d}V(H_{\pm})t(x),$$

form a canonical basis of the three dimensional representation of $SU(2)$ Therefore for $\varphi \in \mathscr{S}(\mathbb{R}^4)$ and $\varphi_R(x) = \varphi(R_x^{-1}), R \in SO(3)$ we have:

$$\begin{aligned} (t_0^1,\varphi) &= D_{0,1}^1(R)(t_1^1,\varphi_R) + D_{00}^1(R)(t_0^1,\varphi_R) + D_{0,-1}^1(R)(t_{-1}^1,\varphi_R), \\ (t_{\pm 1}^1,\varphi) &= D_{\pm 1,1}^1(R)(t_1^1,\varphi_R) + D_{\pm 1,0}^1(R)(t_0^1,\varphi_R) \\ &\quad + D_{\pm 1,-1}^1(R)(t_{-1}^1,\varphi_R). \end{aligned}$$

After integration on $SO(3)$, we get:

(A.11) $(t_0^1,\varphi) = (S,\varphi_0^1), \qquad (t_{\pm 1}^1,\varphi) = (S,\varphi_{\pm 1}^1)$

where $S(x_0,r)$ is a tempered distribution and $\varphi_0^1(x_0,r), \varphi_{\pm 1}^1(x_0,r)$ are fastly decreasing indefinitely differentiable functions, uneven with respect to r, equal to the components of φ on the usual spherical harmonics Y_0^1 and $Y_{\pm 1}^1$.

Now writing the set of equation (A.9) in terms of distributions (A.10), we get;

$$(t_0^1, \mathrm{d}V(F_\pm)\varphi) \pm \sqrt{2}(t_{\pm 1}^1, \mathrm{d}V(F_3)\varphi) = 0$$

With (A.11) we have, after some lengthy calculations:

$$\text{(A.12)} \quad \left(S, \left(\frac{x_0}{r} + \frac{1}{3}\left(x_0\frac{\partial}{\partial r} + r\frac{\partial}{\partial x_0}\right)\right)(\varphi_1^2 + \varphi_{-1}^2)\right) = 0$$

where $\varphi_{\pm 1}^2(\lambda_0, r)$ are the components of φ on the spherical harmonics $Y_{\pm 1}^2$. Therefore, we can set down:

$$\varphi_1^2(x_0, r) + \varphi_{-1}^2(x_0, r) = r\psi(x_0, r)$$

where $\psi(x_0, r)$ is some quickly decreasing function uneven with respect to r. Then, if we discard quantities which do not contribute to (A.11), we derive from (A.12) the following equation:

$$\text{(A.13)} \quad \left(3x_0 - r\left(x_0\frac{\partial}{\partial r} + r\frac{\partial}{\partial x_0}\right)\right)S = 0$$

To solve it, we write:

$$\text{(A.14)} \quad S = r^3 U$$

and we deduce from (A.13) the following equation:

$$\text{(A.15)} \quad \left(x_0\frac{\partial}{\partial r} + r\frac{\partial}{\partial x_0}\right)U = t_0(x_0)\delta(r) + t_1(x_0)\delta'(r) + t_2(x_0)\delta''(r)$$

where $t_0(x_0), t_1(x_0), t_2(x_0)$ are some tempered distributions. But we are interested only by the solutions of (A.14) which are even distributions with respect to r. Therefore we can assume $t_0(x_0) = t_2(x_0) = 0$ in (A.15) and the general solution of the resulting equation can be written:

$$\text{(A.16)} \quad U = u(x_0)\delta(r) + U_0$$

where U_0 verifies the following equation:

$$\text{(A.17)} \quad \left(x_0\frac{\partial}{\partial r} + r\frac{\partial}{\partial x_0}\right)U_0 = 0$$

Now the first term of the right-hand side of (A.16) does not contribute to S, so that finally we have:

$$S = r^3 U_0$$

where U_0 is a solution of (A.17) which is an even distribution with respect to r.

We define now a distribution $Z \in \mathscr{S}'(\mathbb{R}^4)$ by:

$$\text{(A.18)} \quad (Z,\varphi) = (r^2 U_0, \varphi_0^0), \qquad \varphi \in \mathscr{S}(\mathbb{R}^4)$$

where φ_0^0 is the component of φ on Y_0^0.

Suppose now that Z is a Lorentz invariant distribution.

Then our theorem is proved: indeed, since $(x_3\,\varphi)_0^0 = r\varphi_0^1$, (A.11) takes the form:

$$(t,\varphi) = (x_3 Z, \varphi)$$

i.e.

$$t(x) = x_3 Z$$

But, from $\mathrm{d}V(F_3)Z = 0$ and $[\mathrm{d}V(F_3), x_0] = x_3$ we get easily:

$$t(x) = \mathrm{d}V(F_3)x_0 Z$$

where the distribution $x_0 Z$ is invariant by rotation. Then, we get from (A.7):

$$h(x,F_3) = \mathrm{d}V(F_3)(h(x) + x_0 Z).$$

Similarly, (A.6) and the commutation relations of $sl(2,\mathbb{C})$ imply:

$$h(x,F_\pm) = \mathrm{d}V(F_\pm)(h(x), + x_0 Z).$$

Finally, the rotational invariance of $x_0 Z$ allows (A.5) to be written in the form:

$$h(x,X) = \mathrm{d}V(X)(h(x) + x_0 Z), \qquad X \in su(2)$$

Therefore it remains to prove that Z is a Lorentz invariant distribution. Z is obviously invariant by rotation. Thus we have only to prove that $\mathrm{d}V(F_3)Z = 0$. But, we have:

$$(-\,\mathrm{d}V(F_3)\varphi)_0^0 = i\sqrt{\frac{2}{3}}\left(x_0\frac{\partial}{\partial r} + r\frac{\partial}{\partial x_0} + 2\frac{x_0}{r}\right)\varphi_0^1(x_0,r)$$

so that:

$$(\mathrm{d}V(F_3)Z,\varphi)=(Z,-\mathrm{d}V(F_3)\varphi)=(r^2U_0,(-\mathrm{d}V(F_3)\varphi)_0^0)$$
$$=i\sqrt{\frac{2}{3}}\left(r^2U_0,\left(x_0\frac{\partial}{\partial r}+r\frac{\partial}{\partial x_0}+2\frac{x_0}{r}\right)\varphi_0^1\right)$$
$$=i\sqrt{\frac{2}{3}}\left(\left(2\frac{x_0}{r}-\left(x_0\frac{\partial}{\partial r}+r\frac{\partial}{\partial x_0}\right)\right)r^2U_0,\varphi_0^1\right)=0$$

since, by (A.17), the first term in the bracket is equal to zero.

Université Paris VII

References

[1] Strocchi, F., and Wightman, A. S., *J. Math. Phys.* **15**, 2198 (1974).
[2] Rideau, G., in H. Bacry and A. Grossmann (eds.), *Proc. 3rd Int. Coll. on Group Theoretical Methods in Physics*, Marseille, 1974, pp. 210–216.
[3] Hepp, K., *Helv. Phys. Acta* **36**, 355 (1963).
[4] Bayen, F., and Flato, M., *J. Math. Phys.* **17**, 112, (1976).
[5] Tomczak, S. P., and Haller, K., *Nuovo Cimento* **8B**, 1 (1972).
[6] Rideau, G., *J. Math. Phys.* **19**, 1627 (1978).
[7] Rideau, G., 'Extension of representations of Poincaré group', to appear in *Rep. Math. Phys.*
[8] Rideau, G., *Letters Math. Phys.* **2**, 529 (1978).
[9] Pinczon, G., and Simon, J., *Letters Math. Phys.* **1**, 83 (1975).
[10] Naimark, M. A., *Representations linéaires du groupe de Lorentz*, Dunod, Paris.

F. STROCCHI

GAUSS' LAW IN LOCAL QUANTUM FIELD THEORY

ABSTRACT. Symmetry groups generated by charges which obey Gauss' law are discussed as characteristic features of local gauge quantum field theories. The existence of such symmetry groups in a local quantum field theory leads to very important properties. In particular, unbroken charges obeying Gauss' law always define superselection rules and their breaking always leads to the Higgs phenomenon, which has therefore a quite general explanation, related to Gauss' law, without the introduction of Higgs fields.

1. INTRODUCTION

In the last few years there has been increasing support both experimentally and theoretically in favor of gauge quantum field theories (GQFT) as the only QFT relevant to elementary particle physics[1]. The main reasons for such evidence has come from the very peculiar structure properties (asymptotic freedom and good ultraviolet behavior, spontaneous symmetry breaking without massless particles associated to it, unification of various types of interactions, confinement mechanism of elementary constituents, etc.) that GQFT have in contrast with standard (non-gauge) QFT. Since these general features are largely independent from the specific models one is considering, it is natural to ask whether one can understand the reasons of the success of such theories in terms of general physical ideas.

One such explanation was offered in the original paper by Yang and Mills [2] based on the requirement that internal quantum numbers or charges have only a local significance and that their relative identification at different space-time points is physically meaningless. Such a strong localization of internal symmetries is certainly in the spirit of Einstein's principle of general relativity but it is not supported by an analogous equivalence principle, equally forced by physical arguments. To further motivate their requirement, Yang and Mills pointed out that the local

---

Written for an invited talk at the International Symposium on Elementary Particle Physics, organized by JINR (Dubna), Reinhardsbrunn, October 1978.

E. Tirapegui (Ed.), Field Theory, Quantization and Statistical Physics, 227–236.

structure of internal symmetries at the basis of their theory is equivalent to the requirement that the theory be invariant under a group of gauge transformations, which appear as the natural generalization of the familiar quantum electrodynamical case. The gauge invariance requirement actually appears to have become the most popular motivation for gauge theories.

As mentioned at the beginning and as will be discussed later, GQFT represent a strong departure from standard QFT and therefore it is reasonable to ask whether the crucial physical features of gauge theories can be traced back to a more physical property than the invariance under unobservable (gauge) transformations [3]. It is the aim of this talk to emphasize and fully exploit the fact that the gauge invariance requirement is strongly linked to the physical property that the associated internal charges obey *Gauss' law* [4].

$$Q^{(i)} = \int d^3x j_0^{(i)}(x) = \int d^3x \operatorname{div} \boldsymbol{E}^{(i)}(x) \tag{1}$$

(i = internal index). As we will show, most of the general features of GQFT can be traced back to this crucial physical law with no need of mentioning unphysical fields like the (vector) potentials and their gauge transformation properties.

The main output of this analysis is therefore the suggestion [5] that Gauss' law is the basic and primary feature which characterizes elementary particle interactions, rather than gauge invariance, a concept which is more difficult to grasp on physical grounds since it can be given a meaning *only* by introducing unobservable quantities. Gauge invariance can therefore be regarded as a technical tool for constructing Lagrangian functions or equations of motion which guarantee the validity of Gauss' law.

This may be the right track to get an insight into the structure of GQFT and possibly understand why nature seems to choose gauge theories for elementary particle interactions.

As we will see, the general features of GQFT which make them so appealing for elementary particle interactions (Higgs phenomenon, confinement mechanism, superselection rules, etc.) can all be traced back to Gauss' law independently of the specific model and/or the equations of motion.

2. Gauge invariance and Gauss' law

From a technical point of view a Lagrangian $\mathscr{L}$, invariant under an internal symmetry group G, allows a gauge extension of G if it is also invariant under the group $\mathscr{G}$ of gauge transformations of the second kind associated to G. As it stands, this invariance property does not have a direct physical interpretation since gauge transformations act non-trivially only on unobservable quantities. Then one may ask: given an internal symmetry group G, which physical properties should its charges exhibit in order to allow a description in terms of a Lagrangian invariant under the gauge extended group $\mathscr{G}$?

To this purpose we remark that gauge invariance involves an infinite-dimensional group of transformations since the group parameters are functions of space-time. Now, it is well known that in going from a discrete to a continuous group one gains not only a conservation law but also a continuity equation

$$Q = \int d^3x j_0(x) = \text{const}, \qquad \partial^\mu j_\mu = 0, \tag{2}$$

i.e. the conservation law has a local character [6]. It is not difficult to show [7] that when the internal symmetry has a gauge extension then its charge not only gives rise to a continuity equation but it also obeys Gauss' law *in the sense that* the current j_μ is the divergence of an antisymmetric tensor $G_{\mu\nu}$

$$j_\mu^{(i)} = \partial^\nu G_{\nu\mu}^{(i)}, \tag{3}$$

i.e., the current $j_\mu^{(i)}$ is conserved independently of the equations of motion of the fields out of which $j_\mu^{(i)}$ is constructed (*super conservation law*).

In the Abelian case Equations (3) are just Maxwell equations and Equation (1) becomes the familiar Gauss' law. The physical consequences of the validity of Gauss' law will be discussed in the next sections.

3. Gauss' Law and Locality

The introduction of internal (continuous) symmetry groups in local QFT not only has a long history but also it represents a fairly well understood subject even at the rigorous level, especially after the complete clarification of the phenomenon of spontaneous symmetry breaking [9]. *When*, however, *the internal symmetry group has a gauge extension, more precisely its generators obey a Gauss law, the corresponding QFT cannot satisfy locality and positivity at the same time* [10].

To qualify better the above statement we recall that a local QFT is a theory completely formulated in terms of local fields and the property of positivity means that vectors constructed by applying polynomials of local fields to the vacuum have positive 'norm' [11]. One can show [10] that if Gauss law, or equivalently Equation (3), holds as an operator equation in a local QFT [11] then $Q = 0$. To incorporate Gauss' law in a *local* QFT in a non trivial way one has then to allow [10] unphysical vectors (related to longitudinal unobservable modes) and to require the validity of Gauss' law only on the physical states. This is essentially what happens in the familiar Gupta–Bleuler formulation of QED (at least at the level of perturbation theory) and it underlies all the calculations done in QED in the Feynman gauge.

It may be useful at this point to recall the structure which emerges in this case [12, 13]. The whole Hilbert space $\mathscr{H}$ contains a sub-space $\mathscr{H}'$ of physical vector essentially characterized by the validity of Gauss' law:

$$\langle \Psi, (j_\mu^{(i)} - \partial^\nu G_{\nu\mu}^{(i)})\Phi \rangle = 0 \tag{4}$$

for any $\Psi, \Phi \in \mathscr{H}'$, and by the positivity of the scalar product

$$\langle \Psi, \Psi \rangle \geqslant 0 \tag{5}$$

for any $\Psi \in \mathscr{H}'$. The operator $\partial^\nu G_{\nu\mu}^{(i)} - j_\mu^{(i)}$ describes 'longitudinal' unobservable modes (in the QED case this would be equal to $-\partial_\mu \partial^\nu A_\nu$, in the Gupta–Bleuler gauge) and matrix elements of physical states do not change if one adds to a vector, say Φ, of $\mathscr{H}'$ a vector containing such longitudinal modes. (This may be interpreted as a kind of gauge transformation on $\Phi : \Phi \to \Phi + (\partial G - j)\Phi', \Phi' \in \mathscr{H}'$.) This means that when $\partial^\nu G_{\nu\mu}^{(i)} - j_\mu^{(i)}$ is applied to a vector of $\mathscr{H}'$ it gives rise to a vector of $\mathscr{H}'$ which has zero scalar product with all the vectors of $\mathscr{H}'$. The vectors of $\mathscr{H}'$ having such property form a subspace $\mathscr{H}''$.

The features outlined above can be proved [10] to have quite a general character since they occur whenever there is a non-trivial (local) charge which obeys Gauss' law. This result has very strong consequences at a general level since it implies that QFT with charges obeying Gauss' law cannot satisfy all the standard (positive metric) Wightman axioms. One is then facing the embarrassing situation that the deep results of axiomatic QFT (like TCP theorem, Spin Statistics theorem, cluster property and scattering theory, LSZ reduction formulas, proof of dispersion relations, analyticity properties of the scattering amplitudes, etc.) which were regarded as a big success for their phenomenological implications, do not apply straight-forwardly to elementary particle physics since their interactions are described by GQFT (QED seems to be only one of them). The proof of such general properties of elementary particle physics, which seem to be well supported by our present phenomenology require therefore to be investigated anew. The deep reason for these difficulties is that *states corresponding to non-zero eigenvalues of a charge which obeys Gauss' law cannot be obtained as local excitations of the vacuum* [13, 14], i.e. by applying local operators to the vacuum state (more generally by local morphisms). This introduces an element of non-locality in the Haag–Kastler framework based on the local algebra of observables.

4. Gauss' law and superselection rules

A very important physical characterization of internal symmetry groups whose generators obey Gauss' law is that the unbroken generators define charges which are not only conserved but also superselected.

THEOREM 1. *In a local QFT if the $U(1)$ group generated by a (local) charge which obeys Gauss' law is not spontaneously broken, then such charge defines a superselected quantum number* [13].

Proof. We will briefly sketch the simple physical ideas which are the basis of the proof [15]. First we recall that a conserved quantity Q defines a superselection rule if one cannot observe a coherent superposition of states belonging to different eigenvalues of Q. This is equivalent to say that all operators A which describe observable quantities commute with Q. Now, the causality requirement implies that two operators A_1, A_2, localized in some bounded regions of space-time $\mathcal{O}_1, \mathcal{O}_2$, respectively, which are relatively spacelike, must commute, since their measurements cannot interfere [16]. In a local QFT all observable operators can be obtained as suitable limits of localized operators. It is therefore enough to show that Q commutes with all localized operators.

For simplicity we consider the abelian case. By using Gauss' law and Gauss' theorem one then has for any physical vectors Ψ, Φ and an observable operator A

$$
\begin{aligned}
\langle \Psi, [Q, A]\Phi \rangle &= \lim_{R\to\infty} \int_{|x|<R} \mathrm{d}^3x \, \langle \Psi, [j_0(x), A]\Phi \rangle \\
&= \lim_{R\to\infty} \int_{|x|<R} \mathrm{d}^3x \, \langle \Psi, [\operatorname{div} \boldsymbol{E}(x), A]\Phi \rangle \\
&= \lim_{R\to\infty} \int_{S_R} \mathrm{d}S \, \langle \Psi, [\boldsymbol{n}.\boldsymbol{E}(x), A]\Phi \rangle
\end{aligned}
\tag{6}
$$

where $\boldsymbol{n}$ is the normal to the surface S_R of radius R. Now, if A is localized in a bounded region $\mathcal{O}$ of space time, for sufficiently large R the surface S_R will contain only points which are spacelike with respect to $\mathcal{O}$ and therefore the commutator vanishes by causality [17].

An interesting physical application of the above theorem is the *proof of the electric charge superselection rule* in QED. It is important to stress that as the above theorem shows, this important physical law is a consequence of very deep physical ideas, namely Gauss' law (i.e. Maxwell's equations) and causality.

As another application, we remark that the $SU(3)$ color group of Quantum Chromodynamics (QCD) also defines superselection rules, i.e., *all observables are color neutral.* A new interesting feature now arises since $SU(3)$ is non-abelian and therefore its generators define non-abelian superselection rules. This implies that in the labelling of the physical states one cannot use the color charges as good quantum numbers but only the Casimir operators which identify the irreducible representations of $SU(3)$. In particular, *physical states are necessarily described by mixtures with respect to the color quantum numbers.* This also implies that *color charges are not observable*, only combinations like e.g. $\Sigma_a Q_a^2$ may give rise to observable quantum numbers [18].

5. Gauss' law and Higgs phenomenon

Perhaps the most important feature of gauge QFT is that they allow to solve one of the most fundamental problems of elementary particle physics. It is not necessary to emphasize here the importance of approximate symmetries in elementary particle physics and the fact that for a long time they escaped a fully satisfactory explanation and treatment within the framework of QFT. The main reasons being that the introduction of explicit symmetry breaking terms or interactions in the Lagrangian implies that all the symmetry breaking effects cannot be calculated in a sensible way in any conventional renormalizable QFT, since they are divergent (e.g. mass generation, mass differences, etc.); the only way to make them finite is to introduce such breakings at lowest order as ad hoc counterterms. On the other hand, the possibility that approximate symmetries of elementary particle physics could be the result of a spontaneous symmetry breaking, with a completely symmetric Lagrangian and therefore with no need of asymmetric counterterms, was swept out by the Goldstone theorem [9], whose validity was proved under quite general conditions in standard (positive metric) QFT [19].

The discovery of the so called Higgs mechanism in GQFT has opened the way to the solution of this problem [1]. It is first important to remark that the results of Section 2 allow us to understand why this phenomenon is not a counterexample to the general Goldstone theorem, just because internal symmetry groups allowing a gauge extension require that the associated QFT cannot satisfy locality and positivity at the same time and it is not difficult to recognize that both properties play a crucial role in the proof of the Goldstone theorem.

The importance of the Higgs phenomenon demands that one should understand it on a general level without relying on the assumptions or approximations which characterize its treatment in the literature. More precisely, one could reasonably ask whether the discussion of the Higgs phenomenon done at the level of perturbation theory could be improved by exhibiting a non-perturbative explanation of it. It should also be remarked that the standard treatment heavily relies on the existence of the so-called Higgs fields, which enter in the theory at the level of elementary or fundamental fields. This is related to the assumption that the symmetry breaking already appears at the level of a one-point function, whereas one would like to cover the general case in which such breaking shows up only at the level of an n-point function, or more generally as a dynamical effect parametrized by the expectation value of a suitable (possibly composite) operator. Finally, it seems worthwhile to isolate as much as possible the essential features of the Higgs phenomenon from the explicit (perturbative) models in which it has been realized, also with the aim of clarifying the condition of its validity.

The main point of this section is to show in general that the Higgs phenomenon can be made into a theorem and that it only hinges on the validity of Gauss' law, independently of all the other ingredients which enter in the standard treatments. In particular, the following theorem automatically guarantees that the breaking of a generator which obeys Gauss' law does not require the existence of (unwanted) massless particles, quite independently of the specific mechanism of symmetry breaking (whether dynamical or triggered by elementary Higgs fields, etc.).

THEOREM 2. (Higgs phenomenon) [20]. *In a local QFT, the spontaneous breaking of an internal symmetry generated by a current which obeys Gauss' law requires the existence of $\delta\,(p^2)$ singularities, called Goldstone modes, (in suitable expectation values), which however do not show up in the physical spectrum, i.e., they occur as 'longitudinal' unobservable modes of the theory.*

Proof. We emphasize the main physical ideas which enter in the proof leaving aside the mathematical details [21]. First we recall that by definition the symmetry is spontaneously broken if there exists some expectation value, $\langle A \rangle_0$ which is not invariant. (A denotes some local operator (possibly some polynomial of the local fields) and $\langle \ \rangle_0$ denotes the vacuum expectation value). Namely, one has

$$0 \neq \delta \langle A \rangle_0 = \langle \delta A \rangle_0 = i \lim_{R \to \infty} \langle [Q_R, A] \rangle_0$$
$$= i \lim_{R \to \infty} \int_{|x|<R} d^3x \langle [j_0(x), A] \rangle_0. \tag{7}$$

Since by Gauss' theorem (see Section 3)

$$\lim_{R \to \infty} \int_{|x|<R} d^3x [\partial^\nu G_{\nu 0}(x), A] = 0 \tag{8}$$

for any local operator A, one may as well consider the commutator $\langle [Q_R - (\partial G)_R, A] \rangle_0$ instead of $\langle [Q_R, A] \rangle_0$, both giving the same result in the limit $R \to \infty$. Now, by using a Jost–Lehman–Dyson representation for $\langle [j_0(x) - \partial^\nu G_{\nu 0}(x), A] \rangle_0$ one may show [20] that

$$\langle \delta A \rangle_0 = i \lim_{R \to \infty} \int_{|x|<R} d^3x \langle [j_0(x) - \partial^\nu G_{\nu 0}(x), A] \rangle_0 \tag{9}$$

would vanish if the Fourier transform of $\langle [j_0(x) - \partial^\nu G_{\nu 0}(x), A] \rangle_0$ did not contain a $\delta(p^2)$ singularity (*Goldstone mode*). Thus, as in the standard (positive metric) case, spontaneous symmetry breaking is possible only if the theory has infrared singularities. However, since in the presence of a Gauss law, unphysical modes must be present in a local QFT, one has to investigate whether the Goldstone modes occur in the physical or in the unphysical spectrum. This information is provided by Gauss' law.

For suppose that the Goldstone mode corresponds to a physical particle. This would mean that there is a physical vector $\Psi_G \in \mathscr{H}'$, which contributes as intermediate state in the commutator (9), in the limit $R \to \infty$. This is actually impossible since for any $\Psi_G \in \mathscr{H}'$

$$\langle \Psi_0, (j_0(x) - \partial^\nu G_{\nu 0}(x)) \Psi_G \rangle = 0$$

as a consequence of the validity of Gauss' law in matrix elements of physical vectors (Ψ_0 = vacuum vector). In conclusion, only unphysical vectors can contribute to the commutator (9) in the limit $R \to \infty$ and therefore the Goldstone modes are unphysical.

6. Conclusion

We have characterized gauge quantum field theories as QFT invariant under internal symmetry group whose generators obey Gauss' law and we have shown that these internal symmetry groups have very crucial properties: the unbroken generators define quantum numbers which are superselected, whereas the broken generators have Goldstone modes, associated to them, which do not belong to the physical spectrum, i.e., they are unphysical modes.

Scuola Normale Superiore INFN, Pisa, Italy

Notes and References

[1] See e.g. Weinberg, S., *Rev. Mod. Phys.* **46**, 255 (1974).
[2] Yang, C. N., and Mills, R. L., *Phys. Rev.* **96**, 191 (1954).
[3] This is not only a pedagogical question, but also a crucial one if one wants to investigate general properties of GQFT without relying on perturbation theory, Lagrangian functions etc.; one of the main strategies of general QFT is in fact that of extracting basic properties of Lagrangian QFT, so that one may discuss their relations and their consequences in general.
[4] For a more precise meaning of Equation (1) see following sections and Ref. 5.
[5] Strocchi, F., *Phys. Rev.* **D17**, 2010 (1978) and references therein to previous papers.
[6] It need not be stressed here the importance of such a local conservation law (for the charge density), which still retains a meaning even in the case of spontaneously broken symmetry, when the global conservation law fails to exist; the local conservation law also plays a crucial role in understanding symmetries at the level of local fields and the associated Ward identities.
[7] The argument is rather standard at the classical level (see e.g. Strocchi, F., *Nuovo Cimento* **42**, 9 (1966)) and it can be extended to the quantum case by relying on the quantum action principle of Lowenstein and Lam (Ref. 8). (For a more general justification see Note 3). Clearly in the quantization of a classical Lagrangian a certain arbitrariness is involved in specifying the representation of the quantum operators which correspond to the classical fields. For example, in carrying through the renormalization procedure one may completely destroy the group theoretical structures (like Ward identities, etc.) which characterize the classical case and in such a way possibly define a theory which has very little relation with the classical theory. Such relations are instead maintained in the quantum action principle approach of Lowenstein and Lam and it need not be emphasized here that this is the way to define the quantum version of group theoretical structures which exist at the classical level.

[8] Lowenstein, J., *Commun. Math. Phys.* **24**, 1 (1971);
Lam, Y-M. P., *Phys. Rev.* **D6**, 2145, 2161 (1972); **D7**, 2945 (1973).

[9] See, e.g., the excellent review by Swieca, J., 'Goldstone Theorem and Related Topics.' In D. Kastler (ed.). *Cargèse Lectures in Physics*, Vol. 4, New York, Gordon and Breach, 1970.

[10] Ferrari, R., Picasso, L. E., and Strocchi, F., *Commun. Math. Phys.* **35**, 25 (1974).
Strocchi, F., *Phys. Rev.* **D17**, 2010 (1978).

[11] For a more precise definition of locality and positivity in QFT see Streater, R. F., and Wightman, A. S., *PCT, Spin and Statistics and All That*, Benjamin, New York, 1964.

[12] For a more detailed discussion we refer to any standard textbook on quantum electrodynamics, for example Schweber, S., *Introduction to Relativistic Quantum Field Theory*, Harper and Row, New York, 1964; or Mandl, F., *Introduction to Quantum Field Theory*, Interscience, New York, 1966.

[13] Strocchi, F., and Wightman, A. S., *J. Math. Phys.* **15**, 2198 (1974).

[14] Ferrari, R., Picasso, L. E., and Strocchi, F., *Commun. Math. Phys.* **35**, 25 (1974).
Maison, D., and Zwanziger, D., *Nucl. Phys.* **91B**, 425 (1975).
Ferrari, R., Picasso, L. E., and Strocchi, F., *Nuovo Cimento* **39A**, 1 (1977).

[15] For a more detailed and rigorous discussion we refer the reader to Ref. 13.

[16] The importance of this property of causality or locality has been emphasized by Haag, R., and Kastler, D., *J. Math. Phys.* **5**, 848 (1964).

[17] The proof has purposely been oversimplified in order not to obscure the underlying physical ideas. Actually, since by the results discussed in Section 3, Gauss' law can be required to hold only when one takes matrix elements between physical states one should carefully check that this is the case in the above derivation. To this purpose one has to remember that, since 'longitudinal modes' are not observable, matrix elements of observable operators taken between vectors of $\mathscr{H}'$ should not change if one makes a gauge transformation $\Phi \to \Phi + \chi$, χ being a vector of $\mathscr{H}'$ containing longitudinal modes. Hence

$$\langle \Psi, A\Phi \rangle = \langle \Psi, A(\Phi + \chi) \rangle,$$

which implies

$$\langle \Psi, A\chi \rangle = \langle \chi, A\Psi \rangle = 0$$

For a more detailed discussion we refer the reader to Ref. 15.

[18] For a more detailed discussion see Ref. 5, Sect. IV, where also the relevance of this feature for the solution of the infrared problem is emphasized.

[19] Such a theorem requires the existence of massless particles (Goldstone bosons) with the quantum numbers of the broken generators and there does not seem to be any candidate for such particles, the only experimentally observed massless boson being the photon, which has the feature of being associated to the unbroken internal symmetry group generated by the electric charge.

[20] F. Strocchi, *Commun. Math. Phys.* **56**, 57 (1977).

[21] For a more accurate treatment see Ref. 20.

E. C. G. SUDARSHAN

DISCRETE STATES BURIED IN THE CONTINUUM

This paper is dedicated to the memory of Professor B. Jouvet who saw clearly the close connection between renormalizable theories like the Lee model and the nonrenormalizable four-spinor interaction at a time when most physicists would not seriously entertain even the possibility.

ABSTRACT. It is possible to have discrete normalizable eigenstates of energy with eigenvalues overlapping the continuous spectrum. A simple collection of examples exhibiting this behaviour is presented in this paper.

1. Introduction

For most systems of interest in scattering theory there is a continuous spectrum corresponding to those states which represent the scattered particles being far from the scattering center and thus being presumably unaffected by the interaction. In addition to these (usually degenerate) continuum states there may be one or more states with discrete energy eigenvalues which correspond to bound states which represent the component parts which are held together and do not separate outside each other's range of interaction. In general, such states have energies less than the range of the continuous spectrum. For a (local) potential of finite range we can deduce this from the recognition of the phenomenon of quantum tunneling. In such cases we would not have eigenstates of discrete energy but rather a resonance corresponding to eigenstates of complex energy in the analytic continuation of the theory. These considerations appear so natural that we are tempted to consider that discrete eigenstates should not have energies which overlap the continuous spectrum.

There is a trivial sense in which this expectation can be violated: if one or more quantum numbers are conserved by the Hamiltonian. In this case we can consider the various sectors labelled by these quantum

E. Tirapegui (Ed.), Field Theory, Quantization and Statistical Physics, 237–245.

numbers to be distinct families. In the following discussion we shall ignore the possibility of a discrete state in one sector overlapping the continuum in another.

The first example of a local potential with a (normalizable) discrete energy eigenstate buried in the continuum was discovered by Wigner and Von Neumann. The potential in this case is a non-monotonic function that does not vanish at infinity. I know that there are other local potentials known with a single discrete 'bound state' in the continuum.

In this paper, I shall show that one can construct simple Hamiltonians which exhibit one or more discrete states overlapping with the continuum. None of these are local potential systems but are reminiscent of quantum field theory. In all these cases one can compute the scattering amplitude, and at each energy where there is a discrete state, there is a characteristic behaviour of the scattering amplitude.

The plan of the paper is as follows. In Section 2 we outline a Friedrich model theory. To simplify the algebraic manipulations as much as possible we go to the S-wave sector and realize the state space as a function space over the (unperturbed) energy. The spectrum and the (generalized) eigenvectors are obtained in closed form. The question of the existence of discrete energy eigenstates buried in the continuum is posed and solved. Section 3 deals with a system with more than one discrete energy states buried in the continuum. The possibility of repulsive separable potentials generating a discrete state in the continuum is explored in Section 4. With a view to illustrate the essential ideas most simply we have absorbed the phase volume into the normalizations to keep the kinematic complications to a minimum.

2. The Friedrich–Lee Example

Consider a collection of harmonic oscillators with every oscillator mode coupled to a base mode. The Hamiltonian of such a system is of the form

$$H = a_0^+ a_0 m + \sum_j a_j^+ a_j w_j + \sum_j f_j (a_j^+ a_0 + a_0^+ a_j). \tag{1}$$

The number operator

$$N = a_0^+ a_0 + \sum_j a_j^+ a_j \tag{2}$$

is a constant of the motion having nonnegative integer eigenvalues. We shall be interested in the special sector with $N = 1$. The state is then specified by the amplitudes ψ_0, ψ_j. We can now take the limit when the discrete index j takes on continuously many values with the corresponding natural frequencies taking the continuum of values

$$0 < W < \infty. \tag{3}$$

The states are now represented by a vector with a component ψ_0 and a function $\psi(W)$. The scalar product of two vectors ψ, ϕ is given by

$$(\psi, \phi) = \int_0^\infty \mathrm{d}W\, \psi(W)^* \phi(W) + \psi_0^* \phi_0. \tag{4}$$

The action of the Hamiltonian on the vectors is given by the equations.

$$\begin{aligned}(H\psi)_0 &= m\psi_0 + \int_0^\infty \mathrm{d}W f(W)\psi(W), \\ (H\psi)(W) &= f(W)\psi_0 + W\psi(W).\end{aligned} \tag{5}$$

This is the familiar Friedrich–Lee model with the explicit continuum (idealized) eigenvectors

$$\begin{aligned}{}_\lambda\psi_0 &= f(\lambda)/\alpha(\lambda + i\varepsilon), \\ {}_\lambda\psi(W) &= \delta(\lambda - W) + f(W)f(\lambda)/(\lambda - W + i\varepsilon)\alpha(\lambda + i\varepsilon),\end{aligned} \tag{6}$$

with

$$\alpha(Z) = Z - m - \int_0^\infty \mathrm{d}W (Z - W)^{-1} f^2(W). \tag{7}$$

If m is such that

$$\alpha(0) = -m + \int_0^\infty \mathrm{d}W.W^{-1}.f^2(W) > 0 \tag{8}$$

there exists a real negative number M such that

$$\alpha(M) = 0. \tag{9}$$

Then there is a discrete state with the eigenvector:

$$\begin{aligned}{}_M\psi_0 &= (\alpha'(M))^{-1/2}, \\ {}_M\psi(W) &= f(W)(\alpha'(M))^{-1/2}/(M - W).\end{aligned} \tag{10}$$

The set of continuum eigenstates form a complete set if (9) is not satisfied provided $f(W)$ is nonvanishing for $0 < W < \infty$; if (9) is satisfied we will have to include (10) to obtain the complete set:

$$\int_0^\infty d\lambda\, {}_\lambda\psi(W)\,{}_\lambda\psi(W')^* + {}_M\psi(W)\,{}_M\psi^*(W') = \delta(W - W'),$$

$$\int_0^\infty d\lambda\, {}_\lambda\psi(W)\,{}_\lambda\psi_0^* + {}_M\psi(W)\,{}_M\psi_0^* = 0,$$

$$\int_0^\infty d\lambda\, {}_\lambda\psi_0\,{}_\lambda\psi_0^* + {}_M\psi_0\,{}_M\psi_0^* = 1 \tag{11}$$

If the value of m exceeds $\int_0^\infty dW.W^{-1}.f^2(W)$ there would be no discrete state at all. We would, instead have a resonance indicative of an unstable particle in the calculated scattering amplitude:

$$T(W) = f^2(W)/\alpha(W + i\varepsilon) \tag{12}$$

which would correspond to a complex zero of $\alpha(Z)$ near the real axis.

Let us now raise the question: Can we have a discrete (normalizable) eigenstate for an eigenvalue $M > 0$? Such a solution cannot exist if $f(W)$ is nonvanishing for all W in the range $0 < W < \infty$. Therefore we relax this condition and allow for $f(W)$ to have an isolated zero in this domain. This would make it possible for $\alpha(Z)$ to have a zero in this range.

We therefore consider $f(W)$ to be such that the following twin conditions hold:

$$f(M) = 0, \qquad \alpha(M) = 0. \tag{13}$$

Then there is a discrete solution of the form (10) with $M > 0$. Clearly $T(M) = 0$.

There is also a continuum of solutions given by (6). These solutions together with the discrete solution still form a complete set.

It is to be noted that the eigenvalue M is now degenerate: there is, in addition to (10), a (nonnormalizable) solution belonging to the continuum with

$$\psi(W) = \delta(M - W), \qquad \psi_0 = 0 \tag{14}$$

which corresponds to an uncoupled ('plane wave') excitation. We note that if $f(W)$ vanishes faster than $(M-W)$ as $W \to M$, then (14) and (10) are orthogonal.

Since $\alpha(Z)$ defined by (7) is such that the real part of

$$\alpha'(Z) = 1 + \int_0^\infty \mathrm{d}W f^2(W)(W-Z)^{-2} \tag{15}$$

is nonnegative along the real axis, it follows that $\alpha(Z)$ can have at most only one zero. Therefore, there cannot be more than one discrete state whatever be the nature and magnitude of the coupling function $f(W)$.

If we need more than one such discrete state we must generalize the model.

3. Model with many discrete eigenstates

A generalization of the model presented in the previous section is furnished by replacing the Hamiltonian (1) by

$$H = \sum_\alpha a_\alpha^+ a_\alpha m_\alpha + \sum_j a_j^+ a_j W_j + \sum_{\alpha j} f_{\alpha j}(a_j^+ a_\alpha + a_\alpha^+ a_j) \tag{16}$$

and proceeding to the continuum limit

$$0 < W < \infty$$

for W_j. The states are now represented by vectors with discrete components ψ_α and a complex-valued function $\psi(W)$ with the scalar product:

$$(\psi,\phi) = \sum_\alpha \psi_\alpha^* \phi_\alpha + \int_0^\infty \mathrm{d}W \psi(W)^* \phi(W). \tag{17}$$

The action of (16) on a vector leads to

$$(H\psi)_\alpha = m_\alpha \psi_\alpha + \int_0^\infty \mathrm{d}W f_\alpha(W) \psi(W)$$

$$(H\psi)(W) = W\psi(W) + \sum_\alpha f_\alpha(W)\psi_\alpha. \tag{18}$$

The eigenstates can now be worked out. The continuum solutions with

$0 < \lambda < \infty$ are of the form

$$\psi(W) = \delta(\lambda - W) + (\lambda - W + i\varepsilon)^{-1} \sum_\alpha f_\alpha(W)\psi_\alpha,$$

so that

$$\int_0^\infty \mathrm{d}W f_\alpha(W)\psi(W) = \sum_\beta \left\{ \int_0^\infty \mathrm{d}W f_\alpha(W) f_\beta(W)(\lambda - W + i\varepsilon)^{-1} \right\} \psi_\beta + f_\alpha(\lambda)$$
$$= \sum_\beta F_{\alpha\beta}(\lambda + i\varepsilon)\psi_\beta + f_\alpha(\lambda), \tag{19}$$

and

$$(\lambda - m_\alpha)\psi_\alpha = \sum_\beta F_{\alpha\beta}(\lambda + i\varepsilon)\psi_\beta + f_\alpha(\lambda)\theta(\lambda) \tag{20}$$

$$(\lambda - W)\psi(W) = \sum_\beta f_\beta(W)\psi_\beta. \tag{21}$$

Since (20) gives a set of simultaneous inhomogeneous linear equations, we can write down the solutions in terms of the determinants of those coefficients. The essential point is that the function

$$\Delta(Z) = \det\{Z\delta_{\alpha\beta} - m_\alpha\delta_{\alpha\beta} - F_{\alpha\beta}(Z)\} = \det D_{\alpha\beta}(z) \tag{22}$$

plays the role previously assigned to the function $\alpha(Z)$. Unlike $\alpha(Z)$, this new function $\Delta(Z)$ can have multiple zeros atmost equal in number to the range of the index α in (16)–(20).

We now search for discrete solutions of the eigenvalue problem

$$H\psi = \lambda\psi. \tag{23}$$

These may arise if $\Delta(Z) = 0$ for $Z < 0$ and are then analogous to the bound states of the model studied in the last section: the second term on the right-hand side of (20) is then missing and the equations are homogeneous in the ψ_β. As long as the determinant of $D_{\alpha\beta}$ vanishes but none of its principal minors have their determinant vanish the ψ_α are uniquely determined except for a multiplicative constant; this constant can be chosen for each solution by requiring the eigenvectors to be normalized.

If it turns out that for any such solution $\lambda = M > 0$, the coupling functions $f_\alpha(M)$ all vanish then, in complete analogy with (10) we have a discrete solution

$$\sum_\alpha {}_M\psi_\alpha f_\alpha(W)/(M - W) = {}_M\psi(W), \tag{24}$$

with

$$\sum_{\beta} F_{\alpha\beta}(M)_M\psi_\beta = (\lambda - m_\alpha)_M\psi_\alpha. \tag{25}$$

By suitable choice of parameters we can have a number of genuine 'bound state' solutions with $M < 0$, a number of discrete states buried in the continuum and possibly a number of resonances.

4. Repulsive potentials leading to discrete states

Another generalization of the model Hamiltonian (1) is obtained by choosing

$$H = \sum_j a_j^+ a_j W_j \pm \sum_{jk} f_j f_k (a_j^+ a_k + a_k^+ a_j). \tag{26}$$

The $\pm$ signs correspond to repulsive and attractive potentials. Restricting attention to the sector where

$$\Sigma a_j^+ a_j = 1 \tag{27}$$

and proceeding to the continuum limit with $0 < W < \infty$ we get the separable potential model with states represented by functions $\psi(W)$ with scalar product

$$(\psi, \phi) = \int_0^\infty \mathrm{d}W \psi^*(W)\phi(W) \tag{28}$$

The action of the Hamiltonian is

$$(H\psi)(W) = W\psi(W) \pm \int_0^\infty \mathrm{d}W' f(W')\psi(W') f(W). \tag{29}$$

The continuum solutions with eigenvalues

$$0 < \lambda < \infty$$

have the form

$$_\lambda\psi(W) = \delta(\lambda - W) \pm \frac{f(W) \int_0^\infty \mathrm{d}W' f(W')\psi(W')}{(\lambda - W + i\varepsilon)}.$$

Defining

$$D(Z)=1\mp\int_0^\infty \mathrm{d}W f^2(W)/(Z-W). \tag{30}$$

We can write down the complete solution

$$_\lambda\psi(W)=\delta(\lambda-W)\pm f(\lambda)f(W)(\lambda-W+i\varepsilon)^{-1}D^{-1}(\lambda+i\varepsilon). \tag{31}$$

The question naturally arises if there are any discrete solutions. From (29)–(31) we can see that this is possible in general only if

$$D(M)=0, \qquad M<0. \tag{32}$$

Since $\pm D'(Z)>0$ for $Z<0$ and $D(-\infty)=1$, it follows that unless we choose attractive potentials we cannot have a bound state. When we do have a bound state it has the representative wave function

$$_M\psi(W)=\frac{f(W)}{M-W}(-D'(M))^{-1/2}. \tag{33}$$

For a repulsive potential there are no bound states, but there is still the possibility of a discrete state buried in the continuum. The condition for the existence of such a state is

$$D(M)=0, \qquad f(M)=0. \tag{34}$$

The explicit solution is

$$_M\psi(W)=f(W)(M-W)^{-1}(D'(M))^{1/2}. \tag{35}$$

There is also a continuum solution for this energy

$$_\lambda\psi(W)|_{\lambda=M}=\delta(M-W) \tag{36}$$

which will become orthogonal to the discrete solution if $f(W)$ vanishes faster than $M-W$ as $W\to M$.

It is curious that a repulsive potential leads to a discrete normalizable state.

5. Concluding remarks

The discussion in this paper was calculated to show that suitable interactions can lead to discrete normalizable energy eigenstates with energy eigenvalues buried in the continuum. It is possible to design a number of such systems with one or more such states.

In all such cases there is an energy degenerate continuum solution which coincides with the continuum solution for the noninteracting Hamiltonian. Consequently, the scattering amplitude of the interacting system always vanishes at this energy for all the systems considered. This is in marked contrast to the usual case of an unstable state which leads to a reasonance.

Acknowledgement

This work was supported, in part, by the U.S. Department of Energy.

Center for Particle Theory, University of Texas

PART TWO

STATISTICAL PHYSICS

P. COULLET AND C. TRESSER

SOME UNIVERSAL ASPECTS OF THE TRANSITION TO STOCHASTICITY FOR NON-CONSERVATIVE DYNAMICAL SYSTEMS

ABSTRACT. We present several universal aspects of the transition to stochasticity for some one-parameter families of one dimensional difference equations. An analogy is pointed out between this transition for these models and second-order phase transitions. Connections between discrete models and more realistic ones are briefly discussed.

1. Introduction

As early as 1975, de Gennes proposed studying laminar → turbulent (L.T.) transitions in fluids in terms of phase transitions [1]. The model proposed by Lorenz in 1964 in order to modelize convection experiments [2] exhibits a discontinuous transition to chaotic behavior and the questions asked by de Gennes were then [1]:
- Can the L.T. transition be continuous?
- Do the L.T. transitions fall into a certain number of universality class?
- Assuming that a continuous L.T. transition exists, is there a characteristic relaxation time τ which diverges at the transition. If so, is there a scaling law with exponent fixed in one universality class?

In order to throw some light on these questions and related ones for general classes of dynamical systems, we have considered many examples of elementary forced systems like nonlinear oscillators [3] and simple models for interacting biological species [4]. In these cases, numerical studies reveal that τ, defined as the period of oscillations observed in prechaotic régimes, diverges with a power law whose exponent γ is universal. The connexion between these problems and the iteration of diffeomorphisms of $\mathbb{R}^2$ is easily obtained by means of Poincaré's first return map [5]. For a class of diffeomorphisms uniformally area-contracting, turbulent transition is now being studied and reveals also to be universal [6]. The infinite contraction limit of these diffeomorphisms gives non-invertible real maps.

In this paper, we shall focus on this oversimplified case as it displays all the essential features observed in more complicated situations. We shall define various quantities, analogous to correlation times and order

E. Tirapegui (Ed.), Field Theory, Quantization and Statistical Physics, 249–261.

parameters and report observations of typical critical behaviors as power laws and universality properties. Beyond this phenomenological aspect, we shall construct a renormalization group in order to understand this behavior [6–12].

No such techniques have been developed for differential systems but it seems worth noting that the critical exponent ν defined above for oscillators and ecological models appears to be the same as the one we find for the universality class of quadratic real maps ($\nu = \ln 2/\ln 4.669 \ldots$).

2. The models

We consider difference equations

$$x_{i+1} = f_R(x_i) \tag{1}$$

where $f_R (R \in [R_{\min}, R_{\max}])$ is a one-parameter analytic map from the unit interval to itself. Roughly speaking, the maps we shall consider have a unique maximum $M_R = f_R(\bar{x}_R)$ increasing with R, with $f'_R(x) > 0$ when $x < \bar{x}_R, f'_R(x) < 0$ when $x > x_R$ and they verify the relation $f_R(0) = f_R(1) = 0$ (Figure 1). A typical example is $f_R(x) = Rx(1-x)$. These maps have

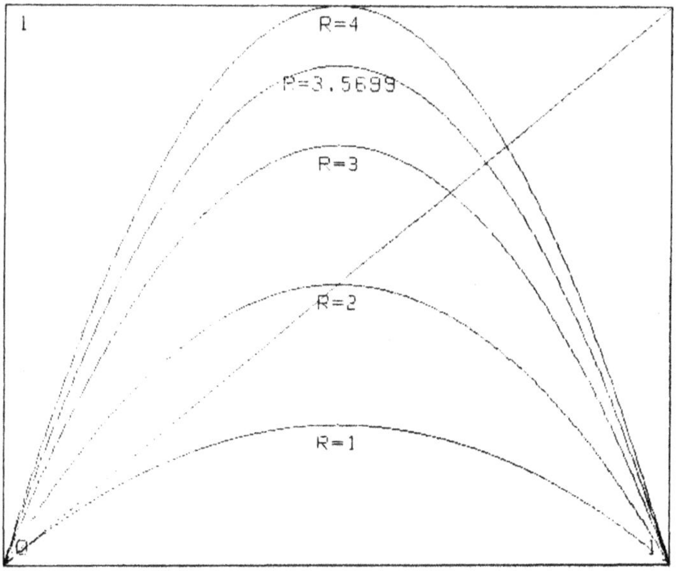

Fig. 1. Graphs of $f_R(x) = Rx(1-x)$ for various values of the parameter R.

been extensively studied [13]. Although very simple, they exhibit stochastic behavior for some values of the parameter.

Let us recall that the orbit of a point is the collection of its successive images by f_R and that a m-periodic cycle is a finite orbit with m elements. Then, the number N_m of period m orbits coexisting when $R = R_{\max}$ grows like 2^m. A theorem due to Sharkovskii [14] give the order of appearance of the first orbit with each period when R grows from $R_{\min}$ to $R_{\max}$; denoting by $n \lhd m$ the fact that the first period n orbit appears before the first one with period m, one gets:

$$1 \lhd 2 \Delta 4 \lhd \ldots \lhd 2^n \lhd 2^{n+1} \lhd \ldots | \ldots | \ldots | \ldots | \ldots \lhd 2^n \times 7 \lhd 2^n \times 5 \lhd 2^n \times 3 |$$
$$\ldots | \ldots \lhd 2^2 \times 7 \lhd 2^2 \times 5 \lhd 2^2 \times 3 \lhd | \ldots 2 \lhd 7 \lhd 2 \times 5 \lhd 2 \times 3 \lhd |$$
$$\ldots \lhd 7 \lhd 5 \lhd 3. \tag{2}$$

In (2), the vertical bars stand for the accumulation points:

$$\lim_{\substack{p \to \infty \\ p \text{ odd}}} 2^n \cdot p. \tag{3}$$

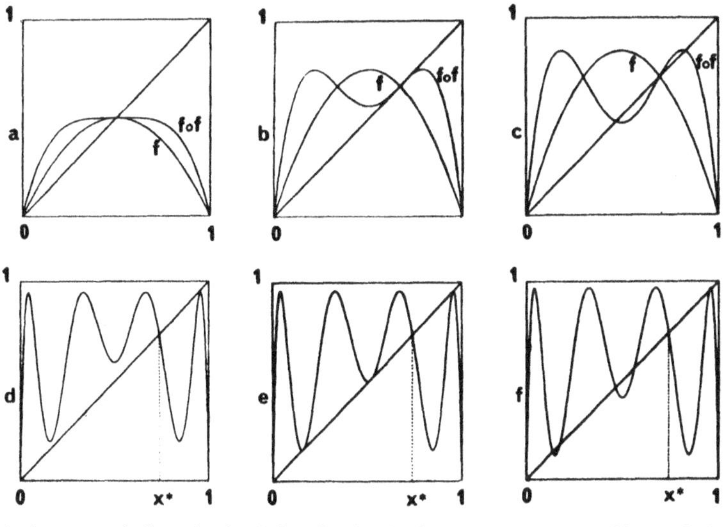

Fig. 2. As a m-periodic point for f_R is a fixed point for f_R^m, we present on Figures 2(a)–2(c) how arise a cycle with period 2 using graphs of f_R and f_R^2 for successive values of R (flip bifurcation) and on Figures 2(d)–2(f) a cycle with period 3 using graphs of f_R^3 (saddle node bifurcation).

We shall denote by $\bar{R}_n$ the corresponding values of R for a given family f_R and by $\tilde{R}_n$ the parameter value corresponding to the appearance of the first 2^n periodic cycle (see Figure 2). The sequences $\{\tilde{R}_n\}_n$ and $\{\bar{R}_n\}_n$ are respectively increasing and decreasing. We shall denote by R_c their common accumulation point. R_c separates two phases, i.e. two parameter domains with different dynamical behaviors. The first one $(R \leqslant R_c)$ is regular and characterized [15] by zero topological entropy [16]. In the second one $(R > R_c)$, one observes stochastic behaviors.

3. Stochastic parameters

The non-invertible character of f_R lead us [12] to define a mean amount of information per iteration, corresponding to the knowledge of a particular obrit of (1) going through a point in [0,1] by

$$h_R(x) = \lim_{n \to \infty} \frac{1}{n} \ln N_n(x, f_R) \tag{3}$$

where $N_n(x, f_R)$ is the number of nth-order inverse images of x. $h_R(x)$ is a local test of stochasticity, easy to handle by numerical methods. Before going further, let us give some examples.

– When $R = R_{\max}$, every point of [0,1] admits two preimages and one gets, for every x in [0,1]:

$$h_{R_{\max}}(x) = \ln 2 \tag{4}$$

– When $R < R_{\max}$, it is reasonable to wait for $h_R(x) < \ln 2$. A situation often considered is the case where f_R admits a stable period three cycle. In this case, according to Sharkovskii' theorem (2), there exist cycles with any period. Furthermore, there exists no point x in [0,1] with a fourth-order inverse such that $f_R(g) > \bar{x}_R, f_R^2(g) < \bar{x}_R$ and $f_R^3(g) < \bar{x}_R$. One can then show that $h_R(x) = \ln((1+\sqrt{5})/2$ if $f_R(M_R) \leqslant x \leqslant M_R$ and that $h_R(x) = 0$ out of this interval.

We have shown [17] that, more generally for each $R > R_c$, [0,1] is constituted of two complementary regions. The first one correspond to $h_R(x) = 0$. In the second one, $h_R(x)$ is constant and equal to the topological entropy [16] of f_R.

In conservative dynamical systems, one usually calls stochastic region the support of the local Kolmogorov entropy [18]. We use here the same terminology for the support S_R of the function h_R. The R-dependence of the geometric structure of S_R has been qualitatively represented on Figure

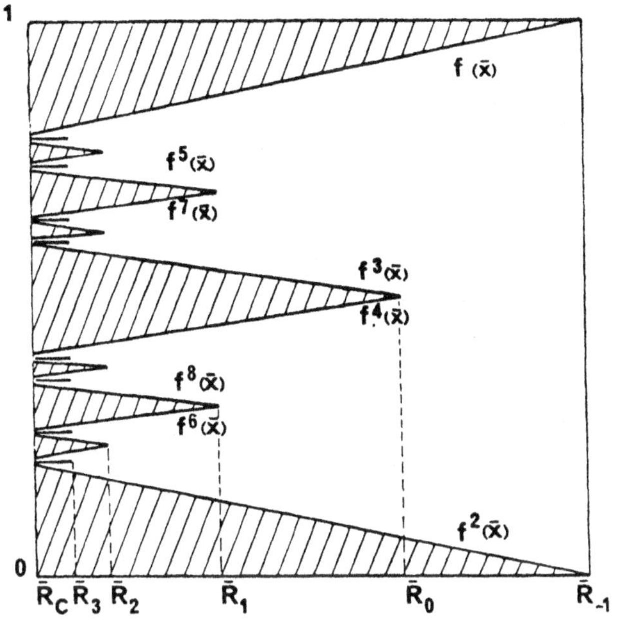

Fig. 3. Rough construction of the stochastic region S_R as a function of the parameter R. The $\bar{R}_n$'s correspond to qualitative change of S_R. The shadowed parts correspond to the complementary of S_R in I. We have indicated the simplest expressions for bounds of S_R.

3. There always exists a $n \geqslant 0$ such that S_R consists of $P'_n = 2^n$ connected components. $\bar{R}_n$ is precisely the parameter value corresponding to a division by a factor 2 of the number of these components.

The topological entropy $h(f_R)$ and Lebesgue's measure $m(S_R)$ of the stochastic region will be considered throughout as stochastic parameters.

4. Transition to Stochasticity as a Phase Transition

In Section 2, we defined R_c as the accumulation point of the sequences $\{\tilde{R}_n\}_n$ and $\{\bar{R}_n\}_n$. R_c plays the role of the critical temperature in second order phase transitions: it marks the transition to stochasticity. We shall now develop this analogy.

4.1. *Scale Invariance*

For the critical value R_c of the parameter, the adherence of the asymptotic

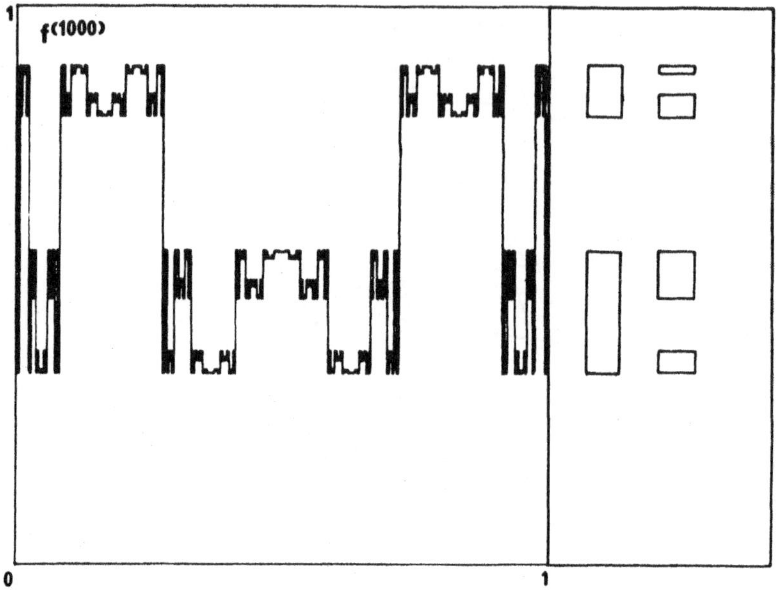

Fig. 4. Graph of f_R^{1000} for $f_R = Rx(1-x)$ with $R = R \simeq 3.5699 \ldots$

orbit of almost all points in $[0,1]$ is a Cantor set (Figure 4)

$$S_{R_c} = \lim_{R \to R_c} S_R, \tag{5}$$

limit, when R decreases to R_C, of the stochastic region defined in the previous section.

For this Cantor set, there does not exist a real scale invariance [8]. Nevertheless, we define a global scaling factor in the following way: denoting by $S_{R_c}^n$ the remaining part of S_{R_c} after the nth step of its construction following the usual way for a Cantor set (Figure 5) and by m Lebesgue's measure, we set

$$\rho = \overline{\lim_{n \to \infty}} \frac{m(S_{R_c}^{n-1})}{m(S_{R_c}^n)}. \tag{6}$$

The global ratio ρ plays the role of the scaling associated to the renormalization of spin blocks in Wilson–Kadanoff's version of critical phenomena theory.

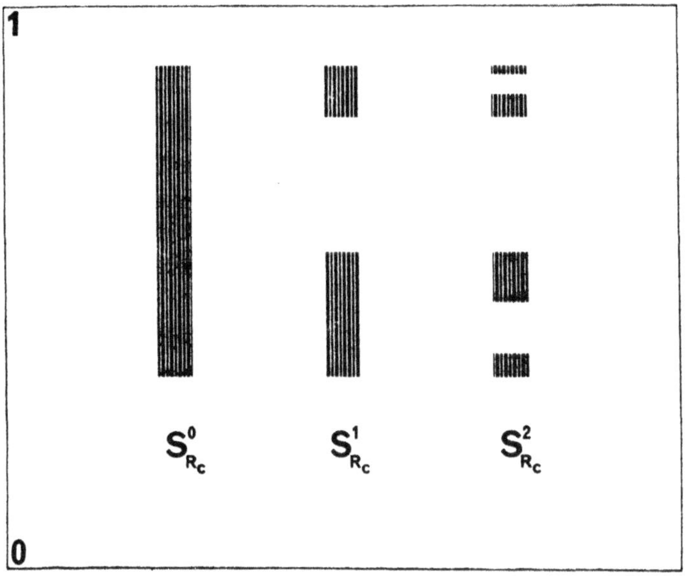

Fig. 5. First approximations of the Cantor set defined by $S_R = \bigcap_{n=0}^{\infty} S_R^n$. The difference of sizes between the connected components at each order does not allow a true scale invariance.

4.2. *Critical Exponents*

We shall define in this section some quantities which are singular in the neighbourhood of R_c.

4.2.1. *The analogous of a correlation time*

We shall denote by $P(\tilde{R}_n) \equiv 2^n$ the period of the stable cycle when $\tilde{R}_n < R < \tilde{R}_{n+1}$, and by $P(\bar{R}_n) \equiv 2^n$ the number of connected components of S_R when $\bar{R}_n < R < R_{n-1}$. At this point, let us remark that this number 2^n of connected components corresponds to a '2^n-periodic noisy motion' [19]. Then, we can define two critical exponents ν and ν' by

$$R < R_c : P(\tilde{R}_n) = 2^n \propto (R_c - \tilde{R}_n)^{-\nu}$$
$$R > R_c : P'(\bar{R}_n) = 2^n \propto (\bar{R}_n - R_c)^{-\nu'}. \tag{7}$$

Numerically, one finds a scaling law as in critical phenomena, namely

$$\nu = \nu' = \ln 2/\ln \lambda \tag{8}$$

where

$$\lim_{n\to\infty}\frac{\bar{\bar{R}}_{n+1}-\bar{\bar{R}}_n}{\bar{\bar{R}}_n-\bar{\bar{R}}_{n-1}}=\lim_{n\to\infty}\frac{\tilde{\bar{R}}_{n+1}-\hat{R}_n}{\tilde{\bar{R}}_n-\tilde{\bar{R}}_{n-1}}=\lambda. \tag{9}$$

In Section 4.4, λ will be redefined in the frame of a renormalization group. For the sample case of quadratic f_R(e.g. $f_R(x)=Rx(1-x)$) one gets $\nu \simeq 0.45$.

4.2.2. *Critical exponents for the stochastic parameters*

Topological entropy. We recall [15] that the topological entropy is zero for $R \leqslant R_c$ and positive for $R > R_c$. On the other hand, it is well known that

$$h(f_{\bar{R}_n})=\frac{1}{2^{n+1}}\ln 2. \tag{10}$$

Using (7) and (8) one then obtains

$$h(f_{\bar{R}_n}) \propto (\bar{R}_n - R_c)^{\nu}, \tag{11}$$

a quite natural result in view of (7) if one keeps in mind that dynamic entropies (metric or topological) are naturally interpreted as inverse of a characteristic time.

4.2.3. *Lebesgue's measure of the stochastic region*

Let us remark that, if n is large enough

$$S_{\bar{R}_n} \approx S^n_{R_n} \tag{12}$$

so that we get

$$m(S_{\bar{R}_n}) \propto \rho^{-n}. \tag{13}$$

This enables us to express the critical exponent for the size of the stochastic region in terms of p and λ. Using (13) and (7)–(9) this becomes

$$m(S_{\bar{R}_n}) \propto (\bar{R}_n - R_c)^{\ln\rho/\ln\lambda}, \tag{14}$$

a result which is corroborated by direct numerical computations.

4.3. *Universality*

The critical exponents we defined depend only on the multiplicity z of the zero of the derivative $f'_R(\bar{x}_R)$. This universality also characterizes the fine

TABLE I
Numerical values of the first approximants of ρ for various functions. Here $\rho(n) = m(S_R^{n-1})/m(S_{R_c}^N)$.

z	1	1	1	3
f_R	$RX(1-X)$	$R\sin\pi X$	$RX(1-X^2)$	$R\sin\left(\frac{\pi}{2}\sin\pi X\right)$
R_c	3.56994567183	.86557926894	2.30228346271	0.94676589611
$\rho(1)$	1.7825	1.7510	1.7042	1.3753
$\rho(2)$	1.7328	1.7391	1.7507	1.3587
$\rho(3)$	1.7434	1.7419	1.7391	1.3624
$\rho(4)$	1.7406	1.7413	1.7419	1.3598
$\rho(5)$	1.7413	1.7413	1.7413	1.3607
$\rho(6)$	1.7413	1.7413	1.7413	1.3602
$\rho(7)$	1.7413	1.7413	1.7413	1.3605
$\rho(8)$	1.7413	1.7413	1.7413	1.3605

structure of the Cantor set [8]. Table I reports results for ρ. Results for λ may be found in [6], [7] and [11] and for the fine structure of S_{R_c} in [8].

4.4. *Renormalization Group*

We shall recall here briefly the construction of the renormalization group which enables an understanding of the origin of the critical exponents and their computation [6–12].

In fact, there are many differences between this renormalization group and those used up to now in physics; here, we study deterministic dynamical systems. We are interested in the variation with some parameter of such quantities as the period of stable cycles. These variations are not continuous and arise for a discrete sequence of values of the parameter. This will impose an irreducibly discrete version of the renormalization procedure. Another fundamental difference comes, as we shall show, from the existence of two generators for this renormalization group. This will generate infinitely many operations which would be necessary to describe the local scale invariance of the Cantor set. At last, let us mention the existence of infinitely many values of the parameter where one can find various critical behaviors [6], but this seems non-relevant to the transition to stochasticity as the topological entropy is non-zero as soon as $R > R_c$.

To achieve greater simplicity in the definition of the renormalization operation, we shall make a translation which will bring the point $[\bar{x}, f(\bar{x})]$, the maximum of the graph, to the point $(0,1)$. This brings the interval $[0,1]$ on some interval $[\alpha,\beta]$. Denoting by g the new function, α is the negative solution of $g(\alpha)=\alpha$ and β verifies $g(\beta)=\alpha$ (Figure 6).

The idea of renormalization is that, for $R=R_c$ and via some simple transformation, the relation between x_{i+2} and x_i is very similar to that between x_{i+1} and x_i. This is suggested by the numerical evidence of a scale invariance for the asymptotic orbit when $R=R_c$ (Cantor like structure) and by the evidence of a self similar order for the iterates of a point on its orbit (the order of the iterates of $\bar{x}_R$ which arise as bounds of the trapping region). Among the transformations we shall have to do, two are similar to those we know in classic renormalization schemes:

- We rename the variables $x_{i+2} \to x_{i+1}$;
- We do as scaling on these variables $x_i \to \xi x_i$.

Then there is one more transformation.

- We do a translation on variables $\xi x_i \to \xi x_i + \eta$.

This leads to study the transformation

$$g(x) \longrightarrow g_{\xi,\eta}(x) = \frac{1}{\xi}[g \circ g(\xi x + \eta) - \eta] \tag{15}$$

This may be written as

$$g_{\xi,\eta}(x) = h_{\xi,\eta}^{-1} \circ g \circ g \circ h_{\xi,\eta},$$

where $h_{\xi,\eta}$ is the affine map $h_{\xi,\eta}(x) = \xi x + \eta$. Then

$$g_{\xi,\eta}(x) = \frac{1}{\xi}[g \circ g(\eta) - \eta] + g'[g(\eta)]g'(\eta)x + O(x^2),$$

where $g'(x) = \partial g(x)/\partial x$,

ξ and η are determined by the conditions

(a) $\eta = 0$ or $\eta > 0$ and $g(\eta) = 0$,
(b) $\xi = [g \circ g(\eta) - \eta]$,

(a) ensures the cancellation of the linear part of $g_{\xi\eta}(x)$ and (b) implies $g_{\xi,\eta}(0) = 1$ (Figure 6).
The two possible choices for η correspond to two distinct operations,

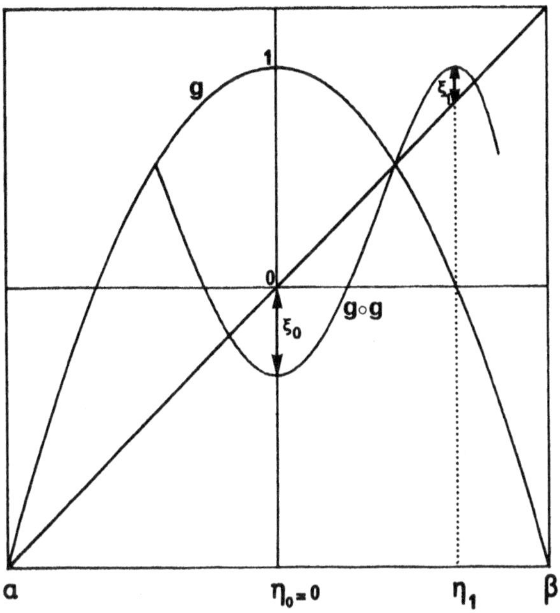

Fig. 6. Geometrical interpretation of the parameters ξ_l and η_l, defined in the main text.

denoted as $\mathscr{R}_1$ and $\mathscr{R}_2$ and which are the generators of the renormalization group. The possibility of finding fixed points for these transformations and the study of their stability is crucial for the explanation of the criticality. This problem has been studied using truncated recursion formula [6] and numerical studies [7–9]. Some exact results have been obtained recently [10].

The conjecture, at least when g is analytic, is then: there exists an unique fixed point g_i^* for the operation $\mathscr{R}_i$ with $g_i^{*\prime\prime}(0) \neq 0$, $i = 1, 2$, such that the spectrum of the linearized operator $D\mathscr{R}_i/g_i^*$ is inside the open unit disk except for a unique positive eigenvalue λ_i such that $\lambda_1 = \lambda_2 = \lambda$.

The picture is then the same that in critical phenomena theory. Let us consider the stable manifold of e.g. g_1^*. In the set of functions g, a sufficiently general one parameter path will get a critical point if precisely it cut this manifold. Then, employing the techniques of the renormalization group we obtain a relation between the critical exponent ν for the period and the eigenvalue λ.

$$\nu = \ln 2/\ln \lambda \tag{16}$$

There is no easy way to compute the other exponent, due to the very existence of two generators for the renormalization group (the existence of an unique operation would led to an exponent $\alpha = \ln \xi / \ln 2$).

Finally, let us mention that in the analytic case, the operations $\mathscr{R}_1$ and $\mathscr{R}_2$ get other fixed points. These will be characterized by $g_i^{*\prime\prime}(0) = 0$. More precisely, we shall find an unique fixed point such that $g_{i(0)}^{*(2n)} \neq 0$, $g_{i(0)}^{*(P)} = 0$ for $p > 2n$. The spectrum of $D\mathscr{R}_i/g_i^*$ will be inside the open unit disk, except n eigenvalues greater than 1. This may be interpreted as a polycritical phenomena for some n parameter families of difference equations. Recently, in a non-analytic case, we constructed with Arneodo, families f_R such that $\mathscr{R}_1$ and $\mathscr{R}_2$ do not admit fixed points but rather order two cycles [11]. The analysis allows then to predict oscillatory behaviors both for the sequences $\{\tilde{R}_n\}_n$ and $\{\bar{R}_n\}_n$ and for the kind of scale invariance on the Cantor set when $R = R_c$.

Conclusion

In this paper, we have considered a kind of transition to stochasticity which may be qualified as 2nd order in analogy with phase transitions. In some dynamical systems that we considered [3,4,21] the transition may be discontinuous, i.e. of 1st order. It is also the case for Lorenz model [2]. These brutal transitions are associated with subcritical bifurcations. Recent experimental results [20] on Rayleigh–Benard transitions reveal 1st- and 2nd-order transitions. In these cases the observed transitions are more complex, as they involve two fundamental frequencies. Nevertheless, it is possible to construct differential models (like forced Van der Pol oscillators) [21] which give qualitatively all the behaviors reported in [20].

The very possibility of modelizing this experiment by a flow which exhibits a transition to strange attractors is, in our opinion, a decisive argument for the ideas of Ruele and Takens [22] on the nature of turbulence.

Acknowledgements

The authors would like to thank A. Chenciner, J. Coste, M. Hénon and R. Thom for encouragement and stimulating discussions.

Laboratoire de Physique de la Matière Condensée, Nice

References

[1] De Gennes, P. G., *Fluctuations, Instabilities and Phase Transitions* in T. Riste (ed.), Proceedings of the Nato Advanced Study Institute, held in Geilo, Norway, April 11–20, 1975, Plenum Press, New York, 1975.

[2] Lorenz, E. N., *J. Atmos. Sci.* **20**, 130 (1963).
For recent works on Lorenz attractor, see for instance the papers of J. E. Marsden, R. Williams and O. E. Landford III in *Turbulence Seminar*, Berkeley 1976–1977, Lectures Notes in Mathematics **615**, Springer Verlag, 1977.
McLaughin, J. B., and Martin, P. C., *Phys. Rev.* **12**, 186 (1975).
Curry, J. H., Preprint, 1978.
Marsden, J. E., and McCracken, M., *The Hopf Bifurcation and its Application*, Springer Verlag, New York, 1976.

[3] Coullet, P., Tresser, C., Arneodo, A., to appear in *Phys. Lett. A* and in preparation.

[4] Coullet, P., and Tresser, C., in preparation.

[5] Hénon, M., *Commun. Math. Phys.* **50**, 69 (1976).

[6] Derrida, B., Gervois, A., and Pomeau, Y., *J. Phys. A* **12**, 269 (1979).

[7] Feigenbaum, M. J., *J. Stat. Phys.* **19**, 25 (1978).

[8] Tresser, C., and Coullet, P., to appear in *Rep. Math. Phys.* (Poland).

[9] Tresser, C., Coullet, P., *C. R. Acad. Sc. Paris* **287**, A-577 (1978).

[10] Collet, P., Eckmann, J. P., and Lanford, III, O. E., Preprint (1978).

[11] Arneodo, A., Coullet, P., Tresser, C., *Phys. Lett.* **70A**, 74 (1979).

[12] Coullet, P., and Tresser, C., Preprint, NPS, 79/1.

[13] E.g. Milnor, J., Thurston, W., Preprint, Princeton (1977).
Guckenheimer, J., Oster, G., and Ipaktchi, A., *J. Math. Biology* **4**, 101 (1977).
May, R., *Nature* **261**, 459 (1976) and references quoted therein.

[14] Sarkovskii, A. N., *Ukranian Math. J.* **16**, 61 (1964).
Stephan, P., *Commun. Math. Phys.* **54**, 237 (1977).

[15] Misiurewicz, M., *Bull. Acad. Pol. Sci.* (to appear).

[16] Denker, M., Grillenberger, C., and Sigmund, K., 'Ergodic Theory on Compact Spaces' *Lecture Notes in Mathematics*, 527, Springer Verlag, 1976.

[17] Coullet, P., and Tresser, C., Preprint NPS 79/2.

[18] E.g. Hénon, M., and Heiles, C., *Astron. J.* **69**, 73 (1964).
Benettin, G., and Strelcyn, G. M., *Phys. Rev. A* **17**, 773 (1978).
Benettin, G., Froeschle, C., and Scheidecker, J. P., to appear in *Phys. Rev. A*.
Benettin, G., Galgani, L., and Strelcyn, G. M., *Phys. Rev. A* **14**, 2338 (1976).
Chirikov, B. V., Researches concerning the theory of nonlinear resonances and stochasticity CERN, Trans. No. 71-40, Geneva, 1971 (unpublished).

[19] Grossmann, S., and Thomae, S., *Z. Naturforsch* **32 a**, 1353 (1977).

[20] Maurer, J., and Libchaber, A., Preprint (1979).

[21] Coullet, P., Tresser, C., and Arneodo A., in preparation.
Gollub, J. P. and Benson, S. V., Preprint (1979).

[22] Ruele, D., and Takens, F., *Commun. Math. Phys.* **20**, 167 (1971).

Added in proof: Many new references may be found in the Proceedings of the International Conference on Nonlinear Dynamics, New York, December 1979.

CHARLES P. ENZ

HYDRODYNAMIC MODELS WITH RANDOM FORCES

ABSTRACT. After a short introduction of the hydrodynamic equations and of the flow patterns leading to instabilities, two specific and equivalent models are studied in the form of Langevin equations, namely a Couette and a Benard flow in two dimensions. Based on a simple and precise definition of 'far away from equilibrium' states, a perturbation formalism applicable in this regime is described. The main result consists in an explicit construction of the unperturbed stationary state which, although it is Gaussian and hence implies the existence of a Wick's theorem, it is nontrivial far away from equilibrium.

1. Introduction

With the development of renormalization group theory and of rigorous methods in mathematical physics, hydrodynamic theory has enjoyed a remarkable revival of interest in the problems both of instabilities and of turbulence. In this paper we address ourselves to the first of these problems with the purpose of showing that, at least in the simplest models of a hydrodynamic instability, methods developed to treat dynamic systems near equilibrium may successfully be carried over to situations far away from equilibrium. The purpose is not to present an actual calculation of diagrams but rather to describe a framework, which here is the Fokker–Planck description of the hydrodynamic equations with and without constraints. This means that random forces simulating the molecular disorder or an external stirring mechanism are included in the equations of motion.

The hydrodynamic equations of motion [1] are just the macroscopic local laws of mass conservation

$$\dot{\rho} + \nabla . \boldsymbol{j} = 0,$$

of momentum balance

$$\dot{\boldsymbol{j}} + \nabla \cdot (\pi + \pi') = \rho \xi$$

and of energy conservation

$$\dot{e} + \nabla \cdot (\boldsymbol{J}_e + \boldsymbol{J}'_e) = 0.$$

E. Tirapegui (Ed.), Field Theory, Quantization and Statistical Physics, 263–275.

Here ρ is the mass density, $\boldsymbol{j}=\rho\boldsymbol{u}$ the momentum density and e the energy density.

$$\boldsymbol{j}, \pi_{ik} = p\delta_{ik} + \rho u_i u_k, \qquad \boldsymbol{J}_e = e\boldsymbol{u} + \boldsymbol{q}$$

are the associated reactive fluxes and

$$\pi'_{ik} = -\nu_0(\nabla_i j_k + \nabla_k j_i - \tfrac{2}{3}\nabla.\boldsymbol{j}\delta_{ik}), \qquad \boldsymbol{J}'_e = \pi'\boldsymbol{u} - \kappa\nabla T$$

their dissipative parts (the dissipative part of $\boldsymbol{j}$ is neglected). $\rho\xi$ is an external force density, $\boldsymbol{u}$ the fluid velocity, p the pressure, $\boldsymbol{q}$ an external heat flux, ν_0 the shear viscosity (the bulk viscosity is neglected) and κ the heat conductivity. These equations have to be supplemented by the local form of the first and second laws of thermodynamics,

$$\mathrm{d}e = T\mathrm{d}s + \mu\mathrm{d}\rho + \boldsymbol{u}.\mathrm{d}\boldsymbol{j}$$

and

$$T\mathrm{d}s = c_v\mathrm{d}T$$

where s is the entropy density, μ the chemical potential and c_v the heat capacity.

In investigations of hydrodynamic instabilities or of turbulence it is customary to introduce the Boussinesq approximation in which ρ, c_v and κ are considered constant and only linear terms in $\boldsymbol{u}$ are kept in the energy conservation law [2]. In the next section these simplified equations are given and supplemented with appropriate destabilizing forces and with constraints describing the proper geometry. The following section introduces random forces and develops the Fokker–Planck description while the remainder is devoted to the analysis of the equations thus obtained.

2. Flow patterns

In the Boussinesq approximation the local mass conservation reduces to the transversality condition

$$\nabla\cdot\boldsymbol{u} = 0 \tag{1}$$

which just expresses incompressibility, and the local momentum balance becomes the Navier–Stokes equation

$$\dot{\boldsymbol{u}} + (\boldsymbol{u}\cdot\nabla)\boldsymbol{u} = -\frac{1}{\rho}\nabla p + \nu_0\nabla^2\boldsymbol{u} + \boldsymbol{\chi} + \boldsymbol{\xi} \tag{2}$$

where we have added a destabilizing force per unit mass, $\boldsymbol{\chi}$.

It should be noted that the incompressibility assumption is not necessarily a simplification because the elimination of the pressure with the help of the transversality (1) introduces the transverse projectors

$$P_{ij} = \delta_{ij} - \nabla^{-2}\nabla_i\nabla_j \tag{3}$$

However, an analogous formalism to the one described below is possible [3]. Also, there are physical situations such as sound propagation or the Benard effect [2] in which density variations $\delta\rho$ are not negligible.

In the Bénard instability the effect of gravity on the density is overcome by an opposite thermal gradient, $T_{up} < T_{down}$, via thermal expansion. This leads to a destabilizing force

$$\boldsymbol{\chi} = (1 + \delta\rho/\rho)\boldsymbol{g} = [1 - \alpha(T - \bar{T})]\boldsymbol{g} \tag{4}$$

where α is the thermal expansion coefficient, $\bar{T}$ and T are the average and local temperatures and $\boldsymbol{g}$ is the acceleration of gravity, $g_x = g_y = 0, g_z = -g$. At the instability the flow pattern changes to rotations on a raw of cylinders (rolls). Temperature enters as a dynamic variable through the local energy conservation, supplemented with the first and second laws and with the Boussinesq approximation,

$$\dot{T} + (\boldsymbol{u}\cdot\nabla)T = \frac{\kappa}{c_v}\nabla^2 T - \frac{1}{c_v}\nabla\cdot q. \tag{5}$$

The flow pattern is only determined if in addition boundary conditions (and, in non-stationary situations, initial conditions) are specified. This fixes the geometry and the stationary local fields $\boldsymbol{u}_0(\boldsymbol{r}), u_{04} = T_0(\boldsymbol{r})$ such that dynamics is described by the fluctuations $v_i = u_i - u_{0i}$ $(i = 1,2,3)$, $v_4 = T - T_0$. Thus in the Bénard effect the lower boundary is a fixed horizontal plane while the upper boundary may be free or fixed.

In the Couette flow [2] an incompressible fluid is confined between two concentric cylinders of radii R_1 and R_2 (and $R_2 > R_1$) rotating with angular velocities Ω_1 and Ω_2, and destabilization is due to the centrifugal force which in cylindrical coordinates ϕ, z, r reads

$$\chi_\phi = \chi_z = 0, \qquad \chi_r = \frac{1}{r}u_\phi^2, \tag{6}$$

and which dominates for $\Omega_2 < \Omega_1$. This is the Taylor instability at which the flow pattern changes to rotations on a raw of rings (toroidal rolls).

The equations of motion for the fluctuations v_i ($i = 1, \ldots 4$) take the form

$$\dot{v}_i = -v_j \nabla_j v_i + v_i \nabla^2 v_i - \frac{1}{\rho} \nabla_i p - S_{ij} v_j + \xi_i \tag{7}$$

where $\boldsymbol{r} = (x_1, x_2, x_3)$, $\nabla_i = \partial/\partial x_i$, $v_i = v_0$ $(i = 1,2,3)$, $\nabla_4 = 0$, $v_4 = \kappa/c_v$, $\xi_4 = -(1/c_v)\nabla \cdot q$ and repeated indiced are summed. The fourth term on the right-hand side of Equation (7) consists of a drag

$$A_{ij} v_j = u_{0j} \nabla_j v_i + (\nabla_j u_{0i}) v_j \tag{8}$$

and of the destabilizing force ($\chi_4 = 0$)

$$\chi_i(u) - \chi_i(u_0) = -B_{ij}(v) v_j \tag{9}$$

such that

$$S_{ij} = A_{ij} + B_{ij}. \tag{10}$$

Of particular interest are models with constant matrix S. An example is the Couette flow in the limit of small differences δR between the radii of the cylinders and $\delta\Omega$ between their angular velocities, as long as the flow is independent of the rotation angle ϕ. Identifying $x_1 = \phi$, $x_2 = z$, $x_3 = r$; $v_1 = v_\phi, v_2 = v_z, v_3 = v_r, v_4 = T - T_0 = 0$, Equations (7) hold for $i = 1, 2, 3$. S_{ij} is determined from the stationary solution of Equation (2) which has the form $u_{0\phi} = ar + b/r$, $u_{0z} = u_{0r} = 0$, where a and b are determined by the boundary conditions $u_{0\phi}(R_i) = \Omega_i R_i (i = 1, 2)$. For $R_i \cong R, \Omega_i \cong \Omega$ it follows from Equation (8) that $A_{13} = \Omega + R\delta\Omega/\partial R \equiv \beta$, and Equations (6) and (9) yield $B_{31} = -2\Omega$. Hence, according to (10),

$$S = \begin{pmatrix} 0 & 0 & \beta \\ 0 & 0 & 0 \\ \gamma & 0 & 0 \end{pmatrix} \tag{11}$$

with $\gamma = -2\Omega$.

The same equations hold for the Bénard flow, as long as it is 2-dimensional, i.e. independent of the coordinate x, and for fixed boundary conditions on both sides [4]. Identifying $x_1 = x$, $x_2 = y$, $x_3 = z$, but relabelling $v_1 = T - T_0$, $v_2 = v_y$, $v_3 = v_z$, $v_4 = v_x = 0$, Equations (7) hold again for $i = 1,2,3$. S_{ij} is determined by the stationary solution of Equations (2) and (5), $u_{01} = T_0 = \bar{T} + \beta x_3$, $u_{02} = u_{0y} = 0$, $u_{03} = u_{0z} = 0$, $u_{04} = u_{0x} = 0$. From (8) and from (4) and (9) one obtains $A_{13} = \beta$ and $B_{31} = -\alpha g$, so that S takes again the form (11), with $\gamma = -\alpha g$.

From the discussion of a similar model in Ref. 5 it follows (see also Ref. 2) that a necessary condition for a soft-mode instability is that $\beta\gamma > 0$ in (11). This implies $\Omega_2 < \Omega_1$ in the Couette flow and $T_{\mathrm{up}} < T_{\mathrm{down}}$ in the Bénard effect.

Eliminating the pressure with the help of (1) and going over to Fourier components according to

$$v_{iq} = V^{-1/2} \int \mathrm{d}^3 r \, \mathrm{e}^{-iq.r} v_i(\boldsymbol{r}),$$

V being the volume, Equations (7) may be written as

$$\dot{v}_{\perp iq} = f^0_{iq} + f'_{iq} + g'_{iq} + \xi_{iq}. \tag{12}$$

Here

$$f^0_{iq} = -V^{-1/2} \sum_k iq_j P_{il}(\boldsymbol{q})\, v_{\perp jk} v_{\perp lq-k} \tag{13}$$

is the non-dissipative mode-coupling force,

$$f'_{iq} = -\nu_0 q^2 v_{\perp iq} \tag{14}$$

the dissipative viscous force (in the Bénard case this implies the assumption $\nu_4 = \nu_0$)

$$g'_{iq} = -P_{il}(\boldsymbol{q}) S_{lj} v_{\perp jq} \tag{15}$$

the drag and destabilizing force, which is also dissipative (hence the prime), i.e. time-reversal violating, and $v_{\perp iq} = P_{ij}(\boldsymbol{q}) v_{jq}$ where $P_{ij}(\boldsymbol{q}) = \delta_{ij} - \hat{q}_i \hat{q}_j$ is the Fourier transform of (3) and $\hat{q} = \boldsymbol{q}/q$. Note that in the above Couette and Bénard models the $v_i(\boldsymbol{r})$ are independent of x_1 so that $v_{iq} \propto \delta_{q_1.0}$.

In the laminar regime without imposed flow, $S_{ij} = 0$, the viscous force f'_{iq} dominates and Equation (12) may be written as

$$\dot{v}_{\perp iq} \cong (-i\omega_0 + \nu_0 q^2) v_{\perp iq}$$

with $\omega_0 = 0$, which means a soft mode. This shows that the laminar regime has an infinite correlation length $\sqrt{\nu_0/\omega_0}$. As a consequence there is scaling in the long wavelength limit $q \to 0$, in analogy to critical phenomena but without any instability. It is therefore possible to study the long-time, long-distance behaviour of the Navier–Stokes equation by dynamical renormalization group (RG) methods [6].

Laminar flow occurs for large viscosity ν_0. In the opposite limit which is dominated by the mode-coupling term f^0_{iq} the flow is in the regime of fully developed turbulence [7, 8]. On the other hand, the instabilities mentioned above are dominated by the linear term g'_{iq} and mark the onset of turbulence. These instabilities are characterized by a soft mode as in continuous phase transitions. They are chaotic in time but not in space, in the sense that time-averages tend to zero for long-time intervals. Fully developed turbulence, on the other hand, is characterized by spatio-temporal chaos due to the non-linearity of f^0_{iq}.

3. Stochastic forces

An important way of influencing flow patterns is by appropriate stochastic forces. Physically this amounts to a random stirring of the fluid. Here, we disregard thermal effects and assume that the external force ξ_q is stochastic with zero mean and Markovian correlation

$$\langle \xi_{iq}(t)\xi_{jk}(t')\rangle = 2C_{iq,jk}\delta(t-t') \tag{16}$$

of the form [6]

$$C_{iq,jk} = D(q)P_{ij}(\boldsymbol{q})\delta_{q+k,0}. \tag{17}$$

The physical meaning of $D(q)$ is given by the expression

$$4D(q) = \int dt \boldsymbol{\xi}_{-q}(t)\cdot\boldsymbol{\xi}_q(0) \sim \boldsymbol{v}_{-q}\cdot\boldsymbol{\xi}_q(0)$$

which is the power injected by the stirring force ξ_q per unit mass at wave-number $\boldsymbol{q}$. It balances dissipation by f'_q and redistribution by f^0_q which is necessary in a stationary state.

In the laminar regime without imposed flow, $S_{ij} = 0$, a forcing spectrum $D(q) \propto q^2$ ensures energy balance at all q-values and leads to a Maxwell distribution, as will be seen below. This means that the fluid is close to global equilibrium and obeys detailed balance. Other functions $D(q)$ have been considered in the literature [6,9–11] and will be discussed below.

In fully developed turbulence a forcing spectrum $D(q)$ peaked at a small q gives rise to the energy cascade of Kolmogorov [12] in which the injected power propagates to flow structures (eddies) of smaller and smaller size and finally dissipates [7]. This looks like scaling in $\boldsymbol{r}$-space and invites again to use RG-techniques (see Ref. 8 for a review).

Equations (12) and (16) are stochastic differential equations and, in analogy to the example of Brownian motion, may be called generalized

Langevin equations. To this 'Heisenberg representation' there is an equivalent 'Schrödinger representation' given by the Fokker–Planck (FP) equation for the probability distribution $P(v,t)$ [13],

$$P = -\partial_\mu J_\mu \tag{18}$$

where $\mu \equiv (i,\boldsymbol{q}), \partial_\mu \equiv \partial/\partial v_{i\boldsymbol{q}}$ and

$$J_\mu = (f_\mu - \partial_\nu C_{\nu\mu})P \tag{19}$$

is the probability flux (repeated indices are summed). Here

$$f_\mu = f_\mu^0 + f'_\mu + g'_\mu \tag{20}$$

and $C_{\mu\nu} = C_{\nu\mu}$ is the correlation matrix of Equation (16). The connection between the two representations may be shown quite generally [14].

Writing the stationary probability distribution in the form

$$P^s(v) = Z^{-1}\mathrm{e}^{-F(v)} \tag{21}$$

the stationary FP equation $\partial_\mu J_\mu^s = 0$ becomes an equation for the 'free energy' F,

$$[\partial_\mu - (\partial_\mu F)][j_\mu + f_\mu^0] = 0, \tag{22}$$

where

$$j_\mu = f'_\mu + g'_\mu + C_{\mu\nu}(\partial_\nu F) - (\partial_\nu C_{\mu\nu}), \tag{23}$$

constitutes the dissipative part, $j_\mu P^s$, of the stationary probability flux J_μ^s.

4. Detailed balance

The flux $j_\mu P^s$ defined in Equation (23) is of particular interest since it represents a current imposed from the outside. $j_\mu = 0$, also called the 'potential conditions' or 'detailed balance' [13], represents a state 'close to global equilibrium', for which the stationary FP equation reduces to

$$(\partial_\mu - (\partial_\mu F))f_\mu^0 = 0. \tag{24}$$

As shown in Ref. 11, the expression (13) implies $\partial_\mu f_\mu^0 = 0$, and

$$F_M = \frac{1}{2T}\sum_{iq} v_{iq} v_{i-q} \tag{25}$$

is the only quadratic solution of Equation (24). Interpreting the constant T as temperature, F_M gives rise to a Maxwell distribution

$P^s_M = Z_M^{-1} \exp(-F_M)$. Inserting Equations (14), (15), (17) and (25) into (23) the condition $j_\mu = 0$ implies $S_{ij} = 0$ and the 'Einstein relation' [13]

$$D(q) = T\nu_0 q^2. \tag{26}$$

This case, already mentioned above, corresponds to model A of Ref. 6. It is characterized by an exact Gaussian stationary state P^s_M, so that the non-linear mode-coupling term f^0_μ affects only dynamics. This particularity gives rise to very general non-renormalization theorems for response functions [11].

5. Gaussian unperturbed stationary state

It is of interest to see how perturbation theory for the non-linear term f^0_μ works beyond detailed balance. A necessary condition for the possibility of a perturbative treatment is the factorization of the unperturbed Green's functions. This is Wick's theorem [15] – which is always connected with a Gaussian unperturbed stationary state $P^s_{un} = Z_{un}^{-1} \exp(-F_{un})$, i.e. with a quadratic free energy

$$F_{un} = \tfrac{1}{2} E_{\mu\nu} v_\mu v_\nu. \tag{27}$$

Therefore, we look for a solution of the stationary FP equation (22) with $f^0_\mu = 0$ and with F given by (27). Writing the linear functions (14) and (15) as

$$f'_\mu + g'_\mu = -\Lambda_{\mu\nu} v_\nu \tag{28}$$

where

$$\Lambda_{iq,jk} = P_{il}(\boldsymbol{q}) \lambda_{lj}(\boldsymbol{q}) \delta_{q,k} \tag{29}$$

and

$$\lambda_{lj}(\boldsymbol{q}) = \nu_0 q^2 \delta_{lj} + S_{lm} P_{mj}(\boldsymbol{q}) \tag{30}$$

this leads to the condition

$$\sigma_{\mu\mu} - E_{\mu\lambda} \sigma_{\lambda\nu} v_\mu v_\nu = 0 \tag{31}$$

valid for all v_μ. Here

$$\sigma_{\mu\nu} = C_{\mu\lambda} E_{\lambda\nu} - \Lambda_{\mu\nu} \tag{32}$$

and the unperturbed flux (23) is $j^{un}_\mu = \sigma_{\mu\nu} v_\nu$.

In the models treated above $C_{iq,jk} \propto \delta_{q+k,0}$ and $\Lambda_{iq,jk} \propto \delta_{q,k}$. Therefore, we can write $E_{iq,jk} = E_{ij}(\boldsymbol{q}) \delta_{q+k,0}$ and $\sigma_{iq,jk} = \sigma_{ij}(\boldsymbol{q}) \delta_{q,k}$. Then follows from

(17), (30) and (32) that

$$\sigma_{ij}(\boldsymbol{q}) = P_{il}(\boldsymbol{q})\left[D(q)E_{lj}(\boldsymbol{q}) - \lambda_{lj}(\boldsymbol{q})\right] \tag{33}$$

and condition (31) becomes, in matrix notation,

$$\sum_{q} \operatorname{Tr}\sigma = 0, \qquad E\sigma + (E\sigma)^T = 0, \tag{34}$$

where the index T means the transposed. Since E is symmetric, the solution of Equations (34) is not trivial, unless λ is also symmetric, in which case

$$\mathrm{DE} = \lambda = \lambda^T \tag{35}$$

and $\sigma = 0$. Hence $j_\mu^{un} = 0$, i.e. the unperturbed stationary state is close to equilibrium. An example of this case is the unconstrained fluid with arbitrary $D(q)$, which is model B or C of Ref. 6: $S = 0$ and $\lambda = v_0 q^2 \mathbb{1}$. Note that $j_\mu^{un} = 0$ does of course not mean that the exact stationary state P^s is also close to equilibrium, and hence there is no contradiction of the present result with the remarks made in the Introduction to Ref. 11. As we shall see below, the full P^s is indeed non-Gaussian and "far from equilibrium" ($j_\mu \neq 0$).

In the non-trivial case when $\lambda \neq \lambda^T$ is given by Equations (30) and (11), the conditions (34) have the unique solution [17]

$$E = \mathbb{1} - P + \frac{v_0 q^2 n}{KD}\{mP + \tfrac{1}{2}\hat{q}_2(\beta + \gamma)v_0 q^2(R + R^T) + \tfrac{1}{2}\hat{q}_2^2(\gamma^2 - \beta^2)Q\} \tag{36}$$

and

$$\sigma = \hat{q}_2(\gamma - \beta)\frac{n}{K}\{mR - m'R^T + \hat{q}_2(\beta + \gamma)v_0 q^2(Q - \tfrac{1}{2}P)\}. \tag{37}$$

Here

$$Q \equiv \begin{pmatrix} 1 & 0 & 0 \\ 0 & 0 & 0 \\ 0 & 0 & 0 \end{pmatrix}, \qquad R = \begin{pmatrix} 0 & -1 & 1 \\ 0 & 0 & 0 \\ 0 & 0 & 0 \end{pmatrix}, \tag{38}$$

and

$$\begin{aligned} m &\equiv v_0^2 q^4 + \tfrac{1}{2}\hat{q}_2^2\beta(\beta - \gamma) \\ m' &\equiv v_0^2 q^4 + \tfrac{1}{2}\hat{q}_2^2\gamma(\gamma - \beta) \\ n &\equiv v_0^2 q^4 - \hat{q}_2^2\beta\gamma \\ K &= mm' - \tfrac{1}{4}\hat{q}_2^2(\beta + \gamma)^2 v_0^2 q^4. \end{aligned} \tag{39}$$

It is clear from (37) that $j_\mu^{un} \neq 0$, i.e. the unperturbed stationary state is far from equilibrium, unless $\hat{q}_2(\gamma - \beta) = 0$, or for a particular value of q.

In an earlier attempt to construct the stationary state P^s for a Couette-type model also determined by Equation (11) but with Einstein relation (26), i.e. for an extension of model A to the case $S \neq 0$, a modification of the model-A solution $P^s{}_m$ was tried where v_{iq} in Equation (25) was replaced by $v_{iq} + w_{iq}$, w_{iq} being independent of $\boldsymbol{v}$ [5]. However, the solution found for S given by (11), namely

$$w_{iq} = -(\beta + \gamma)\delta_{i1} V^{-1/2}(\partial/\partial iq_3)\delta_{q,0},$$

has to be rejected because it is discontinuous at $\boldsymbol{q} = 0$.

The only other far-from-equilibrium Gaussian unperturbed steady state solution produced in the literature is that of Kawasaki and Onuki for the model of a fluid near its critical point subject to a shear flow [18]. This is, in fact, a very natural model to look for a far-from-equilibrium instability since it combines an inherent (thermodynamic) phase-transition with a current imposed from the outside.

6. Perturbation Theory

In the presence of the non-linear term f_μ^0 we look for a solution of Equation (22) in the form of a power series in v_μ,

$$F = F_{un} + \sum_{N=1}^{\infty} F_N \tag{40}$$

with

$$F_N = \frac{1}{N}\phi^{(N)}_{\mu_1 \cdots \mu} \; v_{\mu_N} \cdots v_{\mu_N}. \tag{41}$$

One finds that the solution is determined only up to an external field $\phi_\mu^{(1)}$ which generates non-vanishing terms $\phi^{(N)}$ in all higher orders, while f_μ^0 influences only the terms with $N \geqslant 5$. In the absence of an external field, $\phi_\mu^{(1)} = 0$, the lowest order non-vanishing term is found to be [17]

$$\phi^{(5)}_{\mu_1 \ldots \mu_5} = \tfrac{1}{4}\{E_{\mu_1 \mu_2} \bar{C}^{-1}_{\mu_3 \nu} \bar{\Gamma}_{\nu,\mu_4 \mu_5}\}_{\text{sym}}. \tag{42}$$

Here $\bar{C}_{\mu\nu}$ and $\bar{\Gamma}_{\mu,\nu\lambda}$ are related to $C_{\mu\nu}$ and to the coefficient in $f_\mu^0 = -\Gamma_{\mu,\nu\lambda} v_\nu v_\lambda$ by a factor $P_{ij}(\boldsymbol{q})$,

$$\bar{C}_{iq,jk} \equiv D(q)\delta_{q+k,0} \tag{43}$$

and

$$\bar{\Gamma}_{iq,jklp} \equiv \tfrac{1}{2} V^{-1/2}(iq_m \delta_{in} + iq_n \delta_{im}) P_{mj}(\boldsymbol{k}) P_{nl}(\boldsymbol{p}) \delta_{q,k+p} \tag{44}$$

and the index 'sym' means symmetrization with respect to all free indices. While $\phi^{(6)}$ is again zero, non-vanishing terms are generated in every higher order.

With this result one may analyze the nature of the instability in the above Bénard and Couette flow models by applying static renormalization group methods. On the other hand, for an investigation of the dynamics of these models as well as of models B or C of Ref. 6, the perturbative Green's function formalism of Refs. 11 and 13 turns out again to be very well suited. In this formalism one constructs a 'dynamical matrix'

$$D_{\mu\nu} = D^0_{\mu\nu} - C_{\mu\nu},\ D^0_{\mu\nu} = -D^0_{\nu\mu}, \tag{45}$$

such that

$$f_\mu = (\partial_\nu F) D_{\nu\mu} - (\partial_\nu D_{\nu\mu}). \tag{46}$$

Developing again in powers of v_μ,

$$D_{\mu\nu} = \sum_{N=0}^{\infty} D_{N\mu\nu} \tag{47}$$

with

$$\begin{aligned} D_{N\mu\nu} &= \Delta^{(N)}_{\mu\nu,\lambda_1\ldots\lambda_N} v_{\lambda_1} \cdots v_{\lambda_N}, \qquad N \geqslant 1, \\ D_{0\mu\nu} &= \Delta^{(0)}_{\mu\nu} - C_{\mu\nu}, \end{aligned} \tag{48}$$

one finds for model B that the first non-vanishing term is

$$\Delta^{(3)}_{\mu\nu,\lambda_1\lambda_2\lambda_3} = \left(1 + \sum_\mu 1\right)^{-1} \{\Gamma_{\mu,\lambda_1\lambda_2}\delta_{\nu\lambda_3} - \Gamma_{\nu,\lambda_1\lambda_2}\delta_{\mu\lambda_3}\}_{\text{sym}} \tag{49}$$

while in the case $S \neq 0$ one has in addition to this term [17]

$$\Delta^{(0)}_{iq,jk} = -\sigma_{il}(\boldsymbol{q}) E^{-1}_{lj}(\boldsymbol{q}) \delta_{q+k,0}. \tag{50}$$

Note that according to (33), (34) this has the correct antisymmetry and that in both these cases also $\Delta^{(4)} = 0$.

This result gives rise to an interesting vertex structure of the perturbation theory. In fact, while the only static vertices F_N are of order 5,7 and higher there is a normal third order dynamical vertex $\mathscr{L}_1 = f^0_\mu \partial_\mu$ as in Ref. 11 while 'exceptional vertices' of order 5,7 and higher are generated by the quantities

$$\hat{v}_{N\mu} \equiv D_{N\mu\nu}\partial_\nu \tag{51}$$

for $N = 3,5$ and higher, similarly to Ref. 11. For $N = 0$ this defines the unperturbed conjugate variables $\hat{v}_{0\mu}$ which in the case $S \neq 0$ are modified as compared to Ref. 11. One finds from Equations (17), (48), (50), invoking (33) and (34),

$$\hat{v}_{0iq} = -(E^{-1}\lambda^T P)_{ij}\partial_{j-q}. \tag{52}$$

Note that all the results of this and the last sections are independent of the particular form of the random force spectrum $D(q)$.

7. Conclusion

In giving a complete description of the basic elements of a perturbation theory along the lines of Refs. 11 and 13 we have achieved several goals. First, we have shown that a treatment of models B or C of Ref. 6 is indeed possible and even straightforward in this formalism and allows for a more systematic analysis than was attempted by Foster *et al.* [6]. In particular, the vertex structure found above may be considered as a justification for the use of a third-order vertex in the calculations for model B made in Ref. 6.

Furthermore, the results for the Bénard and Couette flow models show that our Green's function formalism works well in situations far away from equilibrium. Whether the time-dependent Ginzburg–Landau result, which was obtained in the Bénard case by Graham [19] and by Swift and Hohenberg [20] with the help of an iteration procedure of the equations of motion, can be obtained by renormalization of the free energy and of the vertices is not obvious in view of the above vertex structure. As emphasized by Swift and Hohenberg the important question whether the first instability in the Bénard problem is continuous (second-order) or discontinuous (first-order) depends directly on this result which, in turn, is also conditioned by the existence or non-existence of rotational symmetry in the horizontal plane [20]. Both, Graham's treatment [19] and our Bénard flow model lack this symmetry. However, it is easily recovered in our case by going back to the full Equations (7) which show no obvious obstacle in applying the formalism described in this paper [17].

This formalism is a natural generalization along the lines of well-known perturbation schemes using Feynman diagrams or, more exactly,

Dyson's double graphs [15] and a fluctuation-dissipation theorem [11, 13]. A particularly welcome feature is that stationarity appears as the zero-frequency limit of dynamics [13] which is not the case in the other method presently available for treating situations far away from equilibrium [18].

Department of Theoretical Physics, University of Geneva

References

[1] Landau, L. D., and Lifshitz, E. M., *Fluid Mechanics*, Pergamon, Oxford, 1959.
[2] Chandrasekhar, S., *Hydrodynamic and Hydromagnetic Stability*, Clarendon, Oxford, 1961.
[3] Enz, C. P., and Turski, L. A., *Physica* **96A**, 369 (1979).
[4] Scholz, H.-J., *Zur Theorie hydrodynamischer Fluktuationen und Instabilitäten*, Diplomarbeit, Universität Stuttgart, 1975, unpublished.
[5] Droz, M., Enz, C.P., and Malaspinas, A., in L. Garrido, P. Seglar and P. J. Shepherd (eds.), 'Stochastic Processes in Nonequilibrium Systems', Lecture Notes in Physics, Vol. 84, Springer, Berlin–Heidelberg–NewYork, 1978, p. 280.
[6] Forster, D., Nelson, D. R., and Stephen, M. J., *Phys. Rev.* **A16**, 732 (1977).
[7] Rose, H. A., and Sulem, P. L., *J. de Physique* **5**, 441 (1978).
[8] Sulem, P. L., Fournier, J. D., and Pouquet, A., in C. P. Enz (ed.), 'Dynamical Critical Phenomena and Related Topics', Lecture Notes in Physics, Vol. 104, Springer. Berlin–Heidelberg–New York, 1979, p. 320.
[9] Lücke, M., *Phys. Rev.* **A18**, 282 (1978).
[10] de Dominicis, C., and Martin, P. C., *Phys. Rev.* **A19**, 419 (1979).
[11] Enz, C. P., *Physica* **94A**, 20 (1978).
[12] Kolmogorov, A. N., *C. R. Acad. Sci. URSS* **30**, 301 (1941); *Sov. Phys.-Uspeki* **10**, 734 (1968).
[13] Enz, C. P., *Physica* **89A**, 1 (1977).
[14] Graham, R., in L. Garrido, P. Seglar and P. J. Shepherd (eds.), 'Stochastic Processes in Nonequilibrium Systems', Lecture Notes in Physics, Vol. 84, Springer, Berlin–Heidelberg–New York, 1978, p. 83.
[15] Enz, C. P., and Garrido, L., *Phys. Rev.* **A14**, 1258 (1976).
[16] Martin, P. C., Siggia, E. G., and Rose, H. A., *Phys. Rev.* **A8**, 423 (1973).
[17] Enz, C. P., to be published.
[18] Onuki, A., and Kawasaki, K., *Ann. Phys. (N. Y.)*, **121**, 456 (1979).
[19] Graham, R., *Phys. Rev.* **A10**, 1762 (1974).
[20] Swift, J., and Hohenberg, P. C., *Phys. Rev.* **A15**, 319 (1977).

ARNO HOLZ

HIGH-TEMPERATURE STUDY OF PLANAR ROTATOR- AND XY-MODELS FOR DIMENSIONS $D \geqslant 2$

Dedicated to the memory of Bernard Jouvet

ABSTRACT. We establish a relation between discrete excitation models at low temperature and anisotropic rotator- and XY-models at high temperature for dimensions $D \geqslant 2$. It is shown that the correlation functions of these models decay exponentially. The discrete excitation model corresponding to the high-temperature planar rotator model for $D = 2$ is shown to represent the Kac–Ward determinant formulation of the 2D Ising model. A qualitative discussion of the phase transitions occurring in these systems is given.

1. Introduction

In recent publications [1,2] a discrete excitation model has been introduced with a finite number of degrees of freedom per lattice site which can be used for a study of the high-temperature properties of some Heisenberg-type models in dimensions $D \geqslant 2$. The method used is based on the extension of the discrete excitation model introduced by Knops [3] for $D = 2$ to arbitrary dimensions and the notion of monopoles associated with anisotropy fields [4–6].

The present theory makes use of an approximate duality between the low-temperature properties of the discrete excitation models and the high-temperature properties of the Heisenberg-like models. Because the mapping of the Heisenberg-like models onto the discrete excitation models is only reliable in the asymptotic temperature regime it will not be possible to give a satisfactory discussion of the phase transition (PT) occurring in these systems in the context of the present method. In particular, for the case of the 2D planar rotator model where an analytic change of the correlation functions from a low-temperature power law decay to a high-temperature exponential decay law is expected a study of the presumed PT would be highly desirable.

The PT occurring in that system is usually studied using a low-temperature representation as initiated by Kosterlitz and Thouless [7] and further

E. Tirapegui (Ed.), Field Theory, Quantization and Statistical Physics, 277–294.

developed by José *et al.* [5] and many others quoted therein. The phase space of that problem can be represented by a gas of charged vortices with integer valued charge states and interacting via a Coulomb law. The difficulty with such an approach is that with increasing temperature many vortices are present whose statistical mechanics is difficult to handle. Some difficulties, however, can be overcome if one approximates the vortex system by a two-component Coulomb gas with temperature renormalized coupling constants. The high-temperature approach to this problem presented in this work can be considered as complementary to the low-temperature approach described above, except with regard to the PT-region. Here, both methods are faced with severe difficulties being of the same nature respectively. In the high-temperature approach the equivalent to the vortex system is a system of closed current loops with integer-valued current strengths satisfying Kirchhoff's law. In the present approach, this complicated system is approximated by a two-component loop gas, i.e., each loop may have only two orientations. This problem can then be mapped onto the Ising model using the Kac–Ward [8] determinant. It is easy to show then that if the model is extrapolated to lower temperatures a PT occurs into a long-range ordered state signaling the break-down of the two-component loop gas approximation. Accordingly if one wants to penetrate into the low-temperature region using the high-temperature approach all integer-valued current strengths have to be admitted. An approximative treatment of that problem may still be based on a finite-component loop model treating large-order current strengths by perturbation theory. An immediate difficulty which arises then is, however, that the power-law decay of correlations in the low-temperature region cannot be obtained easily or, say, in an obvious mean field approximation.

Similarly if the low-temperature vortex gas approach is used in the high-temperature region then no more correlations seem to exist for an infinitely extended system (see Section 5).

Although we are presently not aware how to overcome these difficulties connected with the PT-region which is certainly the most interesting temperature regime in any model it still seems to be worthwhile to develop a high-temperature approach to the Heisenberg-like models. The reason for that is that the high-temperature behavior of these models can then be studied by means of analytical methods and also some idea of the physics of the system can be obtained which may be useful when extrapolated to the PT region and beyond.

The following work will be devided into various parts. In Section 2 we explain the basic formalism used and in Section 3 the correlation functions for some special cases are investigated. In particular, we study the planar rotator model in a magnetic field and in the presence of an uniaxial anisotropy. Similar formulae as derived for the planar rotator model will shown to hold approximately for the XY-model. In Section 4 we investigate the planar rotator model using the mapping onto the Kac–Ward [8] determinant as mentioned earlier. In Section 5 we give a qualitative discussion of the possible PT occurring in some of the 2D, Heisenberg-like, models from the point of view of the low- and high-temperature approach.

2. Basic formalism

We consider a Hamiltonian of the form

$$H_{DE} = \frac{1}{2} \sum_{\langle i,j \rangle} J^{i}_{DE}(|n_{ij}|) n^2_{i,j} + \sum_{i} J^{a}_{DE}(|n_i|) n^2_i \tag{1}$$

on a D-dimensional simple hypercubic lattice with N lattice sites. In the first term of Equation (1), $\langle i,j \rangle$ indicates that only nearest neighbor sites are included in the summation. Furthermore the following relations have to hold:

$$n_{i,j} = -n_{j,i} \tag{2}$$

and $n_{i,j} \neq 0$ only for nearest neighbors i and j;

$$n_i = 0, \pm np, n = 1,2,3,\ldots, \tag{3}$$

where p is a fixed integer; the conservation law

$$\sum_{j \neq i} n_{i,j} + n_i = 0, \qquad i = 1,2,3,\ldots N. \tag{4}$$

In the following we will study mainly the case

$$n_{ij} = 0, \pm 1, \qquad n = 1. \tag{5}$$

Equations (1)–(5) define the discrete excitation model with finite degrees of freedom per lattice site. It will be shown that this model allows an asymptotically accurate mapping onto the high-temperature states of the planar rotator model. If, however, $n_{ij} = 0, \pm 1, \pm 2, \ldots$ and $n = 0, 1, 2, 3, \ldots$ is used with properly chosen temperature dependent coupling constants $J^{i}_{DE}(|n_{ij}|)$ and $J^{a}_{DE}(|n_i|)$ then a mapping between the two models valid for all temperatures can be achieved.

Using the integral representation for Equation (4)

$$(2\pi)^{-1}\int_0^{2\pi} d\varphi_i \exp\left\{i\varphi_i\left[\sum_{j\neq i} n_{i,j} + n_{i,0}\right]\right\} = \delta \tag{6}$$

where the right-hand side denotes the Kronecker symbol, the partition function to the problem can be written in the form

$$Z_{P,R} = (2\pi)^{-N}\prod_{i=1}^{N}\int_0^{2\pi} d\varphi_i \exp\left\{-\beta_{DE}\left[\sum_{\langle i,j\rangle}\tilde{V}^i(\varphi_i-\varphi_j) + \sum_i \tilde{V}^a(\varphi_i)\right]\right\} \tag{7}$$

where $\langle i,j\rangle$ indicates summation over nearest neighbors. In Equation (7) the abbreviations

$$\tilde{V}^i(\varphi) = -(2\beta_{DE})^{-1}\ln(1+2\cos\varphi e^{-\beta_{DE} j^i_{DE}}), \tag{8}$$

$$\tilde{V}^a(\varphi) = -\beta_{DE}^{-1}\ln(1+2\cos p\varphi e^{-\beta_{DE} p^2 J^a_{DE}}) \tag{9}$$

have been used and β_{DE} indicates the inverse temperature of the discrete excitation model. For

$$\beta_{DE} J^{i(a)}_{DE} \gg 1, \tag{10}$$

Equations (8) and (9) can be Taylor expanded and one obtains approximately

$$Z_{P,R} \sim Z^R_P = (2\pi)^{-N}\prod_{i=1}^{N}\int_0^{2\pi} d\varphi_i \exp\{-\beta_R H_R(\{\varphi_i\})\}. \tag{11}$$

Here the following abbreviations have been introduced

$$H_R(\{\varphi_i\} \equiv -J^i_R\sum_{\langle i,j\rangle}\cos(\varphi_i-\varphi_j) - J^a_R\sum_i \cos p\varphi_i, \tag{12}$$

$$\beta_R J^i_R \equiv e^{-\beta_{DE} J^i_{DE}}, \tag{12b}$$

$$\beta_R J^a_R \equiv 2e^{-\beta_{DE} J^a_{DE}}. \tag{12c}$$

It follows from Equations (10)–(12c) that the low-temperature partition function $Z_{P,R}$ of the discrete excitation model maps onto the high-temperature partition function Z_p^R of the anisotropic rotator model Equation (12a). Here J_R^i, J_R^a represent the coupling constants of the isotropic and anisotropic terms of the rotator model and β_R denotes its inverse temperature. Raising the second subindex of the partition function $Z_{P,X}$ of a discrete excitation model is used in the following always to indicate the partition function Z_p^X of the corresponding continuous model.

Within the same formalism a mapping between a discrete excitation model and the XY-model can be obtained. For that purpose the coupling constants of the discrete excitation models are substituted as follows

$$\beta_{DE} J_{DE}^i(i,j) = \beta_{DE} J_{DE}^i - \ln(\sin\theta_i \sin\theta_j), \tag{13a}$$

$$\beta_{DE} J_{DE}^a(i) = \beta_{DE} J_{DE}^a - \ln f(\sin\theta_i), \tag{13b}$$

where i,j refer to neighboring lattice sites and the function $f(X)$ should satisfy $0 < f(X) < 1$. In addition, the measure of the partition function of the discrete excitation model is supplemented by the integral operator

$$\prod_{i=1}^{N} \frac{1}{2} \int_0^\pi d\theta_i \sin\theta_i. \tag{14}$$

Going through the same routine as earlier and observing Equation (10) one obtains instead of Equation (11)

$$Z_{p,XY} \sim Z_p^{XY} = (4\pi)^{-N} \prod_{i=1}^{N} \int_0^\pi d\theta_i \sin\theta_i \times \int_0^{2\pi} d\varphi_i \exp(-\beta_{XY} H_{XY}(\{\varphi_i\},\{\theta_i\})). \tag{15}$$

Here the following abbreviations have been introduced

$$H_{XY}(\{\varphi_i\},\{\theta_i\}) \equiv -J_{XY}^i \sum_{\langle i,j\rangle} \sin\theta_i \sin\theta_j \cos(\varphi_i - \varphi_j) - J_{XY}^a \sum_i f(\sin\theta_i) \cos p\varphi_i, \tag{16a}$$

$$\beta_{XY} J_{EY}^i \equiv e^{-\beta_{DE} J_{DE}^i}, \tag{16b}$$

$$\beta_{XY} J_{XY}^a \equiv 2e^{-\beta_{DE} J_{DE}^a}. \tag{16c}$$

We consider next the partition function of the discrete excitation model. Equations (4) and (5) imply that each configuration of this model can be represented by an oriented line system carrying a unit current strength. In addition an equal number of sources and sinks of strength p is present which is coupled to the oriented line system such that Kirchhoff's law is satisfied everywhere. Accordingly all lines are oriented and either closed or connected to the sources and sinks. It follows now that $p \leqslant 2D$ has to hold and no more than D oriented lines can cross at a lattice point.

The partition function for $p = 0$ is given by

$$Z_{0,R} = \sum_{\{C_L\}} \exp\left[-\beta_{DE} j^i_{DE} \oint_{C_L} \mathrm{d}s_L \right], \tag{17}$$

where C_L is a collection of oriented and closed loops of total length $c_L = \oint_{C_L} \mathrm{d}s_L$, and $\{C_L\}$ denotes all possible loop collections. We like to point out that $\oint_{C_L} \mathrm{d}s_L$ denotes here the contour integral not only over one loop but a collection of loops. This contour integral is subject to the constraint that no more than D lines may cross at a lattice point. For $J^a_{DE} > 0$ one obtains instead of Equation (17)

$$Z_{p,R} = \sum_{[C_l],\{[C_L]_p\}} \exp\left\{ -\beta_{DE} J^i_{DE} \oint_{C_L} \mathrm{d}s_L - \beta_{DE}\left[J^a_{DE} 2p^2 n([C_s]_p) + J^i_{DE} \sum_{i=1}^{p} \int_{C^i_S} \mathrm{d}s^i_S \right]\right\}. \tag{18}$$

Here $[C_s]_p = (C^1_s, C^2_s, \ldots, C^p_s)$ is a collection of p-fold lines where each constituent of one p-fold line has the same orientation with respect to their common initial point or end point. $n([C_s]_p)$ gives the number of p-fold lines or the number of monopole pairs of strength p. It should be observed that these rules allow the monopole system to be interconnected over an intricate system of oriented lines.

For the XY-model we give the partition function of the discrete excitation model only for the case $f(X) \equiv 0$. One obtains

$$Z_{0,XY} = \sum_{\{C_L\}} \prod_{i=1}^{N} \frac{1}{2} \int_0^{\pi} \mathrm{d}\theta_i \sin\theta_i \prod_{i \in C_l} \sin^2\theta_i \exp\left[-\beta_{DE} j^i_{DE} \oint_{C_l} \mathrm{d}s_L \right] \tag{19a}$$

where $i \in C_L$ means that each point on the lattice which is crossed over by one or more oriented lines is taken once only. Equation (19a) is readily integrated and yields

$$Z_{0,XY} = \sum_{\{C_L\}} \left(\frac{2}{3}\right)^{C_L - m\ (C_L)} \exp\left[- \beta_{DE} j^i_{DE} \oint_{C_L} \mathrm{d}s_L \right] \tag{19b}$$

where $m^*(C_L)$ is proportional to the number of crossings $m(C_L)$ of the collection of loops C_L, where each crossing is counted with its proper multiplicity ranging from 0 to $D-1$ (0: no crossing, 1: two lines cross each other, etc.). for $D = 2$ one obtains $m^*(C_L) = m(C_L)(\log(8/15)/\log(2/3))$.

3. The correlation functions

Consider now the problem that at site $\mathbf{0}$ and site $\boldsymbol{r}$ there is a source and a sink of strength q, respectively, i.e., that

$$\sum_{j \neq i} n_{i,j} + n_i = q\delta_{i,\mathbf{0}} - q\delta_{i,\boldsymbol{r}}, i = 1, 2, \dots N, \tag{20}$$

holds instead of Equation (4). For the case where Equation (10) holds the partition function of that problem when applied to the rotator model is

$$Z_{p,R}(\mathbf{0}, \boldsymbol{r}) \sim Z^R_p(\mathbf{0}, \boldsymbol{r}) = (2\pi)^{-N} \prod_{i=1}^{N} \int_0^{2\pi} \mathrm{d}\varphi_i \, \mathrm{e}^{iq(\varphi_0 - \varphi_r)} * \mathrm{e}^{-\beta_R H_R(\{\varphi_i\})}. \tag{21}$$

Accordingly we get

$$\langle \mathrm{e}^{iq(\varphi_0 - \varphi_r)} \rangle_p = \langle \cos q(\varphi_{\mathbf{0}} - \varphi_r) \rangle_p \sim Z_{p,R}(\mathbf{0}, \boldsymbol{r})/Z_{p,R}, \tag{22}$$

where the subindex p reminds that a p-anisotropy is present. It follows from Equation (20) that $q \leqslant p + 2D$ is necessary in order that $Z_{p,R}(\mathbf{0}, \boldsymbol{r})$ does not vanish. In the following $q = 1$ will be considered mainly.

In a similar way the correlation functions of the XY-model can be constructed. Using the notation

$$\boldsymbol{S}_i = (S^x_i, S^y_i, S^z_i) \equiv (\cos \varphi_i \sin \theta_i, \sin \varphi_i \sin \theta_i, \cos \theta_i)$$

and Equation (20) one obtains

$$\langle S^x_{\mathbf{0}} S^x_r + S^y_{\mathbf{0}} S^y_r \rangle \sim \sum_{\{C_L\}, \{C_S(\mathbf{0}, \boldsymbol{r})\}} \prod_{i=1}^{N} \frac{1}{2} \int_0^{\pi} \mathrm{d}\theta_i \sin \theta *$$

$$* \prod_{i \in C_L} \sin^2\theta_i \prod_{i \in C_S(0,r)} \sin^2\theta_i \exp\left\{ -\beta_{DE} j^i_{DE} \left[\oint_{C_L} ds_L + \right.\right.$$

$$\left.\left. + \int_{\substack{0 \\ C_S(0,r)}}^{r} ds \right]\right\} \Big/ Z_{0,XY}. \qquad (23)$$

Here $i \in C_L$ and $i \in C_s(\mathbf{0}, \boldsymbol{r})$ means that i picks up all points which are crossed once by the collection C_L or by the string $C_s(\mathbf{0}, \boldsymbol{r})$ but each point i is counted only once. Equation (23) can be integrated easily and one obtains

$$\langle S_0 . S_r \rangle \sim \sum_{\{C_L\},\{C_S(0,r)\}} \left(\frac{2}{3}\right)^{C_L + C_S(0,r) + 1 - m^*(C_L, C_S(0,r))}$$

$$* \exp\left\{ -\beta_{DE} j^i_{DE} \left(\oint_{C_L} dS_L + \int_{\substack{0 \\ C_S(0,r)}}^{r} ds \right)\right\} \Big/ Z_{0,XY}, \qquad (24)$$

where use has been made of $\langle S^z_0 S^z_r \rangle \equiv 0$ and where $m^*(C_L, C_s(\mathbf{0}, \boldsymbol{r}))$ is proportional to the total number of crossings $m(C_L, C_s(\mathbf{0}, \boldsymbol{r}))$ of the collection of loops C_L and the string $C_s(\mathbf{0}, \boldsymbol{r})$. For $D = 2$ the same relation holds between m and m^* as given following Equation (19b).

Next we evalute Equation (22) for $q = 1$ and the three cases $p = 0$, $p = 1$, and $p = 2$.

(a) Isotropic planar rotator, $p = 0$. Equation (22) can be written for that case in the form

$$\langle \exp i(\varphi_0 - \varphi_r) \rangle_0$$

$$\sim \sum_{\{C_L\},\{C_S(0r)\}} \exp\left\{ -\beta_{DE} j^i_{DE} \left[\oint_{C_L} dS_L + \int_{\substack{0 \\ C_S(0,r)}}^{r} dS \right] \Big/ Z_{0,R}. \right. \qquad (25)$$

It follows from Equation (17) that for $\beta_{DE} J^i_{DE} \gg 1$ the saddle point of $Z_{0,R}$ consists of a dilute gas of small loops. Similarly the saddle point of $Z_{0,R}(\mathbf{0}, \boldsymbol{r})$ consists of a dilute gas of small loops and an essentially straight string connecting sites $\mathbf{0}$ and $\boldsymbol{r}$. Ignoring the dilute gas of small loops the path integral in Equation (25) can be evaluated by standard methods

[9] yielding for a D-dimensional hypercubic lattice

$$\langle \exp(i\varphi_0 - \varphi_r)\rangle_0 \sim \int_r^\infty \mathrm{d}N g_D(N) f_D(N) \exp(-\gamma_D r^2/N)$$
$$\exp(-\beta_{DE} J_{DE} N). \tag{26}$$

Here $g_D(N)$ is the total number of configurations of a string of length N with one end fixed at the origin. Furthermore $f_D(N)\exp(-\gamma_D r^2/N)$ is the ratio of string conformations with ends attached to $\mathbf{0}$ and $\mathbf{r}$, respectively, to the total number of string conformations $g_D(N)$. γ_D is a factor depending on dimension D. One obtains

$$\begin{aligned} f_2(N) &= 1/\pi N, & \gamma_2 &= 1, \\ f_3(N) &= \left(\frac{3}{2\pi N}\right)^{3/2}, & \gamma_3 &= 3/2, \\ f_4(N) &= 1/\pi N^2, & \gamma_4 &= 2, \ldots. \end{aligned} \tag{27}$$

Taking care of the string in a mean field fashion one may write

$$g_D(N) \sim (z^* - 1)^N, \tag{28}$$

where $z^* \leqslant 2D$ is an effective coordination number and $z^* \to 2D$ for $\beta_{DE} J^i_{DE} \to \infty$. The integral, Equation (26), can now be evaluated by means of saddle point methods. Using the approximation

$$\int_1^\infty \mathrm{d}x\, x^{\nu-1} \mathrm{e}^{-\gamma x - \beta/x} \sim \int_0^\infty \mathrm{d}x\, x^{\nu-1} \mathrm{e}^{-\gamma x - \beta/x} = 2(\beta/\gamma)^{\nu/2} K_\nu(2\sqrt{\beta\gamma}),$$
$$\mathrm{Re}(\beta) > 0,\ \mathrm{Re}(\gamma) > 0, \tag{29}$$

where $K_\nu(z)$ is a Bessel function of imaginary argument, one obtains the exponential decay law

$$\langle \exp i(\varphi_0 - \varphi_r)\rangle_0 \underset{|r| \gg 1}{\sim}$$
$$\times \exp(-2\sqrt{\gamma_D(\beta_{DE} J^i_{DE} - \ln(z^* - 1))r})/r^{(D-1)/2}. \tag{30}$$

Within the present approximation a long-range order develops for

$$\beta_{DE} J^i_{DE} \leqslant \ln(z^* - 1), \tag{31a}$$

whereas from Equation (8) follows that the present approximation breaks down for

$$\beta_{DE} J^i_{DE} \leqslant \ln 2. \tag{31b}$$

In principle it seems therefore to be possible to study even the PT of the planar rotator model within the present scheme for $D \geqslant 2$. To this problem we will come back later. It follows from Equation (30) that the high-temperature correlation function of the isotropic planar rotator model decays exponentially.

(b) Applied magnetic field, $p = 1$. The presence of the magnetic field produces a gas of positive and negative monopoles of unit strength. The string $C_s(\mathbf{0}, \mathbf{r})$ appearing in Equation (18) can now be short circuited by two oppositely charged monopoles. This generates the line conformations $C_s(\mathbf{0})$ and $C_s(\mathbf{r})$, where $C_s(\mathbf{0})$ starts at $\mathbf{0}$ and ends at an arbitrary point and where $C_s(\mathbf{r})$ starts at an arbitrary point and ends at $\mathbf{r}$. It is immediately clear that this effect implies that correlation do not decay anymore exponentially. One obtains approximately, for $v \to \infty$

$$\langle \exp i(\varphi_0 - \varphi_r) \rangle_1 \sim \exp(-2\beta_{DE} J^a_{DE})(\beta_{DE} J^i_{DE} - \ln(z^* - 1))^{-2}. \tag{32}$$

Physically this result means that the gas of monopoles generated by the magnetic field screens the long-range interaction in the system because it short circuits the field lines between the test charges, i.e., the sources and sinks of strength $q = 1$.

(c) Uniaxial anisotropy, $p = 2$. In this case the source and sink of strength $q = 1$ at the sites $\mathbf{0}$ and $\mathbf{r}$, respectively, cannot be screened by the monopoles of strength $q = \pm 2$ as it was the case for $p = \pm 1$. Sources and sinks of strength $q = 2$, however, can be screened implying $\lim_{r \to \infty} \langle \exp 2i(\varphi_0 - \varphi_r) \rangle_2 > 0$.

In order to satisfy the conservation law Equation (20) for the case $p = 2$ one can proceed as follows. In each closed loop or string an even number of alternating source and sink terms is inserted. The partition function Equation (18) can then be written in the form

$$Z_{2,R} = \sum_{\{C_L\}} \prod_{i=1} \left\{ \sum_{n=0}^{C^i_L/2} g_n(C^i_L) \exp\left[-\beta_{DE} J^i_{DE} \oint_{C^i_L} \mathrm{d}S_L - 8\beta_{DE} J^a_{DE} n \right] \right\}. \tag{33}$$

Here C_L represents a collection of closed and oriented loops, i.e.,

$$C_L = \{C^1_L, C^2_L, \ldots C^I_L, \ldots\}$$

and

$$c_L^i = \int_{C_L^i} \mathrm{d}S_L.$$

Furthermore $g_n(C_L^i)$ represents the number of configurations of n pairs along C_L^i starting at one end with say a positive charge always. It is clear that in the presence of loop crossings, $g_n(C_L^i)$ also depends on C_L. In the following we will ignore such crossings altogether and consider a single loop without crossings. In this case we have to determine the quantity

$$\psi_{C_L^i} = \exp\left(- \beta_{DE} J_{DE}^i \oint_{C_L^i} \mathrm{d}S_L \left[\sum_{n=0}^{C_L^i/2} g_n(c_L^i) \exp(-8\beta_{DE} J_{DE}^a n) \right] \right). \tag{34}$$

The first problem is to determine $g_n(c_L^i)$. Within the present approximation this quantity depends only on the length of the loop C_L^i. We determine $g_n(L)$ in such a way that we assume that each charge is represented by a linear piece of unit length. Then the number of states of one particle is

$$\int_0^L \mathrm{d}t = L.$$

For two particles one gets

$$g_1(L) = \int_0^L \mathrm{d}t_1 \int_{t_2}^L \mathrm{d}t_2 F_1(t_2 - t_1), \tag{35}$$

where

$$F_1(t_2 - t_1) = \begin{matrix} 0, |t_2 - t_1| < 1 \\ 1, |t_2 - t_1| \geqslant 1 \end{matrix}. \tag{36}$$

It is now straightforward to apply the method given by Kac *et al.* [10] to obtain

$$g_n(L) \simeq (L - (2n-1))^{2n}/(2n)!. \tag{37}$$

Approximate summation in Equation (34) leads then to

$$\psi_{C_L^i} \simeq \exp\left\{ - \left(\beta_{DE} J_{DE}^i - \exp\left(- 4\beta_{DE} J_{DE}^a \right) \right) \oint_{C_L^i} \mathrm{d}S_L \right\}, c_L^i \gg 1 \tag{38}$$

and

$$Z_{2,R} \simeq \sum_{\{C_L\}} \exp\left\{ -\left(\beta_{DE} J^i_{DE} - \exp\left(-4\beta_{DE} J^a_{DE}\right)\right) \oint_{\{C_L\}} \mathrm{d}S_L \right\}. \quad (39)$$

Within the same approximation used for $Z_{2,R}$ one obtains

$$\langle \exp i(\varphi_0 - \varphi_r)\rangle_2 \simeq \langle \exp i(\varphi_0 - \varphi_r)\rangle_0$$
$$\text{for } \beta_{DE} J^i_{DE} - \exp(-4\beta_{DE} J^a_{DE}), \quad (40)$$

where the right-hand side is given by Equation (30). Accordingly within the present approximation the difference between the high-temperature behavior of the isotropic rotator- and the uniaxial rotator-model amounts to a renormalisation of the coupling constant. It follows from Equations (40) and (30) that, in the presence of uniaxial anisotropy, correlations decay less rapidly as for the isotropic rotator model leading therefore to a higher phase transition temperature. It is clear, however, that the approximation leading to Equation (40) must break down once the PT is approached because the low-temperature phases for the $p = 0$ and $p = 2$ models differ by long-range order.

Finally we like to point out that the XY-model behaves qualitatively similar to the isotropic rotator model as follows from Equation (24). The only difference is that correlations decay more rapidly as follows from

$$\langle \boldsymbol{S}_0 . \boldsymbol{S}_r \rangle \sim \langle \exp i(\varphi_0 - \varphi_r)\rangle_0 \text{ for } \beta_{DE} J^i_{DE} + \ln(3/2), \quad (41)$$

where the left-hand side is evaluated at $\beta_{DE} J^i_{DE}$. This implies that the PT occurs at lower temperature with respect to the isotropic rotator model.

Finally, we would like to study the reasons for the failure of the discrete excitation model to describe the PT region of the planar rotator model in a qualitatively correct way. If an exact mapping between the planar rotator model and the discrete excitation model with $n_{ij} = 0, \pm 1, \pm 2, \ldots$ is made one obtains according to Ref. 5

$$\begin{aligned} \exp(-\beta_{DE} J_{DE}(n) n^2) &= \exp(-2\beta_R J_R) I_n(2\beta_R J_R) \\ &\quad (1/n!)(\beta_R J_R)^n, \text{ for } \beta_R J_R \to 0, \\ &\simeq \exp(-n^2/4\beta_R J_R)/(4\pi\beta_R J_R)^{1/2}, \\ &\quad \text{for } \beta_R J_R \to \infty, \end{aligned}$$

where $I_n(z)$ is a Bessel function of imaginary argument. It follows from this that in the high- and low-temperature phase of the rotator model, $\beta_{DE}J_{DE}(1) \to \infty$ holds. Accordingly $\beta_{DE}J_{DE}(1)$ is not a monotoneously decreasing function of $\beta_R J_R$ but runs through a minimum. The same applies to all other coupling constants $\beta_{DE}J_{DE}(n)n^2$. The difference between the high- and low-temperature phase of the discrete excitation model is therefore that in the high-temperature phase the coupling constants $\beta_{DE}J_{DE}(n)n^2$ for $1 \leqslant n \lesssim 4\beta_R J_R$ are of the same order of magnitude, i.e.,

$$\beta_{DE}J_{DE}(n)n^2 \simeq \ln(4\pi\beta_R J_R)^{1/2}, \text{ for } n \lesssim 4\beta_R J_R.$$

In contrast hereto in the low-temperature phase essentially only $n = 1$ enters.

It is clear that now the arguments which led to Equation (30) cannot apply anymore because then correlations would vanish for $\beta_R J_R \to \infty$. Equation (30) is, however, derived under the assumption that the loop gas is dilute whereas the high-temperature phase of the discrete excitation model allows large numbers of degenerate loop states leading presumably to an infinitely extended loop cluster. Unfortunately, so far we have not been able to show that correlations in such a state exhibit power-law decay when treated from the present point of view.

4. The Kac–Ward[8]Determitant Approach

In the following, we consider the 2D isotropic rotator model approximated by the discrete excitation model defined by Equations (1)–(5) for $p = 0$. This problem can be related to the Kac–Ward cyclic determinant that is equivalent to Onsagers result for the Ising model. Kac and Ward [8] proceed by building up directed graphs on the lattice. These graphs are closed directed line sequences and are parametrised according to their arc length. The Kac–Ward determinant is the sum over all products $W(p)$ of different non-periodical closed directed line sequences p, where no two products have the same directed arc in common. This, however, is exactly the prescription under which Equation (17) has to be evaluated. Accordingly the Kac–Ward determinant is equivalent to $Z_{0,R}$ up to a proportionality factor (see Equation (44) below). This situation is now such that the Kac–Ward determinant enumerates simultaneously (in the thermodynamic limit) the possible unions of closed polygons on two identical lattices. That this is in fact true has been shown rigorously by Sherman [11] based on a conjecture of Feynman (see Ref. 11).

Consider a Ising model with coupling constant $K = J/k_B T$ and Hamiltonian

$$H = -J \sum_{\langle i,j \rangle} \sigma_i \sigma_j \tag{42}$$

on a lattice with $N = nm$ lattice sites. The partition function of the Ising model can be written in the form

$$Z_{n,m} = (\cosh K)^{2N} \sum_{\{\sigma_i\}} \prod_{\langle i,j \rangle} (1 + \sigma_i \sigma_j \tanh K). \tag{43}$$

The second term of Equation (43) represents a system of unoriented polygons. Accordingly one obtains using the same arguments as for the mapping of the Kac–Ward determinant onto the configurations of the two 2D Ising models

$$Z^{DE}_{n,m}(\beta_{DE} J_{DE}) = Z^2_{n,m}(K)/(\cosh K)^{4N}, \tag{44}$$

$$K = \exp(-\beta_{DE} J_{DE}). \tag{45}$$

The correlation function of the Ising model defined on the two n, m lattices is given by

$$\langle \sigma_0 \sigma_r \rangle_{(n,2m)} = \sum_{\{\sigma\}} \sigma_0 \sigma_r \prod_{\langle i,j \rangle} (1 + \sigma_i \sigma_j \tanh K)/Z^2_{n,m}$$

$$= \sum_{\{C'_L\},\{C_S(\mathbf{0},R)\}} \exp\left\{ + \ln(\tanh K)\left[\oint_{C'_L} dS_L + \right.\right.$$

$$\left.\left. + \int_{\substack{\mathbf{0} \\ C_S(\mathbf{0},r)}}^{r} dS \right]\right\} \Big/ \exp\left\{ \ln(\tanh K)\left[\oint_{C'_L} dS_L \right]\right\}, \tag{46}$$

where the prime indicates unoriented loops to be taken. If the sites $\mathbf{0}$ and $\boldsymbol{r}$ are located on one n,m lattice then again all configurations can be mapped onto the discrete excitation model, which is defined on one n,m lattice. Those string conformations $C_s(\mathbf{0}, \boldsymbol{r})$ in Equation (46), which extend onto the second n,m lattice correspond in the discrete excitation model to oriented lines bouncing back from the line separating the two lattices. It should be pointed out that effects connected with the separation line of the two n,m lattices are not taken into account correctly but also do not play a role in the thermodynamic limit. Because the states of the discrete excitation model correspond to those of the Kac-Ward determinant one obtains

$$\langle \exp i(\varphi_0 - \varphi_r)\rangle_0 \simeq \left\langle \exp\left(-\beta_{DE} J_{DE} \int_{\substack{0 \\ C_S(0,r)}}^{r} \mathrm{d}S \right)\right\rangle = \langle \sigma_0 \sigma_r \rangle_K$$

$$\sim \exp(-r/\xi)/r^{1/2}, \qquad K \ll 1, \tag{47a}$$

$$\sim 1/r^{1/4}, \qquad K_c = 0.44069. \tag{47b}$$

For the critical temperature of the discrete excitation model one obtains

$$\beta^c_{DE} J_{DE} = \ln 2.44, \tag{47c}$$

where the last result follows from Equations (45) and (47b). Equations (47a) and (47c) should be compared with Equations (30) and (31a). In both cases a PT occurs into an ordered state which invalidates the present discrete excitation model for that temperature regime. Using Equation (12b) one obtains

$$k_B T^c_R = 2.44 J_R. \tag{48a}$$

This result can be compared with the one obtained from Equation (31a)

$$k_B T^c_R \leqslant (z^* - 1) J_R, \quad z^* \leqslant 4, \tag{48b}$$

and with the low-temperature estimate obtained in Reference 7 for the unbinding of a vortex pair

$$k_B T^c_T = (\pi/\epsilon(0)) J_R, \tag{48c}$$

where $\varepsilon(0) \geqslant 1$ is the dielectric constant of the vortex system. It can be seen that $z^* = 4$ and $\varepsilon = 1$ give almost the same transition temperature. This is not surprising because both estimates neglect any interaction between the objects occurring in these systems. Finally it should be pointed out that using Equation (48c) Kosterlitz and Thouless [7] find Equation (47b) to hold at the PT. Accordingly there is apparently a remarkable agreement between both methods as far as the correlations at the presumed PT are concerned. A close look at this result, however, reveals that the law $\langle \cos(\varphi_0 - \varphi_r)\rangle \overset{r \gg 1}{\sim} 1/r^{1/4}$ at T^c_R as obtained in Eq. (47b) holds for $T \to T^c_R$, whereas in Ref. 7 this law is obtained for $T \to T^c_R$. The jump in $\varepsilon(0)$, $\Delta\varepsilon(0) = \infty$ at T^c_R predicted in Ref. 7 implies then exponential decay of correlations for $T \to T^c_R$. The high- and low-temperature approach therefore do not match when considered from this point of view. With respect to the thermal properties there is also disagreement because the present method would predict a λ-anomaly in the specific heat whereas in Ref. 7 no such heat anomaly is predicted. The deficiencies of the present approach for the PT region, however, have been clearly stated at the end of Section 3 and therefore nothing definite can be concluded from the present treatment with respect to the nature of the PT.

5. Qualitative Discussion of the Phase Transition

In this part we would like to give a qualitative discussion of the PT occurring in some of the models discussed so far.

For $D > 2$ there is in principle no problem to understand the PT occurring in the isotropic rotator- and XY-model because Equations (31a) and (31b) allow the PT to occur in a temperature regime

$$\ln 2 \leqslant \beta_{DE} J^i_{DE} + \ln(3/2) \qquad (\text{for } XY\text{-model}) \leqslant \ln(z^* - 1), \tag{49}$$

where the discrete excitation model is still a good representation of the two models. In particular one expects that for $D \gg 1, z^* \sim 2D$ will hold. In the presence of a magnetic field it has been shown that Equation (32) holds for $\beta_{DE} J^i_{DE} \gg \ln(z^* - 1)$, but breaks down in the temperature regime given by Equation (49). A qualitative argument why the second term of Equation (32) does not blow up and which would imply an infinite extension of the two strings connected to the source and sink is the following. Because the monopole pairs in the system which are connected by strings of unit strength would equally try to 'dissociate', i.e., producing an infinitely extended string. Because for finite dimensions D the system will be able only to accommodate one or a finite number of infinitely extended strings due to excluded volume effects no dissociation will occur at all.

For $D = 2$ the discrete excitation model with $n_{ij} = 0, \pm 1$ leads into an ordered low-temperature phase which is impossible according to the Mermin–Wagner [12] theorem. Further terms in the expression Equation (1) have therefore to be taken into account in order to get realistic results as has been explained at the end of Section 3. The situation has some similarity to the low-temperature vortex gas approach which when used in the high-temperature phase yields no correlations at all in the thermodynamic limit, i.e.,

$$\langle \exp i(\varphi_0 - \varphi_r) \rangle_0 \sim 1/L^\alpha, \tag{50}$$

where $\alpha = \alpha(\beta_R J_R, r) > 0$ and L is the linear dimension of the system. Equation (50) is derived as Equation (10) in Ref. 13, and is based on a Debye ansatz for the dielectric constant $\varepsilon(q)$ and a second-order cumulant expansion. The failure of this approach is either due to a breakdown of the Deye ansatz for the dense vortex system which is rather unlikely but perhaps possible due to cooperative effects, or higher-order cumulants have to be taken into account. In the latter more likely case n-pair 'dielectric functions' $n > 2$ come into play (e.g. for $n = 4, \varepsilon_4(\boldsymbol{q}, \boldsymbol{q}', \boldsymbol{q}''))$ besides the usual 2-pair dielectric function $\varepsilon(\boldsymbol{q})$ (see Equation (6) in Ref. 13) used below T_c. It is clear that in either case pair–pair correlations to all orders must play a decisive role. Such pair–pair correlations play already a considerable role for the evaluation of $\varepsilon(0)$ below T_c and if not properly taken into account lead to topological incompatibilities as can easily be shown [14].

In the authors opinion the PT occurring in the $2D$ rotator-model and XY-model is of the metal-insulator type and is possibly discontinuous as has been shown in Refs. 14 and 15. Although this result has been derived using a Van der Waals equation of state for the vortex-pair gas which may not take all important effects into account it is perhaps true. Because neither the low-temperature nor the high-temperature approach seem to be extendible beyond T_c in a simple fashion, i.e., the method of using instead of bound vortices dissociated vortices and instead of finitely extended loops infinitely extended loops does not work due to the various reasons given. It may therefore be possible that the high- and low-temperature phase are not connected in a continuous fashion. It may well be possible that the discontinuity at the PT may be restricted to a jump in the dielectric constant as predicted by Kosterlitz and Thouless [7] and a jump in higher-order 'dielectric' functions as mentioned earlier, however, this has first to be proven rigorously.

Institute for Theoretical Physics, Universität des Saarlandes, FRG

References

[1] Holz, A., *J. Phys. Lettres* **39**, L331 (1978).
[2] Holz, A., *J. Phys. A: Math. Gen.* **12**, 2099 (1979)
[3] Knops, H. J. F., *Phys. Rev. Lett.* **39**, 766 (1977).
[4] Villian, J., *J. Physique* **36**, 581 (1975).
[5] José, J. V., Kadanoff, L. P., Kirkpatrick, S., and Nelson, D., *Phys. Rev.* **B16**, 1217 (1977).

[6] Kadanoff, L. P., *J. Phys. A: Math. Gen.* **11**, 1399 (1978).
[7] Kosterlitz, J. M., and Thouless, D. J., *J. Phys. C: Solid State Phys.* **6**, 1181 (1973).
[8] Kac, G., and Ward, J. C., *Phys. Rev.* **88**, 1332 (1952).
[9] Feynman, R. P., and Hibbs, A. R., *Quantum Mechanics and Path Integrals*, McGraw-Hill Book Company, 1965.
[10] Kac, M., Uhlenbeck, G. E., and Hemmer, P. C., *J. Math. Phys.* **4**, 53 (1963).
[11] Sherman, S., *J. Math. Phys.* **1**, 202 (1960).
[12] Mermin, N. D., and Wagner, H., *Phys. Rev. Lett.* **22**, 1133 (1966).
[13] Holz, A., *Annals of Israel Phys. Soc.* **2**, *Stat. Phys.* **13**, 641 (1978).
[14] Holz, A., 'On the phase transitions in 2-dimensional planar rotator systems', unpublished (1977).
[15] Holz, A., *Phys. Lett.* **61A**, 399 (1977).

F. LANGOUCHE, D. ROEKAERTS AND E. TIRAPEGUI

GENERAL LANGEVIN EQUATIONS AND FUNCTIONAL INTEGRATION

ABSTRACT. We study functional integral representations for the transition probability density of Markov processes defined by general nonlinear Langevin equations. We give a review of our work covering the definition through discretization, the relation to ordering problems in the operator formalism and the ambiguities of perturbation expansions. We also derive the functional integral representations directly from the stochastic differential equation without using an operator formalism.

1. Introduction

We study here Markov processes defined by general nonlinear Langevin equations. These processes are extremely useful in statistical mechanics in general, and especially to model systems far from thermodynamic equilibrium [1–3].

The general Langevin equation we consider is $(\gamma,\alpha = 1,2,\dots M)$

(1.1) $\dot{q}^{\gamma} + a^{\gamma}(\boldsymbol{q}(t)) = \sqrt{\eta}\,\sigma^{\gamma}_{\alpha}(\boldsymbol{q}(t)) f^{\alpha}(t),$

where the Gaussian white noise $f^{\alpha}(t)$ has correlation $\{f^{\alpha}(t)f^{\beta}(t')\} = \delta^{\alpha\beta}\delta(t-t')$, and $\boldsymbol{q} = (q^1,\dots,q^M)$.

In Section 2, we start by deriving the Fokker–Planck equation for the transition probability density $P(\boldsymbol{Q},t\,|\,\boldsymbol{Q}_0,t_0)$ of the Markovian process $\boldsymbol{q}(t)$ defined by (1) and then we proceed to construct an operator formalism [4–8]. We show, starting from this formalism, that $P(\boldsymbol{Q},t\,|\,\boldsymbol{Q}_0,t_0)$ admits an infinity of functional integral representations which have to be specified introducing the notion of discretization [9–12]. Once the discretization is specified, the functional integral becomes a well defined object and all ambiguities disappear. The relation of discretization to ordering problems is studied in detail. The ambiguities of the perturbation expansions starting from the functional integral representation for the generating functional of correlation and response functions are also explained following our work [13, 14] and solved.

*Onderzoeker IIKW, Belgium
**Aspirant NFWO, Belgium

E. Tirapegui (Ed.), Field Theory, Quantization and Statistical Physics, 295–319.

Finally, as an example, and for future use in Section 3, we construct explicitly a special class of functional integral representations for $P(\boldsymbol{Q},t|\boldsymbol{Q}_0,t_0)$ depending on two numbers (r,s), which correspond to a discretization that we note $\gamma(r,s)$.

In Section 3, we treat the Langevin equations directly replacing the badly defined white noise $f^{\alpha}(t)\,dt$ by the well-defined differential of the Wiener process dw^{α}. Without using an operator formalism but, rather, working directly with (1) and using the usual probabilistic methods we derive again the functional integral representations for $P(\boldsymbol{Q},t|\boldsymbol{Q}_0,t_0)$ with their respective discretizations. This is done by three different methods and we find complete agreement with the results of Section 2. Thus the fact that functional integral representations are not unique and that they become well defined only when the discretization is specified is confirmed directly, as well as the fact that different interpretations are possible for the stochastic differential Equation (1.1) (Stratonovich and Ito).

Finally in Section 4 we comment briefly on (and give references to) other definitions of path integrals (approaches of Feynman, Graham and Dekker), on the problem of the most probable path and on WKB-type expansions.

2. Operator formalism and functional integral representation

Let $P(\boldsymbol{Q},t|\boldsymbol{Q}_0,t_0)$ be the transition probability density for the Markovian process $\boldsymbol{Q}(t)$ determined by the Langevin equation (1.1.) It satisfies the Fokker–Planck equation

$$(2.1)\quad P(\boldsymbol{Q},t|\boldsymbol{Q}_0,t_0)=\frac{\partial}{\partial Q^{\mu}}\left\{A^{\mu}(\boldsymbol{Q})+\frac{1}{2}\eta\frac{\partial}{\partial Q^{\nu}}g^{\mu\nu}(\boldsymbol{Q})\right\}P(\boldsymbol{Q},t|\boldsymbol{Q}_0,t_0)$$

with

$$(2.2)\quad A^{\mu}(\boldsymbol{Q})\equiv a^{\mu}(\boldsymbol{Q})-\tfrac{1}{2}\eta\sigma^{\nu}_{\alpha}(\boldsymbol{Q})\partial_{\nu}\sigma^{\mu}_{\alpha}(\boldsymbol{Q})$$
$$g^{\mu\nu}(\boldsymbol{Q})=\sigma^{\mu}_{\alpha}(\boldsymbol{Q})\sigma^{\nu}_{\alpha}(\boldsymbol{Q})$$

(Sum over repeated indices is to be understood)

Equation (2.1) is obtained from (1.1.) in an unambiguous way when the two point correlation of the white noise is interpreted as

$$\lim_{\tau\to 0}\delta^{\alpha\beta}\Delta_{\tau}(t-t')$$

where $\Delta_{\tau}(t)=\Delta_{\tau}(-t)\to\delta(t)$, $\tau\to 0$ (See [4, 15]). We shall construct an operator formalism [4–8] and introduce the operators $\hat{q}^\mu, \hat{p}_\nu, [\hat{q}^\mu, \hat{p}_\nu] = i\delta^\mu_\nu, [\hat{q}^\mu, \hat{q}^\nu] = [\hat{p}_\mu, \hat{p}_\nu] = 0$ and the 'Hamiltonian' $\hat{H}(\hat{\boldsymbol{p}}, \hat{\boldsymbol{q}})$ by

$$(2.3)\quad \hat{H}(\hat{\boldsymbol{p}}, \hat{\boldsymbol{q}}) = -\frac{i\eta}{2}\hat{p}_\mu \hat{p}_\nu g^{\mu\nu}(\hat{\boldsymbol{q}}) - \hat{p}_\mu A^\mu(\hat{\boldsymbol{q}}).$$

The evolution operator

$$i\partial_t U(t,t') = \hat{H} U(t,t'), U(t,t) = 1$$

allows us to write

$$(2.4)\quad P(\boldsymbol{Q},t|\boldsymbol{Q}_0,t_0) = \langle \boldsymbol{Q}|U(t,t_0)|\boldsymbol{Q}_0\rangle.$$

Setting

$$t_j = t_0 + j\varepsilon, \quad t_{n+1} = t, \quad \boldsymbol{q}_{n+1} = \boldsymbol{Q}, \quad \boldsymbol{q}_0 = \boldsymbol{Q}_0$$

and inserting N times the completeness relation

$\int \mathrm{d}\boldsymbol{q}_i |\boldsymbol{q}_i\rangle\langle \boldsymbol{q}_i| = 1, i = 1, 2, \ldots . N$, one obtains from (4) that

$$(2.5)\quad P(\boldsymbol{Q},t|\boldsymbol{Q}_0,t_0) = \int \prod_{i=1}^{N} \mathrm{d}\boldsymbol{q}_i \prod_{j=1}^{N+1} \langle \boldsymbol{q}_j|U(t_j,t_{j-1})|\boldsymbol{q}_{j-1}\rangle.$$

We shall compute (5) in the limit $N \to \infty$, consequently we need $U(t_j, t_{j-1})$ only up to $O(\varepsilon)$ (the validity of this has been discussed by us in [12] and [16], see also [9]). Then

$$(2.6)\quad \langle \boldsymbol{q}_j|U(t_j,t_{j-1})|\boldsymbol{q}_{j-1}\rangle = \langle \boldsymbol{q}_j|\boldsymbol{q}_{j-1}\rangle - i\varepsilon\langle \boldsymbol{q}_j|\hat{H}|\boldsymbol{q}_{j-1}\rangle.$$

The point now is that $\langle \boldsymbol{q}'|\hat{H}|\boldsymbol{q}\rangle$ admits representations of the form $(\boldsymbol{\Delta} \equiv \boldsymbol{q}' - \boldsymbol{q})$

$$(2.7)\quad \langle \boldsymbol{q}'|\hat{H}|\boldsymbol{q}\rangle = \int \frac{\mathrm{d}\boldsymbol{p}}{(2\pi)^M} \exp[i\boldsymbol{p}.\boldsymbol{\Delta}] h^\gamma(\boldsymbol{p}, \boldsymbol{q}', \boldsymbol{q})$$

where h^γ is not unique. The different h^γ that can be used in (7) are in correspondence with the different orderings in (3) of the noncommuting operators $\hat{p}_\mu$ and $\hat{q}^\nu$. From (6) and (7) we have $(\Delta\boldsymbol{q}_j \equiv \boldsymbol{q}_j - \boldsymbol{q}_{j-1})$

$$(2.8)\quad \langle \boldsymbol{q}_j|U(t_j,t_{j-1})|\boldsymbol{q}_{j-1}\rangle = \int \frac{\mathrm{d}\boldsymbol{p}}{(2\pi)^M} \exp[i\boldsymbol{p}.\Delta\boldsymbol{q}_j](1 - i\varepsilon h^\gamma(\boldsymbol{p}, \boldsymbol{q}_j \boldsymbol{q}_{j-1})).$$

Up to $O(\varepsilon)$ we can put the last term in (2.8) in exponential form and replacing in (2.5):

$$(2.9)\quad P(Q,t|Q_0,t_0) = \lim_{N\to\infty} \int \prod_{i=1}^{N} \mathrm{d}q_i \prod_{j=1}^{N+1} \frac{\mathrm{d}p_j}{(2\pi)^M} \times \exp\left\{ i\varepsilon \sum_{j=1}^{N+1} [p_j.\frac{\Delta q_j}{\varepsilon} j - h^\gamma(p_j, q_j, q_{j-1})] \right\}.$$

The function h^γ is of the form (note that γ parametrizes the different h^γ in (7))

$$(2.10)\quad h^\gamma(p,q',q) = -\frac{i\eta}{2} G^{\mu\nu}_{[\gamma]}(q',q) p_\mu p_\nu - p_\mu B^\mu_{[\gamma]}(q',q) - iC_{[\gamma]}(q',q)$$

where $G^{\mu\nu}_{[\gamma]}(q',q)|_{q'=q} = g^{\mu\nu}(q)$. Setting $H^\gamma(p,q) = h^\gamma(p,q',q)|_{q'=q}$ the limit (9) is written as the phase space functional integral

$$(2.11)\quad P(Q,t|Q_0,t_0) = \int_\gamma \mathrm{D}q\mathrm{D}p \exp\left\{ i \int \mathrm{d}\tau [p.q - H^\gamma(p,q)] \right\} \times \delta(q(t) - Q)\delta(Q(t_0) - Q_0)$$

where γ stands for the discretization involved in the definition of the functional integral (see also Appendix A), and which here is specified by (2.9).

We illustrate this with an example of h^γ that we note h_r. We write $\hat{H}$ given by (2.3) doing a commutator as ($[\hat{p}_\mu, A^\nu(\hat{q})] = -i\partial_\mu A^\nu(\hat{q})$)

$$(2.12)\quad \hat{H} = -\frac{i\eta}{2} \hat{p}_\mu \hat{p}_\nu g^{\mu\nu}(\hat{q}) - (1-r)\hat{p}_\mu A^\mu(\hat{q}) - rA^\mu(\hat{q})\hat{p}_\mu + ir\partial_\mu A^\mu(\hat{q})$$

where r is a number. We compute now (2.7). Note that introducing

$$\int \mathrm{d}p |p\rangle\langle p| = 1, \quad \hat{p}_\mu |p\rangle = p_\mu |p\rangle$$

$$(2.13)\quad \langle q'|\hat{p}_\mu \hat{p}_\nu g^{\mu\nu}(\hat{q})|q\rangle = \int \mathrm{d}p \langle q'|\hat{p}_\mu \hat{p}_\nu|p\rangle\langle p|g^{\mu\nu}(\hat{q})|q\rangle = \int \frac{\mathrm{d}p}{(2\pi)^M} \exp[ip.\Delta] p_\mu p_\nu g^{\mu\nu}(q).$$

Calculating the other terms of (2.12) as in (2.13) we obtain

$$(2.14)\quad h_r(p,q',q) = -\frac{i\eta}{2} p_\mu p_\nu g^{\mu\nu}(q) - (1-r)p_\mu A^\mu(q) - rp_\mu A^\mu(q') + ir\partial_\mu A^\mu(q)$$

and $H_r(\boldsymbol{p},\boldsymbol{q}) = h_r(\boldsymbol{p},\boldsymbol{q},\boldsymbol{q})$ is

$$H_r(\boldsymbol{p},\boldsymbol{q}) = -\frac{i\eta}{2} p_\mu p_\nu g^{\mu\nu}(\boldsymbol{q}) - p_\mu A^\mu(\boldsymbol{q}) + ir\partial_\mu A^\mu(\boldsymbol{q}).$$

This simple calculation illustrates the relation between ordering of operators and discretizations. Note that for $r = 0$ one has antistandard ordering ($\hat{p}_\mu$ at the left-hand side of $\hat{q}^\nu$), and that this corresponds to discretizing all functions of $\boldsymbol{q}$ in $\boldsymbol{q}_{j-1}$ (see [14]). Discretization problems and their relation to ordering have been studied e.g. by Berezin [9], Mayes and Dowker [10, 17], Leschke and Schmutz [11] and ourselves [12–16].

There is a systematic way of constructing the different h^γ based on the relation between operators and c-number functions in phase space which has been used by the above-mentioned authors. The problems of covariance of the different functional integral representations (2.11) were studied by us in [12] and the detailed analysis of the perturbation expansions of (2.11) was done in [13, 14] where we show that the perturbation series is the same, as it must be, for any discretization γ. This is obvious since $P(\boldsymbol{Q},t;\boldsymbol{Q}_0,t_0)$ has nothing to do with γ, and the mechanism that works here is a cancellation between tadpoles which have discretization dependent values and extra vertices generated by the different discretizations. It is rather simple to understand this: perturbation expansions (and also WKB-type expansions [18–20]) are obtained splitting $H^\gamma = H_0^\gamma + H_1^\gamma$, where H_0^γ is quadratic in $(\boldsymbol{p},\boldsymbol{q})$, and writing (2.11) formally as

$$P(\boldsymbol{Q},t;\boldsymbol{Q}_0,t_0) = \exp\left[-i\int \mathrm{d}\tau\, H_1^\gamma\left(\frac{1}{i}\frac{\delta}{\delta \boldsymbol{J}^*(\tau)}, \frac{1}{i}\frac{\delta}{\delta \boldsymbol{J}(\tau)}\right)\right] \times \tag{2.15}$$

$$Z_0[\boldsymbol{J},\boldsymbol{J}^*]\big|_{J=J^*=0}$$

with

$$Z_0[J,J^*] = \int_\gamma \mathrm{D}\boldsymbol{q}\,\mathrm{d}\boldsymbol{p} \exp\left\{i\int \mathrm{d}\tau[\boldsymbol{p}.\dot{\boldsymbol{q}} - H_0^\gamma(\boldsymbol{p},\boldsymbol{q}) + \boldsymbol{J}.\boldsymbol{q} + \boldsymbol{J}^*.\boldsymbol{p}]\right\}\delta(\boldsymbol{q}(t) - \boldsymbol{Q})\delta(\boldsymbol{q}(t_0) - \boldsymbol{Q}_0). \tag{2.16}$$

The functional integral (2.16) is Gaussian and can be computed exactly. Its value is of the form ($\bar{N}$ a normalization factor)

$$(2.17)\quad Z_0[\boldsymbol{J},\boldsymbol{J}^*] = \bar{N}\exp\left\{-\int dt'\,dt''\left[\tfrac{1}{2}J_\mu(t')G^{\mu\nu}(t',t'')J_\nu(t'') + J^{*\mu}(t')S^\nu_\mu(t',t'')J_\nu(t'') + \tfrac{1}{2}J^{*\mu}(t')\Delta_{\mu\nu}(t',t'')J^{*\nu}(t'')\right]\right\}$$

with $G^{\mu\nu}(t',t'')$ and $\Delta_{\mu\nu}(t',t'')$ continuous at $t'=t''$, while $S^\nu_\mu(t',t'')$ has a jump at $t'=t''$ (this is of course just a direct consequence of the fact that $\hat{p}_\mu$ and $\hat{q}^\nu$ do not commute).

The perturbation series is obtained developing $\exp[-i\int d\tau\, H^\gamma_1]$ in (2.15) and operating on Z_0. We see immediately that when in H^γ_1 there is a product $p_\mu(\tau)q^\nu(\tau)$ this will give

$$(2.18)\quad \frac{1}{i}\frac{\delta}{\delta J^{*\mu}(\tau)}\frac{1}{i}\frac{\delta}{\delta J_\nu(\tau)}Z_0[\boldsymbol{J},\boldsymbol{J}^*]|_{J=J^*=0} = S^\nu_\mu(\tau,\tau)$$

an undefined tadpole due to the discontinuity of $S^\nu_\mu(t',t'')$. But the solution of this ambiguity is simple: one has to remember that (2.15) was obtained from (2.11) which is the limit of the discretized expression in (2.9) and that the splitting of H^γ induces a corresponding splitting of h^γ as $h^\gamma = h^\gamma_0 + h^\gamma_1$.

Consequently $\int d\tau\, H^\gamma_1$ in (2.15) has to be interpreted as

$$(2.19)\quad \lim_{\varepsilon\to 0}\int d\tau\, h^\gamma_1\left(\frac{1}{i}\frac{\delta}{\delta \boldsymbol{J}^*(\tau)}, \frac{1}{i}\frac{\delta}{\delta \boldsymbol{J}(\tau+\varepsilon)}, \frac{1}{i}\frac{\delta}{\delta \boldsymbol{J}(\tau-\varepsilon)}\right).$$

This gives at each order an unambiguous meaning to the expansion fixing the values of the tadpoles $S^\nu_\mu(\tau,\tau)$. The same expansion is then obtained starting from any H^γ due to the cancellation mechanism mentioned before [13, 14] (an operator formalism analysis confirming this conclusion has been done by Leschke, Hirshfeld and Suzuki [21]).

In the case we are treating here the simplest thing is to use the prepoint discretization, i.e. to take for H^γ the function H_r for $r=0$ (see (2.14)) which is just what is obtained from the original $\hat{H}$ in (4) replacing $\hat{\boldsymbol{p}}\to\boldsymbol{p}$, $\hat{\boldsymbol{q}}\to\boldsymbol{q}$, then in the perturbation expansion for the generating functional of correlation and response functions all $S(\tau,\tau)$ – tadpoles vanish when (2.19) is used [15]. We write explicitly the representation

$$(2.20)\quad P(\boldsymbol{Q},t;\boldsymbol{Q}_0,t_0) = \int_{\gamma(0)} D\boldsymbol{q}\,D\boldsymbol{p}\exp\left\{i\int d\tau\left[\boldsymbol{p}\cdot\dot{\boldsymbol{q}} + \frac{i\eta}{2}g^{\mu\nu}(\boldsymbol{q})p_\mu p_\nu + p_\mu A^\mu(\boldsymbol{q})\right]\right\}\delta(q(t)-Q)\delta(Q(t_0)-Q_0)$$

where the right-hand side is lim $I_N^{\gamma(0)}, N \to \infty$, with

$$(2.21)\quad I_N^{\gamma(0)} = \int_{q_0 = Q_0} \left(\prod_{i=1}^{N} \mathrm{d}\boldsymbol{q}_i \prod_{j=1}^{N+1} \frac{\mathrm{d}\boldsymbol{p}_j}{(2\pi)^M} \right) \times \exp\left\{ i\varepsilon \sum_{j=1}^{N+1} \left[p_{j\mu} \frac{\Delta q_j^\mu}{\varepsilon} + \frac{i\eta}{2} g^{\mu\nu}(\boldsymbol{q}_{j-1}) p_{j\mu} p_{j\nu} + p_{j\mu} A^\mu(\boldsymbol{p}_{j-1}) \right] \right\}.$$

We shall need in the next section a representation (2.11) for a discretization $\gamma(r,s)$ parametrized by two numbers r and s that we derive now. We recall following Agarwal and Wolf [22] and [13, 14] that to an operator $\hat{D}(\hat{\boldsymbol{p}}, \hat{\boldsymbol{q}})$ we can associate through a function $\Omega(\boldsymbol{u}, \boldsymbol{v})$ a phase space function $\mathrm{D}^\Omega(\boldsymbol{p}, \boldsymbol{q})$ and a function $\mathrm{d}^\Omega(\boldsymbol{p}, \boldsymbol{q}', \boldsymbol{q})$ defined by ($\bar{\boldsymbol{q}} \equiv \frac{1}{2}(\boldsymbol{q} + \boldsymbol{q}')$)

$$(2.22)\quad \mathrm{D}^\Omega(\boldsymbol{p}, \boldsymbol{q}) = \bar{\Omega}(i\partial/\partial \boldsymbol{q}, i\partial/\partial \boldsymbol{p}) \mathrm{D}_{\mathrm{AS}}(\boldsymbol{p}, \boldsymbol{q}),$$

$$(2.23)\quad \mathrm{d}^\Omega(\boldsymbol{p}, \boldsymbol{q}', \boldsymbol{q}) = \Omega(-i\partial/\partial \bar{\boldsymbol{q}}, -\boldsymbol{\Delta}) \mathrm{D}^\Omega(\boldsymbol{p}, \bar{\boldsymbol{q}}),$$

where

$$\bar{\Omega}(\boldsymbol{u}, \boldsymbol{v}) = \tilde{\Omega}(\boldsymbol{u}, \boldsymbol{v}) \exp\left[\frac{i}{2} \boldsymbol{u}.\boldsymbol{v} \right], \qquad \tilde{\Omega}(\boldsymbol{u}, \boldsymbol{v}) = [\Omega(-\boldsymbol{u} - \boldsymbol{v})]^{-1},$$

and the function $\mathrm{D}_{\mathrm{AS}}(\boldsymbol{p}, \boldsymbol{q})$ is defined by $\langle \boldsymbol{p} | \hat{\mathrm{D}} | \boldsymbol{q} \rangle = \mathrm{D}_{\mathrm{AS}}(\boldsymbol{p}, \boldsymbol{q}) \langle \boldsymbol{p} | \boldsymbol{q} \rangle$, where the subscript AS stands for antistandard ordering. One shows that

$$(2.24)\quad \langle \boldsymbol{q}' | \hat{\mathrm{D}} | \boldsymbol{q} \rangle = \int \frac{\mathrm{d}\boldsymbol{p}}{(2\pi)^M} \exp[i\boldsymbol{p}.\boldsymbol{\Delta}] \mathrm{d}^\Omega(\boldsymbol{p}, \boldsymbol{q}', \boldsymbol{q}),$$

and different Ω functions correspond to different orderings of the operator $\hat{\mathrm{D}}(\hat{\boldsymbol{p}}, \hat{\boldsymbol{q}})$. Using formulas (2.22) to (2.24) with our operator $\hat{H}$ instead of $\hat{\mathrm{D}}$ we obtain the representations

$$(2.25)\quad P(\boldsymbol{Q}, t; \boldsymbol{Q}_0, t_0) = \int_{\gamma(\Omega)} \mathrm{D}\boldsymbol{q} \mathrm{D}\boldsymbol{p} \exp\left\{ i \int_{t_0}^{t} \mathrm{d}\tau [\boldsymbol{p}.\dot{\boldsymbol{q}} - H^{\gamma(\Omega)}(\boldsymbol{p}, \boldsymbol{q})] \right\} \times \delta(\boldsymbol{q}(t) - \boldsymbol{Q}) \delta(\boldsymbol{q}(t_0) - \boldsymbol{Q}_0)$$

with $H^{\gamma(\Omega)}(\boldsymbol{p}, \boldsymbol{q}) = h^\Omega(\boldsymbol{p}, \boldsymbol{q}, \boldsymbol{q})$. Choosing now two functions Ω_1 and Ω_2 and splitting $\hat{H}$ as $\hat{H} = \hat{H}_1 + \hat{H}_2$ we can compute $h_i^{\Omega_i}, i = 1,2$, and from

(2.24) we have

$$(2.26)\quad \langle q'|\hat{H}|q\rangle = \int \frac{\mathrm{d}p}{(2\pi)^M} \exp[ip.\Delta](h_1^{\Omega_1}(p,q',q) + h_2^{\Omega_2}(p,q',q)).$$

Putting

$$H_{12} = H_1^{\gamma(\Omega_1)} + H_2^{\gamma(\Omega_2)}, \qquad H_i^{\gamma(\Omega_i)} = h_i^{\Omega_i}(p,q,q),$$

we have the representations

$$(2.27)\quad P(Q,t;Q_0,t_0) = \int_{\gamma_{12}} \mathrm{D}q\mathrm{D}p \exp\left\{i\int_{\Omega_{12}} \mathrm{d}\tau[p.\dot{q} - H_{12}(p,q)]\right\} \times \delta(q(t) - Q)\delta(q(t_0) - Q_0),$$

where γ_{12} discretizes H_{12} as $(h_1^{\Omega_1} + h_2^{\Omega_2})$ according to (2.26). The representations we shall need are obtained by taking

$$(2.28)\quad \hat{H}_1 = -\frac{i\eta}{2}\hat{p}_\mu\hat{p}_\nu g^{\mu\nu}(\hat{q}) + \frac{\eta}{2}\hat{p}_\mu\sigma_\alpha^\nu(\hat{q})\,\partial_\nu\sigma^\mu(\hat{q}),$$

$$(2.29)\quad \hat{H}_2 = -\hat{p}_\mu a^\mu(\hat{q}),$$

and

$$\Omega_1(u,v) = \Omega_s(u^\mu v_\mu), \qquad \Omega_2(u,v) = \Omega_r(u^\mu v_\mu),$$

with

$$\Omega_r(x) = \exp[i(\tfrac{1}{2} - r)x].$$

One has $(i = 1,2)$

$$(2.30)\quad h^{\Omega_i}(p,q',q) = \exp[(r - \tfrac{1}{2})\Delta^\mu\,\partial/\partial\bar{q}^\mu]\,H_i^{\Omega_i}(p,\bar{q}) = H_i^{\Omega_i}(p,q^{(r)}),$$

where

$$q^{(r)} \equiv q + r\Delta.$$

Putting

$$H^{[r,s]} = H_1^{\Omega_r} + H_2^{\Omega_s}, h^{[r,s]} = h_1^{\Omega_r} + h_2^{\Omega_s}$$

one has that

$$(2.31)\quad h^{[r,s]}(p,q',q) = H_1^{\Omega_s}(p,q^{(s)}) + H_2^{\Omega_r}(p,q^{(r)})$$

and obviously $H^{[r,s]}(\boldsymbol{p},\boldsymbol{q}) = h^{[r,s]}(\boldsymbol{q}=\boldsymbol{q}')$. Explicitly one obtains

$$\begin{aligned}(2.32)\quad h^{[r,s]}(\boldsymbol{p},\boldsymbol{q}',\boldsymbol{q}) = &-\frac{i\eta}{2}g^{\mu\nu}(\boldsymbol{q}^{(s)})p_\mu p_\nu + \frac{\eta}{2}p_\mu\sigma^\nu_\alpha(\boldsymbol{q}^{(s)})\partial_\nu\sigma^\mu_\alpha(\boldsymbol{q}^{(s)})\\ &-\eta s p_\nu\partial_\mu g^{\mu\nu}(\boldsymbol{q}^{(s)}) - \frac{i\eta s}{2}\partial_\mu[\sigma^\mu_\alpha(\boldsymbol{q}^{(s)})\partial_\nu\sigma^\mu_\alpha(\boldsymbol{q}^{(s)})]\\ &+\frac{i\eta s}{2}\partial_{\mu\nu}g^{\mu\nu}(\boldsymbol{q}^{(s)}) - p_\mu a^\mu(\boldsymbol{q}^{(r)}) + ir\partial_\alpha a^\alpha(\boldsymbol{q}^{(r)}).\end{aligned}$$

Calling $\gamma(r,s)$ the discretization we have from (2.32)

$$\begin{aligned}(2.33)\quad P(\boldsymbol{Q},t;\boldsymbol{Q}_0,t_0) = &\int_{\gamma(r,s)} \mathrm{D}\boldsymbol{q}\,\mathrm{D}\boldsymbol{p}\exp\left\{i\int_{t_0}^{t}\mathrm{d}\tau[\boldsymbol{p}.\dot{\boldsymbol{q}} - H^{[r,s]}(\boldsymbol{p},\boldsymbol{q})]\right\}\\ &\times\delta(\boldsymbol{q}(t)-\boldsymbol{Q})\delta(\boldsymbol{q}(t_0)-\boldsymbol{Q}_0)\end{aligned}$$

As we have explained, (2.33) is defined as lim $I_N^{[r,s]}, N\to\infty$, with $I_N^{[r,s]}$ given by the right-hand side of (2.9) with h^γ replaced by $h^{[r,s]}$.

Let us see now how it can be shown that all these different functional integrals have the same value, namely $P(\boldsymbol{Q},t;\boldsymbol{Q}_0,t_0)$. This is, of course, obvious by construction but we want to see how it can be checked directly in the functional integral. In order to see this we do in (2.9) the Gaussian integrals over $\mathrm{d}\boldsymbol{p}_j$($h^\gamma$ is given by (2.10)) using the formula

$$\begin{aligned}(2.34)\quad &\int_{-\infty}^{\infty}\frac{\mathrm{d}x^1\ldots.\mathrm{d}x^n}{(2\pi)^{n/2}}\exp\left[-\frac{1}{2}\sum_{i,j}x^iA_{ij}x^j + \sum_i b_ix^i\right]\\ &= \frac{1}{\sqrt{\det A}}\exp\left[\frac{1}{2}\sum b_i(A^{-1})_{ij}b_j\right].\end{aligned}$$

One obtains (calling I_N^γ the multidimensional integral in the right-hand side of (2.9))

$$\begin{aligned}(2.35)\quad I_N^\gamma = &\int_{q_0=Q_0}^{q_{n+1}=Q}\left(\prod_{i=1}^{N}\mathrm{d}\boldsymbol{q}_i\prod_{j=1}^{N+1}\{(2\pi\varepsilon\eta)^M\det G^{\mu\nu}_{[\gamma]}(\boldsymbol{q}_j,\boldsymbol{q}_{j-1})\}^{-1/2}\right)\\ &\times\exp\left\{-\varepsilon\sum_{j=1}^{N+1}\left[\frac{1}{2\eta}G_{[\gamma]\mu\nu}(\boldsymbol{q}_j,\boldsymbol{q}_{j-1}\left(\frac{\Delta q^\mu_j}{\varepsilon} + B^\mu_{[\gamma]}(\boldsymbol{q}_j,\boldsymbol{q}_{j-1})\right)\right.\right.\\ &\left.\left.\times\left(\frac{\Delta q^\nu_j}{\varepsilon} + B^\nu_{[\gamma]}(\boldsymbol{q}_j,\boldsymbol{q}_{j-1})\right) + C_{[\gamma]}(\boldsymbol{q}_j,\boldsymbol{q}_{j-1})\right]\right\}\end{aligned}$$

where $G_{[\gamma]\mu\nu}$ is the inverse matrix of $G^{\mu\nu}_{[\gamma]}$. In the case of $\gamma(0)$ (see Equation (2.21)) one obtains ($g_{\mu\nu}$ is the inverse matrix of $g^{\mu\nu}$)

$$(2.36)\quad I_N^{\gamma(0)} = \int_{q_0=Q_0}^{q_{n+1}=Q} \left(\prod_{i=1}^{N} \mathrm{d}\boldsymbol{q}_i \prod_{j=1}^{N+1} \{(2\pi\varepsilon\eta)^M \det g^{\mu\nu}(\boldsymbol{q}_{j-1})\}^{-1/2}\right) \times \exp\left\{-\frac{\varepsilon}{2\eta}\sum_{j=1}^{N+1} g_{\mu\nu}(\boldsymbol{q}_{j-1})\left(\frac{\Delta q_j^\mu}{\varepsilon} + A^\mu(\boldsymbol{q}_{j-1})\right) \times \left(\frac{\Delta q_j^\nu}{\varepsilon} + A^\nu(\boldsymbol{q}_{j-1})\right)\right\}$$

and we want to show that when $N \to \infty$ the limit of (2.35) for any γ is the same as the limit of (2.36). The argument of the exponential in (2.35) can be written $\Sigma_{j=1}^{N+1} E_j$, with

$$\begin{aligned}(2.37)\quad E_j = &- G_{[\gamma]\mu\nu}(\boldsymbol{q}_j, \boldsymbol{q}_{j-1})(\Delta q_j^\mu \Delta q_j^\nu/2\eta\varepsilon)\\ &- (1/\eta)G_{[\gamma]\mu\nu}(\boldsymbol{q}_j, \boldsymbol{q}_{j-1})\Delta q_j^\mu B^\nu_{[\gamma]}(\boldsymbol{q}_j, \boldsymbol{q}_{j-1})\\ &- (\varepsilon/2\eta)G_{[\gamma]\mu\nu}(\boldsymbol{q}_j, \boldsymbol{q}_{j-1})B^\mu_{[\gamma]}(\boldsymbol{q}_j, \boldsymbol{q}_{j-1})B^\mu_{[\gamma]}(\boldsymbol{q}_j, \boldsymbol{q}_{j-1})\\ &- \varepsilon C_{[\gamma]}(\boldsymbol{q}_j, \boldsymbol{q}_{j-1}).\end{aligned}$$

When $N \to \infty$ $(\varepsilon = (t - t_0)/(N+1) \to 0)$ the first term $(-\Delta \boldsymbol{q}_j^2/2\varepsilon)$ is dominant, and as it appears in the exponential $\exp[-\Delta \boldsymbol{q}_j^2/2\varepsilon]$ we see that $\Delta \boldsymbol{q}_j$ is effectively of order $\varepsilon^{1/2}$. Let us develop now E_j around $\boldsymbol{q}_{j-1}$ keeping only terms up to $O(\varepsilon)$ (which are the only terms that contribute in the limit).

The result is of the form (taking into account $\Delta \boldsymbol{q}_j = O(\varepsilon^{1/2})$)

$$\begin{aligned}(2.38)\quad E_j = &- g_{\mu\nu}(\boldsymbol{q}_{j-1})(\Delta q_j^\mu \Delta q_j^\nu/2\eta\varepsilon) + a_\mu(\boldsymbol{q}_{j-1})\Delta q_j^\mu\\ &+ a_{\mu\nu\rho}(\boldsymbol{q}_{j-1})(\Delta q_j^\mu \Delta q_j^\nu \Delta q_j^\rho/\varepsilon)\\ &+ a_{\mu\nu\rho\sigma}(\boldsymbol{q}_{j-1})(\Delta q_j^\mu \Delta q_j^\nu \Delta q_j^\rho \Delta q_j^\sigma/\varepsilon) + \varepsilon b(\boldsymbol{q}_{j-1}) + O(\varepsilon^{3/2}).\end{aligned}$$

Note that I_N^γ in (35) is of the form $\int \prod_i \mathrm{d}\boldsymbol{q}_i \prod_j F_j^\gamma$, with

$$\begin{aligned}(2.39)\quad F_j^\gamma = &[(2\pi\varepsilon\eta)^M \det G^{\mu\nu}_{[\gamma]}(\boldsymbol{q}_j, \boldsymbol{q}_{j-1})]^{-1/2}\\ &\times \exp[-g_{\mu\nu}(\boldsymbol{q}_{j-1})(\Delta q_j^\mu \Delta q_j^\nu/2\varepsilon\eta) + \tilde{E}_j].\end{aligned}$$

Where $\tilde{E}_j$ can be read from (2.38), and that we only need each F_j^γ up to $O(\varepsilon)$. What we want to show is how all the different F_j^γ coming from different discretizations coincide up to $O(\varepsilon)$. The determinant in (2.39) has an expansion of the form

$$(2.40)\quad [\det G^{\mu\nu}_{[\gamma]}(\boldsymbol{q}_j, \boldsymbol{q}_{j-1})]^{-1/2} = [\det g^{\mu\nu}(\boldsymbol{q}_{j-1})]^{-1/2} \times [1 + c_\mu(\boldsymbol{q}_{j-1})\Delta q^\mu_j + c_{\mu\nu}(\boldsymbol{q}_{j-1})\Delta q^\mu_j \Delta q^\nu_j + O(\varepsilon^{3/2})].$$

Let us develop now $\exp \tilde{E}_j = 1 + \tilde{E}_j + \frac{1}{2}\tilde{E}_j^2 + O(\varepsilon^{3/2})$ in (2.39), after using (2.40) we have

$$\begin{aligned}(2.41)\quad F^\gamma_j = {} & [(2\pi\varepsilon\eta)^M \det g^{\mu\nu}(\boldsymbol{q}_{j-1})]^{-1/2} \\ & \times \exp[-(1/\eta) g_{\mu\nu}(\boldsymbol{q}_{j-1})(\Delta q^\mu_j \Delta q^\nu_j / 2\varepsilon)] \\ & \times [1 + (a_\mu(\boldsymbol{q}_{j-1}) + c_\mu(\boldsymbol{q}_{j-1}))\Delta q^\mu_j + a_{\mu\nu\rho}(\boldsymbol{q}_{j-1})(\Delta q^\mu_j \Delta q^\nu_j \Delta q^\rho_j/\varepsilon) \\ & + (c_{\mu\nu}(\boldsymbol{q}_{j-1}) + \tfrac{1}{2} a_\mu(\boldsymbol{q}_{j-1}) a_\nu(\boldsymbol{q}_{j-1}))\Delta q^\mu_j \Delta q^\nu_j \\ & + (a_\mu(\boldsymbol{q}_{j-1}) a_{\nu\rho\sigma}(\boldsymbol{q}_{j-1}) + a_{\mu\nu\rho\sigma}(\boldsymbol{q}_{j-1}))(\Delta q^\mu_j \Delta q^\nu_j \Delta q^\rho_j \Delta q^\sigma_j/\varepsilon) \\ & + \tfrac{1}{2} a_{\mu\nu\rho}(\boldsymbol{q}_{j-1}) a_{\sigma\alpha\beta}(\boldsymbol{q}_{j-1})(\Delta q^\mu_j \Delta q^\nu_j \Delta q^\rho_j \Delta q^\sigma_j \Delta q^\alpha_j \Delta q^\beta_j/\varepsilon^2) \\ & + \varepsilon b(\boldsymbol{q}_{j-1}) + O(\varepsilon^{3/2})].\end{aligned}$$

One has then from (35) that

$$\begin{aligned}(2.42)\quad I^\gamma_N = {} & \int \prod_i \mathrm{d}\boldsymbol{q}_i \prod_j [(2\pi\varepsilon\eta)^M \det g^{\mu\nu}(\boldsymbol{q}_{j-1})]^{-1/2} \\ & \times \exp\left[-\frac{1}{2\eta\varepsilon} g_{\mu\nu}(\boldsymbol{q}_{j-1})\Delta q^\mu_j \Delta q^\nu_j\right][1 + \tilde{F}^\gamma_j + O(\varepsilon^{3/2})]\end{aligned}$$

where $\tilde{F}^\gamma_j$ can be read from (2.41). In (2.42) one can change $\tilde{F}^\gamma_j$ as far as the difference does not contribute when $N \to \infty$ ($\varepsilon \to 0$). We use the symbol $\doteq$ introduced by B. S. de Witt [23] to indicate equivalence under the N-dimensional integral (2.42) when $N \to \infty$. One has the following relations for the terms in $\tilde{F}^\gamma_j$

$$(2.43)\quad \text{(a)}\ \Delta q^\mu_j \Delta q^\nu_j \doteq \varepsilon\eta g^{\mu\nu}(\boldsymbol{q}_{j-1}),$$

$$\text{(b)}\ \frac{1}{\varepsilon}\Delta q^\mu_j \Delta q^\nu_j \Delta q^\sigma_j \doteq \eta[\Delta q^\mu_j g^{\nu\sigma} + \Delta q^\nu_j g^{\mu\sigma} + \Delta q^\sigma_j g^{\mu\nu}]|_{\boldsymbol{q}_{j-1}},$$

$$\text{(c)}\ \frac{1}{\varepsilon}\Delta q^\mu_j \Delta q^\nu_j \Delta q^\zeta_j \Delta q^\sigma_j \doteq \varepsilon\eta^2[g^{\mu\nu}g^{\zeta\sigma} + g^{\mu\zeta}g^{\nu\sigma} + g^{\mu\sigma}g^{\nu\zeta}]|_{\boldsymbol{q}_{j-1}}$$

$$\begin{aligned}\text{(d)}\ \frac{1}{\varepsilon^2}\Delta q^\mu_j \Delta q^\nu_j \Delta q^\zeta_j \Delta q^\sigma_j \Delta q^\kappa_j \Delta q^\lambda_j \doteq {} & \varepsilon\eta^3[g^{\mu\nu}(g^{\zeta\sigma}g^{\kappa\lambda} + g^{\zeta\kappa}g^{\sigma\lambda} + g^{\zeta\lambda}g^{\sigma\kappa}) \\ & + g^{\mu\zeta}(g^{\nu\sigma}g^{\kappa\lambda} + g^{\nu\kappa}g^{\sigma\lambda} + g^{\nu\lambda}g^{\sigma\kappa}) + g^{\mu\sigma}(g^{\nu\zeta}g^{\kappa\lambda} + g^{\nu\kappa}g^{\zeta\kappa} \\ & + g^{\nu\lambda}g^{\zeta\kappa}) + g^{\mu\kappa}(g^{\nu\zeta}g^{\sigma\lambda} + g^{\nu\sigma}g^{\zeta\lambda} + g^{\nu\lambda}g^{\zeta\sigma}) + g^{\mu\lambda}(g^{\nu\zeta}g^{\sigma\kappa} \\ & + g^{\nu\sigma}g^{\zeta\kappa} + g^{\nu\kappa}g^{\zeta\sigma})]|_{\boldsymbol{q}_{j-1}}.\end{aligned}$$

These relations are simple to prove. They are discussed in [16, 23–27]. The first one is (when there is only one variable) a direct consequence of the formula

$$(2.44)\quad \int dq_j \exp[-\Delta q_j^2/2\varepsilon D]\Delta q_j^2 f(q_j) = \varepsilon D f(q_{j-1}) \times \int dq_j \exp[-\Delta q_j^2/2\varepsilon D] + O(\varepsilon^2)$$

which says that $\Delta q_j^2 \doteq \varepsilon D$. The others are proved similarly. In fact it can also be shown, as we have done in detail in [16], that all relations (43) are a rather direct consequence of the commutation relations $[\hat{q}^\mu, \hat{p}_\nu] = i\delta^\mu_\nu$. In order to see this for Δq_j^2 we start from the equality

$$(2.45)\quad \int dq_j \exp[-(\Delta q_j^2/2\varepsilon D) + \Delta q_j^2 + O(\varepsilon^{3/2})] = \int dq_j dp_j (\varepsilon D/2\pi)^{1/2} \exp\{i\varepsilon[p(\Delta q_j/\varepsilon) + (iD/2)p^2 - Dp\Delta q_j]\}.$$

Comparing the right-hand side of (2.45) with (2.9) we see that the terms there would have been found computing $\langle Q|U(t,t_0)|Q_0\rangle$ with $U(t',t)$ determined by (see (2.5) and the calculation in formula (2.14)).

$$(2.46)\quad \hat{H} = -(iD/2)\hat{p}^2 + D(\hat{q}\hat{p} - \hat{p}\hat{q}) = -(iD/2)\hat{p}^2 + D[\hat{q},\hat{p}].$$

But $[\hat{q},\hat{p}] = i$, and $\hat{H} = -(iD/2)\hat{p}^2 + iD$. Using $\hat{H}$ written in this last form one can see that the right-hand side of (2.45) can be effectively replaced by

$$(2.47)\quad \int dq_j dp_j (\varepsilon D/2\pi)^{1/2} \exp i\varepsilon[p(\Delta q_j/\varepsilon) + (iD/2)p^2 - iD] = \int dq_j \exp[-(\Delta q_j^2/2\varepsilon D) + \varepsilon D],$$

which, when compared with (2.45) indeed shows that $\Delta q_j^2 \doteq \varepsilon D$.

Let us go back now to (2.41). Using the relations (2.43) one can eliminate under the sign $\doteq$ all powers $\Delta \boldsymbol{q}^n, n \geqslant 2$, and remain with an expression of the form

$$(2.48)\quad F_j^\gamma \doteq [(2\pi\varepsilon\eta)^M \det g^{\mu\nu}(\boldsymbol{q}_{j-1})]^{-1/2} \exp\left[-\frac{1}{2\eta\varepsilon} g_{\mu\nu}(\boldsymbol{q}_{j-1})\Delta q_j^\mu \Delta q_j^\nu\right] \times [1 + \bar{a}_\mu(\boldsymbol{q}_{j-1})\Delta q_j^\mu + \varepsilon\bar{b}(\boldsymbol{q}_{j-1})].$$

The right-hand side of (48) must be equal to $F_j^{\gamma(0)}$ computed from (2.36) which is (after using (2.43 (a))

$$(2.49)\quad F_j^{\gamma(0)} = [(2\pi\varepsilon\eta)^M \det g^{\mu\nu}(\boldsymbol{q}_{j-1})]^{-1/2} \exp\left[-\frac{1}{2\eta\varepsilon} g_{\mu\nu}(\boldsymbol{q}_{j-1})\Delta q_j^\mu \Delta q_j^\nu\right]$$
$$\times\left[1 - \frac{1}{\eta} g_{\mu\nu}(\boldsymbol{q}_{j-1}) A^\nu(\boldsymbol{q}_{j-1})\Delta q_j^\mu + O(\varepsilon^{3/2})\right].$$

We see then that in (48) one must find $\bar{b}(\boldsymbol{q}_{j-1}) = 0$ and $\bar{a}_\mu(\boldsymbol{q}_{j-1}) = -(1/\eta)$ $g_{\mu\nu}(\boldsymbol{q}_{j-1})A^\nu(\boldsymbol{q}_{j-1})$. This can be checked explicitly once γ, and hence h^γ, is specified.

We remark that (2.35) is written in the limit $N \to \infty$ as the configuration space path integral ($g = \det g_{\mu\nu}$)

$$(2.50)\quad P(\boldsymbol{Q},t;\boldsymbol{Q}_0,t_0) = \int_\gamma D(\sqrt{g}\,\boldsymbol{q})$$
$$\times \exp\left\{-\int \mathrm{d}\tau\left[\frac{1}{2\eta} g_{\mu\nu}(\boldsymbol{q})(\dot{q}^\mu + \bar{B}^\mu(\boldsymbol{q}))(\dot{q}^\nu + \bar{B}^\nu(\boldsymbol{q}))\right.\right.$$
$$\left.\left. + \bar{C}(\boldsymbol{q})\right]\right\}\delta(\boldsymbol{q}(t) - \boldsymbol{Q})\delta(\boldsymbol{q}(t_0) - \boldsymbol{Q}_0),$$

where the discretization γ is specified by (2.35) which defines (2.50) and $\bar{B}^\mu(\boldsymbol{q}) = B^\mu_{[\gamma]}(\boldsymbol{q},\boldsymbol{q})$, $\bar{C}(\boldsymbol{q}) = C_{[\gamma]}(\boldsymbol{q},\boldsymbol{q})$. We finally note that in a term that is already of order ε, such as $\varepsilon C_{[\gamma]}(\boldsymbol{q}_j,\boldsymbol{q}_{j-1})$ in (35), one can of course also write instead $\varepsilon\bar{C}(\boldsymbol{q}_{j-1})$ or $\varepsilon\bar{C}(\boldsymbol{q}_j)$ since the difference is $O(\varepsilon^{3/2})$. We shall use this repeatedly in what follows, in the configuration and also in the phase space functional integrals.

3. Path Integral and Stochastic Differential Calculus

We write now the Langevin equation (1.1) in the form

$$(3.1)\quad \mathrm{d}q^\nu + a^\nu(\boldsymbol{q}(t))\,\mathrm{d}t = \sqrt{\eta}\,\sigma^\nu_\alpha(\boldsymbol{q}(t))\,\mathrm{d}w^\alpha$$

where $\mathrm{d}w^\alpha$ is the differential of the Wiener process (formally $\mathrm{d}w^\mu = f^\mu(t)\mathrm{d}t$). Let us consider the interval (t_0,t), putting $t_j = t_0 + j\varepsilon$, $t_{N+1} = t$, the probability P_1 that $\{\boldsymbol{w}(t_0) = 0$,

$$\boldsymbol{w}_1 \leqslant \boldsymbol{w}(t_1) \leqslant \boldsymbol{w}_1 + \mathrm{d}\boldsymbol{w}_1, \boldsymbol{w}_2 \leqslant \boldsymbol{w}(t_2) \leqslant \boldsymbol{w}_2 + \mathrm{d}\boldsymbol{w}_2, \ldots,$$
$$\boldsymbol{w}_{N+1} \leqslant \boldsymbol{w}(t) \leqslant \boldsymbol{w}_{N+1} + d\boldsymbol{w}_{N+1}\}$$ is given by

$$(3.2)\quad P_1 = \prod_{j=1}^{N+1} \frac{1}{\sqrt{(2\pi\varepsilon)^M}} \prod_{\mu=1}^{M} \mathrm{d}w_j^\mu \times \exp\left[-\frac{1}{2\varepsilon} \sum_{j=1}^{N+1} \sum_{\mu=1}^{M} (w_j^\mu - w_{j-1}^\mu)^2 \right], \qquad \boldsymbol{w}_0 = 0.$$

Let us make the change of variables $\boldsymbol{Z}_j = \boldsymbol{w}_j - \boldsymbol{w}_{j-1}, j = 1, 2, \ldots, N+1$. One has det $(\partial Z_j^\mu / \partial w_i^\nu) = 1$ and consequently we can write P_1 as •

$$(3.3)\quad P_1 = \left(\prod_{j=1}^{N+1} [(2\pi\varepsilon)^M]^{-1/2} \prod_{\mu=1}^{M} \mathrm{d}Z_j^\mu \right) \exp\left[-\frac{1}{2\varepsilon} \sum_{j=1}^{N+1} \sum_{\mu=1}^{M} (Z_j^\mu)^2 \right].$$

Now, through the stochastic differential equation (3.1) one can transform P_1 in the probability for paths in $\boldsymbol{q}$-space. In order to do this we discretize (1) putting $\boldsymbol{q}(t_j) = \boldsymbol{q}_j, \boldsymbol{q}_0 = \boldsymbol{Q}_0, t_j - t_{j-1} = \varepsilon, \boldsymbol{w}(t_j) = \boldsymbol{w}_j$. One has

$$(3.4)\quad q_j^\nu - q_{j-1}^\nu + \varepsilon a^\nu(\boldsymbol{q}_{j-1}^{(r)}) = \sqrt{\eta}\, \sigma_\alpha^\nu(\boldsymbol{q}_{j-1}^{(s)})(w_j^\alpha - w_{j-1}^\alpha),$$

where $\boldsymbol{q}_{j-1}^{(r)} \equiv \boldsymbol{q}_{j-1} + r\Delta\boldsymbol{q}_j, \qquad \Delta\boldsymbol{q}_j \equiv \boldsymbol{q}_j - \boldsymbol{q}_{j-1}$.
We shall see that the stochastic process $\boldsymbol{q}(t)$ defined by (3.1) depends on s in (3.4), but not on r.

The dependence on r is just the form in which appears here the δ-function ambiguity of the white noise in the previous section (see after (2.2)), and we shall see that for $s = 1/2$ one has the process of Section 2. The value $s = 1/2$ corresponds to the Stratonovich interpretation of (1), while $s = 0$ corresponds to the Ito interpretation [28].

Defining $\sigma_{\beta\mu}$ by $\sigma_\alpha^\mu(\boldsymbol{q})\sigma_{\beta\mu} = \delta_{\alpha\beta}$ one has $g_{\mu\nu} = \sigma_{\alpha\mu}\sigma_{\alpha\nu}$ and $\sigma_{\beta\mu} = g_{\mu\nu}\sigma_\beta^\nu$. Recalling that $Z_j^\rho = w_j^\rho - w_{j-1}^\rho$ one obtains from (3.4) that

$$(3.5)\quad Z_j^\rho = \frac{1}{\sqrt{\eta}} \sigma_{\rho\nu}(\boldsymbol{q}_{j-1}^{(s)}) [\Delta q_j^\nu + \varepsilon a^\nu(\boldsymbol{q}_{j-1}^{(r)})].$$

We have to replace (3.5) in (3.3) and compute of course the Jacobian J, $\prod_j \mathrm{d}\boldsymbol{Z}_j = J \prod_j \mathrm{d}\boldsymbol{q}_j$, $J = \det\,(\partial Z_j^\rho / \partial q_i^\alpha)$. Since $\partial Z_j^\rho / \partial q_k^\alpha = 0, k > j$, one has $J = \prod_{j=1}^{N+1} \det\,(\partial Z_j^\rho / \partial q_j^\alpha)$ the last determinant being of the $M \times M$ matrix of indices $(\rho\alpha)$. From (3.5) one has

$$(3.6)\quad \partial Z_j^\rho / \partial q_j^\alpha = (1/\sqrt{\eta})\sigma_{\rho\nu}(\boldsymbol{q}_{j-1}^{(s)})[\delta_{\nu\alpha} + \varepsilon r \partial_\alpha a^\nu(\boldsymbol{q}_{j-1}^{(r)}) + s\sigma_\mu^\nu(\boldsymbol{q}_{j-1}^{(s)})\partial_\alpha \sigma_{\mu\gamma}(\boldsymbol{q}_{j-1}^{(s)})(\Delta q_j^\gamma + \varepsilon a^\gamma(\boldsymbol{q}_{j-1}^{(r)}))].$$

Consequently $\det\,(\partial Z_j^\rho / \partial q_j^\alpha) = \eta^{-M/2} \det\,[\sigma_{\rho\nu}(\boldsymbol{q}_{j-1}^{(s)})] \det(1 + X)$

where X is obtained from (3.6). We have $\det\,(1 + X) = \exp \operatorname{tr} \ln\,(1 + X) = \exp \operatorname{tr}\,(X - (X^2/2) + 0(\varepsilon^{3/2}))$, where we have used $\Delta q_j = O(\varepsilon^{1/2})$ since the dominant term in the argument of the exponential in (3.3) will again be $(-\Delta \boldsymbol{q}_j^2/2\varepsilon)$ when (3.5) is replaced. Explicitly

$$
\begin{aligned}
(3.7)\quad \det(1+X) = \exp\Big\{ & s\sigma^\nu_\mu(\boldsymbol{q}^{(s)}_{j-1})\partial_\nu\sigma_{\mu\gamma}(\boldsymbol{q}^{(s)}_{j-1})(\Delta q^\gamma_j + \varepsilon a^\gamma(\boldsymbol{q}^{(r)}_{j-1})) \\
& - \frac{\varepsilon}{2}s^2\eta\sigma^\nu_\mu(\boldsymbol{q}_{j-1})\sigma^\beta_{\mu'}(\boldsymbol{q}_{j-1})g^{\gamma\gamma'}(\boldsymbol{q}_{j-1})\partial_\beta\sigma_{\mu\gamma}(\boldsymbol{q}_{j-1}) \\
& \times \partial_\nu\sigma_{\mu'\gamma'}(\boldsymbol{q}_{j-1}) + \varepsilon r\partial_\alpha a^\alpha(\boldsymbol{q}_{j-1}) + O(\varepsilon^{3/2})\Big\}
\end{aligned}
$$

where we have used $\Delta q^\gamma_j \Delta q^{\gamma'}_j \doteq \varepsilon\eta g^{\gamma\gamma'}(\boldsymbol{q}_{j-1})$ (see (2.43)) and we have put $\boldsymbol{q}_{j-1}$ instead of $\boldsymbol{q}^{(s)}_{j-1}$ or $\boldsymbol{q}^{(r)}_{j-1}$ in terms of $O(\varepsilon)$.
One has $\det\sigma_{\rho\nu} = (\det g^{\mu\nu})^{-1/2}$. The Jacobian J is finally

$$
\begin{aligned}
(3.8)\quad J = \left(\prod_{j=1}^{N+1}[\eta^M \det g^{\mu\nu}(\boldsymbol{q}^{(s)}_{j-1})]^{-1/2}\right)\exp\Big\{ & \sum_{j=1}^{N+1}[s\sigma^\nu_\mu(\boldsymbol{q}^{(s)}_{j-1})\partial_\nu\sigma_{\mu\gamma}(\boldsymbol{q}^{(s)}_{j-1}) \\
& \times(\Delta q^\gamma_j + \varepsilon a^\gamma(\boldsymbol{q}^{(r)}_{j-1})) - \frac{\varepsilon}{2}s^2\eta\partial_\beta\sigma^\nu_\mu(\boldsymbol{q}_{j-1})\partial_\nu\sigma^\beta_\mu(\boldsymbol{q}_{j-1}) \\
& + \varepsilon r\partial_\alpha a^\alpha(\boldsymbol{q}_{j-1})]\Big\},
\end{aligned}
$$

since

$$
\sigma^\nu_\mu\sigma^\beta_{\mu'}g^{\gamma\gamma'}(\partial_\beta\sigma_{\mu\gamma})(\partial_\nu\sigma_{\mu'\gamma'}) = (\partial_\beta\sigma^\nu_\mu)(\partial_\nu\sigma^\beta_\mu).
$$

Replacing (3.5) in (3.3) and using (3.8) for the Jacobian one obtains $(\boldsymbol{q} = \boldsymbol{Q}_0)$

$$
\begin{aligned}
(3.9)\quad P_1 = & \prod_{j=1}^{N+1}[(2\pi\varepsilon\eta)^M \det g^{\mu\nu}(\boldsymbol{q}^{(s)}_{j-1})]^{-1/2}\prod_{\mu=1}^{M} \mathrm{d}q^\mu_j \\
& \times\exp\Big\{-\frac{\varepsilon}{2\eta}\sum_{j=1}^{N+1} g_{\mu\nu}(\boldsymbol{q}^{(s)}_{j-1})((\Delta q^\mu_j/\varepsilon) + a^\mu(\boldsymbol{q}^{(r)}_{j-1}) \\
& \times((\Delta q_j/\varepsilon) + a^\nu(\boldsymbol{q}^{(r)}_{j-1})) + s\sigma^\nu_\mu(\boldsymbol{q}^{(s)}_{j-1})\partial_\nu\sigma_{\mu\gamma}(\boldsymbol{q}^{(s)}_{j-1})(\Delta q^\gamma_j \\
& + \varepsilon a^\gamma(\boldsymbol{q}^{(r)}_{j-1}) + \varepsilon r\partial_\alpha a^\alpha(\boldsymbol{q}_{j-1}) - \frac{\varepsilon}{2}s^2\eta\partial_\beta\sigma^\nu_\mu(\boldsymbol{q}_{j-1})\partial_\nu\sigma^\beta_\mu(\boldsymbol{q}_{j-1})\Big\}.
\end{aligned}
$$

Using formula (2.34) we can introduce integrations $\int\sum_{j=1}^{N+1}\mathrm{d}\boldsymbol{p}_j$ and write (3.9) in the form

$$(3.10)\quad P_1 = \left(\prod_{j=1}^{N+1}\prod_{\mu=1}^{M} \mathrm{d}q_j\right)\int\left(\prod_{j=1}^{N+1}\frac{\mathrm{d}p_j}{(2\pi)^M}\right)\exp\left\{i\varepsilon\sum_{j=1}^{N+1}\left[p_j\cdot\frac{\Delta q_j}{\varepsilon}+\right.\right.$$
$$+\frac{i\eta}{2}g^{\alpha\beta}(q_{j-1}^{(s)})p_{j\alpha}p_{j\beta}+p_{j\mu}(a^\mu(q_{j-1}^{(r)})+\eta s\partial_\alpha g^{\alpha\mu}(q_{j-1}^{(s)})$$
$$-\eta s\sigma_\alpha^\nu(q_{j-1}^{(s)})\partial_\nu\sigma_\alpha^\nu(q_{j-1}^{(s)}))-ir\partial_\alpha a^\alpha(q_{j-1})+\frac{i}{2}\eta s^2$$
$$\left.\left.\times(\partial_\beta\sigma_\mu^\nu(q_{j-1})\partial_\nu\sigma_\mu^\beta(q_{j-1})-\partial_\nu\sigma_\mu^\nu(q_{j-1})\partial_\beta\sigma_\mu^\beta(q_{j-1}))\right]\right\}.$$

Fixing now $q_{N+1}=Q$, i.e. integrating $\delta(q_{N+1}-Q)P_1$ over $\mathrm{d}q_i$, $i=1,2,\ldots$ $N+1$, one obtains from (3.10) $P(Q,t;Q_0,t_0)$ as

$$(3.11)\quad P(Q,t;Q_0,t_0)=\lim_{N\to\infty}\int_{q_0=Q_0}^{q_{N+1}=Q}\left(\prod_{j=1}^{N+1}\mathrm{d}q_i\prod_{j=1}^{N+1}\frac{\mathrm{d}p_j}{(2\pi)^M}\right)$$
$$\times\exp\left\{i\varepsilon\sum_{j=1}^{N+1}\left[p_j\cdot\frac{\Delta q_j}{\varepsilon}-\bar{h}^{[r,s]}(p,q_j,q_{j-1})\right]\right\}$$

with

$$(3.12)\quad \bar{h}^{[r,s]}(p,q',q)=-\frac{i\eta}{2}g^{\mu\nu}(q^{(s)})p_\mu p_\nu-p_\mu a^\mu(q^{(r)})$$
$$+p_\mu(\eta s\sigma_\alpha^\nu(q^{(s)})\partial_\nu\sigma_\alpha^\mu(q^{(s)})-\eta s\partial_\alpha g^{\alpha\mu}(q^{(s)})+ir\partial_\alpha a^\alpha(q)$$
$$-\frac{i\eta s^2}{2}[(\partial_\beta\sigma_\mu^\nu(q))(\partial_\nu\sigma_\mu^\beta(q))-(\partial_\nu\sigma_\mu^\nu(q))(\partial_\beta\sigma_\mu^\beta(q))].$$

We want to compare (3.12) with $h^{[r,s]}(p,q',q)$ as given by (2.32). We remark that one can change in (2.32) $q^{(s)}$ and $q^{(r)}$ in q in all terms without p since these terms are already $O(\varepsilon)$ (see the last paragraph of Section 2).

In general in phase space, and as a consequence of $\Delta q_j=O(\varepsilon^{1/2})$, one must count p_j as being effectively $O(\varepsilon^{-1/2})$ when computing the order in ε in (3.12). We shall see now that $h^{[r,1/2]}=\bar{h}^{[r,1/2]}$, i.e. they coincide for $s=1/2$. From (2.32) and using

$$2\partial_\mu(\sigma_\alpha^\nu\partial_\nu\sigma_\alpha^\mu)-\partial_{\mu\nu}g^{\mu\nu}=(\partial_\nu\sigma_\alpha^\mu)(\partial_\nu\sigma_\alpha^\nu)-(\partial_\mu\sigma_\alpha^\mu)(\partial_\nu\sigma_\alpha^\nu)$$

one obtains

$$(3.13)\quad h^{[r,1/2]}(p,q',q)=-\frac{i\eta}{2}g^{\mu\nu}(q^{(1/2)})p_\mu p_\nu$$

$$+\frac{\eta}{2}p_\mu(\sigma^\nu_\alpha(\boldsymbol{q}^{(1/2)})\partial_\nu\sigma^\mu_\alpha(\boldsymbol{q}^{(1/2)})-\partial_\nu g^{\nu\mu}(\boldsymbol{q}^{(1/2)})$$

$$-p_\mu a^\mu(\boldsymbol{q}^{(r)})+ir\partial_\alpha a^\alpha(\boldsymbol{q})-\frac{i\eta}{8}[(\partial_\nu\sigma^\mu_\alpha(\boldsymbol{q}))(\partial_\mu\sigma^\nu_\alpha(\boldsymbol{q}))$$

$$-(\partial_\mu\sigma^\mu_\alpha(\boldsymbol{q}))(\partial_\nu\sigma^\nu_\alpha(\boldsymbol{q}))]$$

which indeed coincides with (3.12) when one puts $s = 1/2$ there. The conclusion is then that for $s = 1/2$, i.e. when (1) is interpreted in the Stratonovich sense, one obtains the same stochastic process as in Section 2, and this independently of r. This Markovian process $\boldsymbol{q}(t)$ is characterized by the Fokker–Planck equation (2.1) for $P(\boldsymbol{Q},t;\boldsymbol{Q}_0,t_0)$. When $s \neq 1/2$, eq. (3.1) defines a different stochastic process whose Fokker–Planck equation can easily be computed from (3.11) and (3.12) using the methods of Section 2 (see [13], section on the inverse problem). We give now a more direct method to treat (1). It is based on some simple rules for the differentials $\mathrm{d}t$, $\mathrm{d}w^\alpha$, appearing in (1) (called the Ito rules), and which say that terms of higher order than $O(\mathrm{d}t)$ should be dropped with $\mathrm{d}w = O(\mathrm{d}t^{1/2})$, and that $\mathrm{d}w^\mu\,\mathrm{d}w^\nu$ should be replaced by $\delta^{\mu\nu}\,dt$ (note that these rules are just what one obtains from the dominating term $(-\Delta w_j^2/2\varepsilon)$ in (3.2)). We write (3.1) as $(\boldsymbol{q}^{(r)} = \boldsymbol{q} + r\,\mathrm{d}\,\boldsymbol{q})$

$$\text{(3.14)}\quad \mathrm{d}q^\nu + a^\nu(\boldsymbol{q}^{(r)})\,\mathrm{d}t = \sqrt{\eta}\,\sigma^\nu_\mu(\boldsymbol{q}^{(s)})\,\mathrm{d}w^\mu$$

and we see immediately that we can put $\boldsymbol{q}$ instead of $\boldsymbol{q}^{(r)}$. Developing $\sigma^\nu_\mu(\boldsymbol{q}^{(s)}) = \sigma^\nu_\mu(\boldsymbol{q} + s\,\mathrm{d}\boldsymbol{q})$ around $\boldsymbol{q}$, using (14) for $\mathrm{d}\boldsymbol{q}$ and also $\mathrm{d}w^\mu\,\mathrm{d}w^\nu = \delta^{\mu\nu}\mathrm{d}t$, one has

$$\text{(3.15)}\quad \sqrt{\eta}\,\sigma^\nu_\mu(\boldsymbol{q}^{(s)})\mathrm{d}w^\mu = \sqrt{\eta}\,\sigma^\nu_\mu(\boldsymbol{q})\,\mathrm{d}w^\mu + \eta s\,\sigma^\rho_\mu(\boldsymbol{q})\partial_\rho\sigma^\nu_\mu(\boldsymbol{q})\mathrm{d}t.$$

Replacing (3.15) in (3.14) one obtains

$$\text{(3.16)}\quad \mathrm{d}q^\nu + \bar{A}^\nu(\boldsymbol{q})\mathrm{d}t = \sqrt{\eta}\,\sigma^\nu_\mu(\boldsymbol{q})\mathrm{d}w^\mu$$

where $\bar{A}^\nu(\boldsymbol{q}) = a^\nu(\boldsymbol{q}) - \eta s\,\sigma^\rho_\mu(\boldsymbol{q})\partial_\rho\sigma^\nu_\mu(\boldsymbol{q})$.
We can now discretize (3.16) as we have just done in formula (3.4) of this section, to obtain $(\boldsymbol{q}_0 = \boldsymbol{Q}_0)$

$$\text{(3.17)}\quad w_j^\beta - w_{j-1}^\beta = \frac{1}{\sqrt{\eta}}\sigma_{\beta\nu}(\boldsymbol{q}_{j-1})(q_j^\nu - q_{j-1}^\nu + \varepsilon A^\nu(\boldsymbol{q}_{j-1})).$$

The calculation of P_1 (formulas (2) to (9)) is now much simpler since

$$\text{(3.18)}\quad \det(\partial w_j^\beta/\partial q_i^\mu) = \prod_{j=1}^{N+1}\det\frac{\partial w_j^\beta}{\partial q_j^\mu} = \prod_{j=1}^{N+1}\eta^{-M/2}[\det g^{\mu\nu}(\boldsymbol{q}_{j-1})]^{-1/2}$$

and then from (3.2), (3.17) and (3.18) one immediately obtains

$$(3.19)\quad P_1 = \left(\prod_{j=1}^{N+1} \mathrm{d}q_j[(2\pi\varepsilon\eta)^M \det g^{\mu\nu}(\boldsymbol{q}_{j-1})]^{-1/2}\right)\exp\left\{-\frac{\varepsilon}{2\eta}\sum_{j=1}^{N+1} g^{\mu\nu}(\boldsymbol{q}_{j-1})\right.$$
$$\left.\times\,((\Delta q_j^\mu/\varepsilon) + \bar{A}^\mu(\boldsymbol{q}_{j-1}))((\Delta q_j^\nu/\varepsilon) + \bar{A}^\nu(\boldsymbol{q}_{j-1}))\right\}.$$

Integrating $\delta(\mathrm{q}_{N+1} - \mathrm{Q})P_1$ over $\mathrm{d}\boldsymbol{q}_j$, $j = 1,2,\ldots.N+1$, we have

$$(3.20)\quad P(\boldsymbol{Q},t;\boldsymbol{Q}_0,t_0) = \lim_{N\to\infty}\int_{q_0=Q_0}^{q_{N+1}=Q_N}\left(\prod_{i=1}^{N}\mathrm{d}\boldsymbol{q}_i\prod_{j=1}^{N+1}\right.$$
$$\times\,[(2\pi\varepsilon\eta)^M \det g^{\mu\nu}(\boldsymbol{q}_{j-1})]^{-1/2}\exp\left\{-\frac{\varepsilon}{2\eta}\sum_{j=1}^{N+1} g_{\mu\nu}(\boldsymbol{q}_{j-1})\right.$$
$$\left.\left.\times\left(\frac{\Delta q_j^\mu}{\varepsilon} + \bar{A}^\mu(\boldsymbol{q}_{j-1})\right)\left(\frac{\Delta q_j^\nu}{\varepsilon} + \bar{A}^\nu(\boldsymbol{q}_{j-1})\right)\right\}\right).$$

For $s = 1/2$ one has $\bar{A}^\mu(\boldsymbol{q}) = A^\mu(\boldsymbol{q})$ (see (3.16) and (2.2)) and comparing (3.20) with (2.36) we arrive again at the conclusion that the process defined by (3.14) when $s = 1/2$ is the same we studied in Section 2.

We shall give in this section a third method to obtain a functional integral representation for $P(\boldsymbol{Q},t;\boldsymbol{Q}_0,t_0)$. This method is used in Refs. [29–31] and was also discussed by us in relation to discretization problems in [15].

Formally it consists in solving (2.1) for fixed $f^\mu(t)$ and to do the average over $f^\mu(t)$ afterwards using the formula [1,15]

$$(3.21)\quad \{F[\boldsymbol{f}]\} = N^{-1}\int D\boldsymbol{f}F[\boldsymbol{f}]\exp\left[-\frac{1}{2}\int \mathrm{d}\tau\sum_{\mu=1}^{M}(f^\mu(\tau))^2\right]$$

for the average $\{F[\boldsymbol{f}]\}$ of a functional $F[\boldsymbol{f}]$(N is such that $\{1\} = 1$). Let $\boldsymbol{q}_f(t;\boldsymbol{Q}_0,t_0)$ be the solution of (1.1) for fixed $\boldsymbol{f}(t)$ and initial condition $\boldsymbol{Q}_0$ at t_0. Then obviously the probability density $P(\boldsymbol{Q},t;\boldsymbol{Q}_0,t_0)$ is $\{P_f(\boldsymbol{Q},t;\boldsymbol{Q}_0,t_0)\}$, where $P_f = \delta(\boldsymbol{Q} - \boldsymbol{q}_f(t;\boldsymbol{Q}_0,t_0))$ is the probability density of the deterministic process with $f(t)$ fixed. We indicate now the formal steps of the calculation. One has for any functional $F[\boldsymbol{q}_f(\tau)]$ that

$$(3.22)\quad F[\boldsymbol{q}_f(\tau)] = \int D\boldsymbol{q}(\tau)F[\boldsymbol{q}(\tau)]\delta[\boldsymbol{q}(\tau) - \boldsymbol{q}_f(\tau)]|_{t_0}^{t}.$$

But

$$\delta[\boldsymbol{q}(\tau) - \boldsymbol{q}_f(\tau)]|_{t_0}^{t} = \prod_i \delta(\boldsymbol{q}(t_i) - \boldsymbol{q}_f(t_i))$$

just expresses that the original differential equation (2.1) is satisfied and consequently we can write

$$(3.23)\quad \delta[\boldsymbol{q}(\tau)-\boldsymbol{q}_f(\tau)]|_{t_0}^{t}=\delta[\dot{q}^{\nu}(\tau)+a^{\nu}(\boldsymbol{q})(\tau)) -\sqrt{\eta}\sigma^{\nu}_{\mu}(\boldsymbol{q}(\tau))f^{\mu}(\tau)]|_{t_0}^{t}\,\delta(\boldsymbol{q}(t_0)-\boldsymbol{Q}_0)J$$

where J is a Jacobian due to the change in the argument of the δ-functions. Using $\delta[\boldsymbol{K}(\tau)]|_{t_0}^{t}=\int D\boldsymbol{p}\exp\{i\int \mathrm{d}\tau \boldsymbol{p}(\tau).\,\boldsymbol{K}(\tau)\}$ one obtains from (3.22) and (3.23) if $F[\boldsymbol{q}_f]=\delta(\boldsymbol{Q}-\boldsymbol{q}_f(t))$ that

$$(3.24)\quad \delta(\boldsymbol{Q}-\boldsymbol{q}_f(t))=\int D\boldsymbol{q}D\boldsymbol{p}\exp i\int_{t_0}^{t}\mathrm{d}\tau\, p_{\mu}(\tau) \times[\dot{q}^{\mu}+a^{\mu}(\boldsymbol{q})-\sqrt{\eta}\sigma^{\mu}_{\nu}(\boldsymbol{q})f^{\nu}(\tau)]J\delta(\boldsymbol{q}(t)-\boldsymbol{Q})\delta(\boldsymbol{q}(t_0)-\boldsymbol{Q}_0).$$

Now $P(\boldsymbol{Q},t;\boldsymbol{Q}_0,t_0)$ is obtained as $\{\delta(\boldsymbol{Q}-\boldsymbol{q}_f(t))\}$ from (24) and using (21) to compute the average over $\boldsymbol{f}(\tau)$ (the integral over $\boldsymbol{f}(\tau)$ being a Gaussian it can always be done exactly). Let us do now this calculation carefully in order to account for all the discretization problems.

Formula (22) for $F[\boldsymbol{q}_f(\tau)]=\delta(\boldsymbol{Q}-\boldsymbol{q}_f(t))$ reads in discretized version $(\boldsymbol{q}_j=\boldsymbol{q}(t_j), t_j=t_0+j\varepsilon, t_{N+1}=t)$

$$(3.25)\quad \delta(\boldsymbol{Q}-\boldsymbol{q}_f(t))=\int\prod_{i=1}^{N+1}\mathrm{d}\boldsymbol{q}_i\,\delta(\boldsymbol{q}_{N+1}-\boldsymbol{Q})\prod_{j=1}^{N+1}\delta(\boldsymbol{q}_j-\boldsymbol{q}_f(t_j)).$$

We now use the discretized form (3.4) of the stochastic differential equation to replace in (3.25) the product of δ-functions as follows

$$(3.26)\quad \prod_{j=1}^{N+1}\delta(\boldsymbol{q}_j-\boldsymbol{q}_f(t_j))=\det\left\{\frac{\partial}{\partial q^{\mu}_k}[((q^{\nu}_j-q^{\nu}_{j-1})/\varepsilon)+a^{\nu}(\boldsymbol{q}^{(r)}_{j-1}) -\sqrt{\eta}\sigma^{\nu}_{\alpha}(\boldsymbol{q}^{(s)}_{j-1})(w^{\alpha}_j-w^{\alpha}_{j-1}/\varepsilon)]\right\}\prod_{j=1}^{N+1}\prod_{\nu=1}^{M}\delta(((q^{\nu}_j-q^{\nu}_{j-1})/\varepsilon) +a^{\nu}(\boldsymbol{q}^{(r)}_{j-1})-\sqrt{\eta}\sigma^{\nu}_{\alpha}(\boldsymbol{q}^{(s)}_{j-1})(w^{\alpha}_j-w^{\alpha}_{j-1})/\varepsilon).$$

This formula is the discretized version of (3.23) and J is the determinant on the right-hand side of (26). We have (see the similar calculation of formula (6))

$$(3.27)\quad J=\prod_{j=1}^{N+1}\frac{1}{\varepsilon^M}\det\left\{\delta^{\nu}_{\mu}+\varepsilon r\partial_{\mu}a^{\nu}(\boldsymbol{q}^{(r)}_{j-1})-\sqrt{\eta}s\partial_{\mu}\sigma^{\nu}_{\alpha}(\boldsymbol{q}^{(s)}_{j-1})\Delta w^{\alpha}_j\right\}$$

where $\Delta\boldsymbol{w}_j=\boldsymbol{w}_j-\boldsymbol{w}_{j-1}$ and the last determinant applies to the $M\times M$

matrix of indices $(\nu\mu)$ in (27). Using det $(1+X) = \exp \operatorname{tr} \ln (1+X) = \exp \operatorname{tr} (X - X^2/2) + O(\varepsilon^{3/2}))$ since $\Delta w_j = 0(\varepsilon^{1/2})$ one obtains after replacing $\Delta w_j^\alpha \Delta w_j^\beta = \varepsilon\delta^{\alpha\beta}$ in tr X^2 (this replacement is valid since we shall average over $d\boldsymbol{w}_j$ with the Brownian motion weight (2) with dominating term $(-\Delta w_j^2/2\varepsilon)$) the expression

$$(3.28)\quad J = \left(\prod_{j=1}^{N+1} \frac{1}{\varepsilon^M}\right) \exp\left\{\sum_{j=1}^{N+1} \left[\varepsilon r \partial_\mu a^\mu(\boldsymbol{q}_{j-1}^{(r)}) - \sqrt{\eta} s \partial_\mu \sigma_\alpha^\mu(\boldsymbol{q}_{j-1}^{(s)}) \Delta w_j^\alpha \right.\right.$$
$$\left.\left. -\varepsilon \frac{\eta s^2}{2} (\partial_\mu \sigma_\alpha^\rho(\boldsymbol{q}_{j-1}))(\partial_\rho \sigma_\alpha^\mu(\boldsymbol{q}_{j-1}))\right]\right\}$$

We replace now (3.26) in (3.25) using (3.28) for the determinant, and average over Brownian motion (this corresponds to the average (3.21) in terms of the white noise) in order to obtain $P(\boldsymbol{Q},t;\boldsymbol{Q}_0,t_0) = \{\delta(\boldsymbol{Q} - \boldsymbol{q}_f(t))\}$. At the same time we write each of the $(N+1)$ δ-functions $\delta(K_j)$ in (3.26) as $\int (d\boldsymbol{p}_j/(2\pi)^M) \exp[i\boldsymbol{p}_j.\boldsymbol{K}_j]$. The result is (the limit $N \to \infty$ is to be understood)

$$(3.29)\quad P(\boldsymbol{Q},t;\boldsymbol{Q}_0,t_0) = \int \left(\prod_{j=1}^{N+1} \frac{1}{\sqrt{(2\pi\varepsilon)^M}} \prod_{\mu=1}^{M} dw_j^\mu\right)$$
$$\times \exp\left[-\frac{1}{2\varepsilon}\sum_{j=1}^{N+1}\sum_{\mu=1}^{M}(\Delta w_j^\mu)^2\right] \int \left(\prod_{j=1}^{N+1} d\boldsymbol{q}_i\right) \delta(\boldsymbol{q}_{N+1} - \boldsymbol{Q})$$
$$\times \int \left(\prod_{j=1}^{N+1} \frac{d\boldsymbol{p}_j}{(2\pi\varepsilon)^M}\right) \exp\left\{i \sum_{j=1}^{N+1}\left[p_{j\mu}\left(\frac{\Delta q_j^\mu}{\varepsilon} + a^\mu(\boldsymbol{q}_{j-1}^{(r)})\right.\right.\right.$$
$$\left.\left.\left. -\sqrt{\eta}\sigma_\alpha^\mu(\boldsymbol{q}_{j-1}^{(s)})\frac{\Delta w_j^\alpha}{\varepsilon}\right)\right]\right\} \exp\left\{\sum_{j=1}^{N+1}\left[\varepsilon r \partial_\mu a^\mu(\boldsymbol{q}_{j-1}^{(r)})\right.\right.$$
$$\left.\left. -\sqrt{\eta} s \partial_\mu \sigma_\alpha^\mu(\boldsymbol{q}_{j-1}^{(s)})\Delta w_j^\alpha - \varepsilon\frac{\eta s^2}{2}(\partial_\mu \sigma_\alpha^\rho(\boldsymbol{q}_{j-1}^{(s)}))(\partial_\rho \sigma_\alpha^\mu(\boldsymbol{q}_{j-1}^{(s)}))\right]\right\}.$$

Doing the change of variables $Z_j^\mu = \Delta w_j^\mu$ of Jacobian equal to unity the Gaussian integral over $d\boldsymbol{Z}_j$ is easily done in (3.29), and after changing $p_{j\mu} \to \varepsilon p_{j\mu}$ one obtains from (3.29) the expression

$$(3.30)\quad P(\boldsymbol{Q},t;\boldsymbol{Q}_0,t_0) = \int_{q_0 = Q_0}^{q_{N+1} = Q} \left(\prod_{i=1}^{N} d\boldsymbol{q}_i \prod_{j=1}^{N+1} \frac{d\boldsymbol{p}_j}{(2\pi)^M}\right)$$
$$\times \exp\left\{i\varepsilon \sum_{j=1}^{N+1}\left[p_{j\mu}(\Delta q_j^\mu/\varepsilon) + \frac{i\eta}{2} g^{\mu\nu}(\boldsymbol{q}_{j-1}^{(s)}) p_{j\mu} p_{j\nu} + p_{j\mu}(a^\mu(\boldsymbol{q}_{j-1}^{(r)})\right.\right.$$

$$+\eta s\sigma^{\mu}_{\alpha}(\boldsymbol{q}^{(s)}_{j-1})\partial_{\nu}\sigma^{\nu}_{\alpha}(\boldsymbol{q}^{(s)}_{j-1}) - ir\partial_{\mu}a^{\mu}(\boldsymbol{q}_{j-1}) + i\frac{\eta s^2}{2}(\partial_{\mu}\sigma^{\rho}_{\alpha}(\boldsymbol{q}_{j-1})$$

$$\times\, \partial_{\rho}\sigma^{\mu}_{\alpha}(\boldsymbol{q}_{j-1}) - \partial_{\mu}\sigma^{\mu}_{\alpha}(\boldsymbol{q}_{j-1})\partial_{\nu}\sigma^{\nu}_{\alpha}(\boldsymbol{q}_{j-1})\Big]\Big\}$$

where in terms of $O(\varepsilon)$ we have replaced $\boldsymbol{q}^{(r)}_{j-1}, \boldsymbol{q}^{(s)}_{j-1}$, by $\boldsymbol{q}_{j-1}$.

Noting that $\sigma^{\mu}_{\alpha}\partial_{\nu}\sigma^{\nu}_{\alpha} = \partial_{\nu}g^{\mu\nu} - \sigma^{\nu}_{\alpha}\partial_{\nu}\sigma^{\mu}_{\alpha}$ we see that (3.30) is just formula (3.11) since the argument of the exponential in (3.30) is $i\sum_j[p_{j\mu}\Delta q^{\mu}_j - \varepsilon\bar{h}^{[r,s]}(\boldsymbol{p}_j, \boldsymbol{q}_j, \boldsymbol{q}_{j-1})]$ with $\bar{h}^{[r,s]}$ given by (3.12).
Thus this third method gives the same result as the two previous ones.

4. Other definitions of path integrals

Apart from the definition through discretizations that we have been using we can mention the definition of Feynman [32] (the case of Langevin equations in this frame is discussed in [33, 34]) and the approaches of Graham [35–37] and Dekker [38].

These approaches are formally covariant under point transformations in $\boldsymbol{q}$-space, but if one wants to obtain perturbation expansions one has previously to determine the corresponding discretizations. In [39] we presented a critical examination of Dekker's approach.

We have not discussed here the problem of the most probable path in $\boldsymbol{q}$-space, this problem is solved in flat space [40–43], but the answer is missing in curved space.

Let us finally point out that the problem of obtaining a WKB-type expansion [44, 45] for the probability density of the process $\boldsymbol{q}(t)$ defined by (1.1) (i.e. an expansion in powers of η) has been solved in [20] and all higher-order corrections explicitly computed.

Appendix A: Definition of functional integrals through discretization

We first define phase space functional integrals of the form ($\boldsymbol{p} = (p_1, \ldots, p_M)$, $\boldsymbol{q} = (q^1, \ldots, q^M)$, $\quad \boldsymbol{p}\cdot\boldsymbol{q} = \sum_{\mu=1}^{M} p_{\mu}q^{\mu}$)

$$\text{(A1)}\quad I = \int_{\gamma} D\boldsymbol{q}D\boldsymbol{p}\exp\left\{i\int_{t_0}^{t} \mathrm{d}\tau[\boldsymbol{p}\cdot\dot{\boldsymbol{q}} - H^{\gamma}(\boldsymbol{p}(\tau), \boldsymbol{q}(\tau))]\right\}\delta(\boldsymbol{q}(t_0) - \boldsymbol{Q}_0)$$
$$\delta(\boldsymbol{q}(t) - \boldsymbol{Q}).$$

The symbol γ stands for the discretization, and it means that one is given a function $h^\gamma(\boldsymbol{p},\boldsymbol{q}',\boldsymbol{q})$ such that $h^\gamma(\boldsymbol{p},\boldsymbol{q},\boldsymbol{q}) = H^\gamma(\boldsymbol{p},\boldsymbol{q})$, in terms of which I is defined as lim $I_N, N \to \infty$, with $(\varepsilon = (t-t_0)/(N+1), \Delta\boldsymbol{q}_j \equiv \boldsymbol{q}_j - \boldsymbol{q}_{j-1})$

$$\text{(A2)}\quad I_N = \int_{q_0 = Q_0}^{q_{N+1} = Q} \left(\prod_{i=1}^{N} \mathrm{d}\boldsymbol{q}_i \prod_{j=1}^{N+1} (\mathrm{d}\boldsymbol{p}_j/(2\pi)^M)\right) \times \exp\left\{ i\varepsilon \sum_{j=1}^{N+1} [p_{j\mu}(\Delta q_j^\mu/\varepsilon) - h^\gamma(\boldsymbol{p}_j,\boldsymbol{q}_j,\boldsymbol{q}_{j-1})]\right\}.$$

One speaks of equivalent discretizations if one is given two functions $h_i^\gamma(p,q',q), i = 1,2$, such that $h_i^\gamma(q' = q) = H^\gamma(p,q)$, $i = 1,2$ and such that the limit of I_N, $N \to \infty$, is the same in both cases. It is clear that the equivalence classes of discretizations are the useful notion here. For details about this approach, see [16].

We define now the configuration space path integral

$$\text{(A3)}\quad I = \int_\gamma D(\sqrt{g}\,\boldsymbol{q}) \exp\left[-\int_{t_0}^{t} \mathrm{d}\tau L(\boldsymbol{q},\dot{\boldsymbol{q}})\right] \delta(\boldsymbol{q}(t) - \boldsymbol{Q})\delta(\boldsymbol{q}(t_0) - \boldsymbol{Q}_0)$$

where $L(\boldsymbol{q},\dot{\boldsymbol{q}})$ is of the form

$$\text{(A4)}\quad L = \tfrac{1}{2} g_{\mu\nu}(\boldsymbol{q})\dot{q}^\mu \dot{q}^\nu + \dot{q}^\mu A_\mu(\boldsymbol{q}) + V(\boldsymbol{q})$$

and $g(\boldsymbol{q}) = \det g_{\mu\nu}(\boldsymbol{q})$. The discretization γ here means that one is given functions $G_{[\gamma]\mu\nu}(\boldsymbol{q}',\boldsymbol{q})$ and $\alpha_\mu^{[\gamma]}(q',q)$, such that they reduce to $g_{\mu\nu}(\boldsymbol{q})$ and $A_\mu(\boldsymbol{q})$, respectively, for $\boldsymbol{q}' = \boldsymbol{q}$, and in terms of which I is defined as lim $I_N, N \to \infty$ ($G_{[\gamma]}^{\mu\nu}$ is the inverse matrix of $G_{[\gamma]\mu\nu}$, and ε, $\Delta\boldsymbol{q}_j$, are as before)

$$\text{(A5)}\quad I_N = \int_{q_0 = Q_0}^{q_{N+1} = Q} \left(\prod_{i=1}^{N} \mathrm{d}\boldsymbol{q}_i \prod_{j=1}^{N+1} [(2\pi\varepsilon)^M \det G_{[\gamma]}^{\mu\nu}(\boldsymbol{q}_j,\boldsymbol{q}_{j-1})]^{-1/2}\right) \times \exp\left\{ -\varepsilon \sum_{j=1}^{N+1} [\tfrac{1}{2} G_{[\gamma]\mu\nu}(\boldsymbol{q}_j,\boldsymbol{q}_{j-1})(\Delta q_j^\mu/\varepsilon)(\Delta q_j^\nu/\varepsilon) + a_\mu^{[\gamma]}(\boldsymbol{q}_j,\boldsymbol{q}_{j-1})(\Delta q_j^\mu/\varepsilon) + V(\boldsymbol{q}_{j-1})]\right\}.$$

There is of course a notion of equivalent discretizations here also. For details see [13, 16]. The relation between these two types of functional integrals has been explained here in the text.

REFERENCES

[1] Graham, R., *Springer Tracts on Modern Physics* **66**, (1973) Berlin.
[2] Graham, R., in 'Stochastic Processes in Nonequilibrium Systems', *Springer Lecture Notes in Physics* **84**, (1978) Berlin.
[3] Van Kampen, N. G., *Phys. Rep.* **24C**, 171 (1976).
[4] Langouche, F., Roekaerts, D. and Tirapegui, E., *Progr. Theor. Phys.* **61**, 1617 (1979).
[5] Martin, P. C., Siggia, E. D. and Rose, M. A., *Phys. Rev.* **A8**, 423 (1973).
[6] Phythian, R., *J. Phys.* **A8**, 1423 (1975).
[7] Garrido L., and San Miguel, M., *Progr. Theor. Phys.* **59**, 40, 55 (1978).
[8] Enz, C. P., *Physica* **89A**, 1 (1977).
[9] Berezin, F. A., *Theor. Math. Phys.* **6**, 194 (1971).
[10] Mayes, I. W., and Dowker, J. S., *Proc. R. Soc. London* **A327**, 131 (1972).
[11] Leschke, H. and Schmutz, M., *Z. Phys.* **B27**, 85 (1977).
[12] Langouche, F., Roekaerts, D., and Tirapegui, E., *Nuov. Cim.* **53B**, 135 (1979).
[13] Langouche, F., Roekaerts, D., and Tirapegui, E., *Phys. Rev.* **D20**, 419 (1979).
[14] Langouche, F., Roekaerts, D., and Tirapegui, E., *Phys. Rev.* **D20**, 433 (1979).
[15] Langouche, F., Roekaerts, D., and Tirapegui, E., *Physica* **95A**, 252 (1979).
[16] Langouche, F., Roekaerts, D., and Tirapegui, E., 'Functional Integrals in Phase space', (to be published).
[17] Dowker, J. S., *J. Math. Phys.* **17**, 1873 (1976).
[18] Langouche, F. Roekaerts, D., and Tirapegui, E., *Lett. Nuov. Cim.* **25**, 307 (1979).
[19] Langouche, F., Roekaerts, D., and Tirapegui, E., *Phys. Lett.* **72A**, 413 (1979).
[20] Langouche, F., Roekaerts, D., and Tirapegui, E., WKB-type Expansion for Langevin Equations, *Physica* **A 101**, 301 (1980)
[21] Leschke, H., Hirshfeld, A. C., and Suzuki, T., *Phys. Rev.* **D18**, 2834 (1978).
[22] Agarwal, B. S., and Wolf, E., *Phys. Rev.* **D2**, 2161 (1970).
[23] De Witt, B. S., *Rev. Mod. Phys.* **29**, 377 (1957).
[24] Mc. Laughlin, D., Schulman, L. S., *J. Math. Phys.* **12**, 2520 (1971).
[25] Cheng, K. S., *J. Math. Phys.* **13**, 1723 (1972).
[26] Gervais, J. L., and Jevicki, A., *Nucl. Phys.* **B110**, 93 (1976).
[27] Hirshfeld, A. C., *Phys. Lett.* **67A**, 5 (1978).
[28] Mortensen, R. E., *J. Stat. Phys.* **1**, 271 (1969).
[29] Jouvet, B., *Quantum Aspects of Classical and Statistical Fields*, preprint University of Paris XI, September 1977.
[30] Jouvet, B., and Phythian, R., *Phys. Rev.* **A19**, 1350 (1979).
[31] Phythian, R., *J. Phys.* **A10**, 777 (1977).
[32] Feynman, R. P., *Rev. Mod. Phys.* **20**, 267 (1948).
[33] Weiss, U., *Z. Phys.* **B30**, 429 (1978).
[34] Langouche, F. Roekaerts, D., and Tirapegui, E., *J. Phys.* **A13**, 449 (1980).
[35] Graham, R., *Z. Phys.* **B26**, 281 (1977).
[36] Graham, R., *Ann. Isr. Phys. Soc.* **2**, 948 (1978).
[37] Deininghaus, U., and Graham, R., *Z. Phys.* **B34**, 211 (1979).
[38] Dekker, H., *Phys. Rev.* **A19**, 2101 (1979).
[39] Langouche, F., Roekaerts, D., and Tirapegui, E., *Phys. Rev.* **A21**, 1344 (1980).
[40] Langouche, F., Roekaerts, D., and Tirapegui, E., *Physica* **A97**, 195 (1979).

[41] Bach, A., and Dürr, D., *Comm. Math. Phys.* **60**, 153 (1978).
[42] Ito, H., *Progr. Theor. Phys.* **59**, 725 (1978).
[43] Langouche, F., Roekaerts, D., and Tirapegui, E., *J. Phys.* **A11**, L263 (1978).
[44] De Witt-Morette, C., Maheshwari, A., and Nelson, B., *Phys. Rep.* **50**, 255 (1979).
[45] Elworthy, K. D., and Truman, A., *Classical Mechanics, the Diffusion (Heat) Equation, and the Schrödinger Equation on Riemannian Manifolds*, preprint, Edinburgh (1979).

AUTHOR AND SUBJECT INDEX

MATHEMATICAL PHYSICS AND
APPLIED MATHEMATICS

Editors:

M. FLATO *(Université de Dijon, Dijon, France)*
R. RĄCZKA *(Institute of Nuclear Research, Warsaw, Poland)*

1. M. Flato, Z. Maric, A. Milojevic, D. Sternheimer, and J. P. Vignier (Eds.), *Quantum Mechanics, Determinism, Causality, and Particles.* 1976, x+252 pp.
2. K. Maurin and R. Rączka (Eds.), *Mathematical Physics and Physical Mathematics.* 1976, xviii+504 pp.
3. M. Cahen and M. Flato (Eds.), *Differential Geometry and Relativity.* 1976, xii+304 pp.
4. M. Flato, C. Fronsdal, and K. A. Milton (Eds.), *Selected Papers (1937–1976) of Julian Schwinger.* 1979, xxviii+413 pp.
5. J. A. Wolf, M. Cahen, and M. De Wilde (Eds.), *Harmonic Analysis and Representations of Semisimple Lie Groups.* 1980, viii+495 pp.

GPSR Compliance

The European Union's (EU) General Product Safety Regulation (GPSR) is a set of rules that requires consumer products to be safe and our obligations to ensure this.

If you have any concerns about our products, you can contact us on ProductSafety@springernature.com

In case Publisher is established outside the EU, the EU authorized representative is:

Springer Nature Customer Service Center GmbH
Europaplatz 3
69115 Heidelberg, Germany

Batch number: 10761599

Printed by Printforce, the Netherlands